Energy Technology Handbook

Energy Technology Handbook

Edited by Lucas Collins

☐SYRAWOOD
PUBLISHING HOUSE
New York

Published by Syrawood Publishing House,
750 Third Avenue, 9th Floor,
New York, NY 10017, USA
www.syrawoodpublishinghouse.com

Energy Technology Handbook
Edited by Lucas Collins

International Standard Book Number: 978-1-68286-493-7 (Hardback)

Cataloging-in-Publication Data

Energy technology handbook / edited by Lucas Collins.
 p. cm.
Includes bibliographical references and index.
ISBN 978-1-68286-493-7
1. Power (Mechanics). 2. Power resources. 3. Energy development--Technological innovations.
4. Energy policy. 5. Renewable energy sources. I. Collins, Lucas.
TJ163.9 .E54 2017
621.042--dc23

Printed in the United States of America.

TABLE OF CONTENTS

Permissions

List of Contributors

Index

PREFACE

Energy technology is an umbrella discipline that concerns itself with the diverse aspect of energy such as its storage, distribution, efficiency, etc. The ever growing need of advanced technology is the reason that has fueled the research in the field of energy technology in recent times. Some of the diverse topics covered in the book address the varied branches that fall under this category. Different approaches, evaluations, methodologies and advanced study on energy technology have been included in this book.

After months of intensive research and writing, this book is the end result of all who devoted their time and efforts in the initiation and progress of this book. It will surely be a source of reference in enhancing the required knowledge of the new developments in the area. During the course of developing this book, certain measures such as accuracy, authenticity and research focused analytical studies were given preference in order to produce a comprehensive book in this area of study.

This book would not have been possible without the efforts of the authors and the publisher. I extend my sincere thanks to them. Secondly, I express my gratitude to my family and well-wishers. And most importantly, I thank my students for constantly expressing their willingness and curiosity in enhancing their knowledge in the field, which encourages me to take up further research projects for the advancement of the area.

Editor

Analysis of Design of Technologies, Tariff Structures, and Regulatory Policies for Sustainable Growth of the Smart Grid

ZAHEERUDDIN and MUNISH MANAS*

Department of Electrical Engineering, Jamia Millia Islamia (A Central University), New Delhi, India

Abstract: Deployment of the smart grid technologies, regulatory policies, and tariff designs in an efficient, cost-effective, and scalable manner is necessary for sustainable growth of the smart grid. This can be achieved by bringing together all enabling technologies and all key stakeholders under one roof. Technological innovations, utility restructuring, power sector regulatory policies and environmental concerns are a vital basis for smart grid development and promotion of renewable energy integration in the modern power sector. Utility restructuring and favorable regulatory policies open the energy market. These market dynamics will promote the small-scale power generation technology. As the renewable energy sources (RESs) can respond well to market signals by control actions, they are chosen for small-scale generation. A preferential transmission tariff to promote RESs integration is also a practical need of the present power sector. In this article, we try to analyze the technological base development, development of preferential tariff, and vital regulatory policies that help in sustainable growth of the smart grid in the modern power sector.

Keywords: Smart grid, renewable energy, advanced metering infrastructure, demand response, transmission tariff, regulatory policies, distributed generation, cyber security

1. Introduction

The sustainable growth of the smart grid is based on three important aspects, which include the development of a sound technological base, design of favorable regulatory policies, and preferential tariff designs. In this article, all three aspects are discussed in detail. Case studies of Indian power sector are presented in various sections because India's power usage has increased 16 times and the installed electricity capacity by 84 times in the last 6 decades.[1] Considering the rapid urbanization and improving standards of Indian households, the demand is likely to grow further. The demand for energy in India is estimated to increase up to 900 GW by 2032. This could be met by increasing the renewable energy generation of India to 183 GW by 2032.[2,3] Therefore, India needs a smart power grid to ensure a reliable and quality electric supply to its huge number of customers. Also, India needs a serious upgrade of its old electricity infrastructure and allied technologies from transmission to distribution in order to meet certain challenges such as grid losses, poor reliability,

*Address correspondence to: Munish Manas, Department of Electrical Engineering, Jamia Millia Islamia (A Central University), New Delhi, India. Email: msd.gkg85@gmail.com

power theft, and poor accessibility of power in rural areas. The smart grid is a strong solution to all these issues. Here, we try to analyze how the modern power sector is building a strong foundation for the smart grid technologies such as advanced metering infrastructure, demand response, distributed generation, home area network, communication, cyber security, electric vehicles, and electric energy storage (EES).

The installed capacity in India was 30,000 MW in 1981 and rose to over 143,000 MW by 2009.[4] In the recent past, this has put a burden on the conventional energy-based power generation sector of India. Three fifths of India's power generation depends on coal reserves, which are depleting at a fast rate. With natural gas being the next production fuel option, its demand has risen rapidly in the last 10 years. India is dependent on other countries for its natural gas supply to a large extent. On the other hand, India has a huge potential for renewal energy sources (RES). The commercial harnessing of RESs will help in the growth of the Indian economy in general and that of the Indian power sector in particular. In India, renewable sources such as wind, solar, small hydro, and biomass are producing 32,425 MW of energy, which is about 12.8% of the total installed capacity of 2, 53, 389 MW.[5,6] The importance of this green and clean energy is amplified by the fact that more than 33% of the coal-based thermal power plants produce large amounts of ash and pollutants, such as carbon dioxide, carbon monoxide, and sulfur dioxide. With the further growth of industry in India, the difference between demand and supply of electric power will also increase. Thus, the integration of renewable energy in the Indian power sector has a larger role to play in meeting the ever-increasing power demand in a clean way. This would also curtail the pollution from thermal electric power generation. To promote renewable energy integration in the Indian power sector, the Ministry of Power (MoP) introduced various policies such as National Electricity Policy, National Tariff Policy, and Electricity Act 2003. State Electricity Regulatory Commissions (SERCs) are directed by the Electricity Act 2003 to increase investment in renewable energy development and fix a percentage of RESs generated power in the total distribution system energy.[7] In addition, the policies such as Electricity Act 2003, National Electricity Policy, and National Tariff also encourage preferential tariffs for RESs. In these acts, policies such as renewable portfolio standards (RPS) and renewable purchase obligation (RPO) make it compulsory for a state to derive a fixed percentage of electricity from RESs. A 5-year tax holiday is given to companies as an incentive to set up RES projects.[8] MNRES plans to increase the renewable integration up to 35 GW in the 12th five-year plan (2012–2017).[9] Other policies such as cost reduction policy, competitive bid, and renewable resources obligation are also vital in promoting renewable energy integration in the Indian smart grid sector.

This article is organized as follows: Section 2 describes the technologies being employed for sustainable growth of the smart grid; section 3 deals with preferential transmission tariff for renewable generation; section 4 describes the regulatory policies for renewable integration, and the last section 5 provides the concluding remarks.

2. Technologies for Sustainable Growth of the Smart Grid

In this section we have discussed various technologies incorporated in the smart grid setup. The combination of these technologies in the operational framework of the smart grid makes the entire system more sustainable and reliable.

2.1 Advanced Metering Infrastructure and Their Benefits in the Smart Grid

The term *advanced metering infrastructure* (AMI) is used to describe the complete infrastructure, comprising smart meters, bilateral communication network, and control center instruments. It also includes the application software, which empowers the collection and transfer of real-time energy data. The tasks of AMI are remote meter reading of accurate data, identification of grid problem, load profiling, energy audit, and partial load shedding.

The various components of AMI in the smart grid projects include smart meter, meter data acquisition system, meter data management system (MDMS), communication network, and home area network (HAN). A smart meter is an electronic device, which measures electric energy consumption in predefined intervals and sends the reading back to the utility for monitoring and billing purposes.[10] A meter data acquisition system is the combination of various application software that empowers the hardware of data concentrator units (DCUs) and the control center to share the information between smart meters and MDMS through the communication network. MDMS is the control center's host system, which imports the metering data and authenticates and processes it before the final billing and analysis. The communication network is the network that

empowers bilateral information exchange between smart meters and utilities. Examples of communication networks are power line carrier communication, fixed radio frequency, fiber optic, and broadband over power Line (BPL). Home area network (HAN) is an expansion of AMI installed at the consumer end to facilitate the communication of home appliances with the smart meter. This will provide a better load management between utility and consumer. The IEEE standard P1888.4 (Green Smart Home and Residential Quarter Control Network Protocol) provides communication protocols and management solutions for optimized functioning of HAN.

AMI empowers the power distributors in the identification and automatic load dispatch of power demand, which improves the system reliability. The improvement in the system reliability, serviceability, reduction in power outages, and smooth billing infrastructure would reduce the cost of maintenance and operation of the grid. This will lead to a reduction in power tariff rates. The real-time monitoring of power usage through AMI will make the system more transparent, which will reduce the power theft.

2.2 Demand Response Technologies and Their Advantages in the Smart Grid

Demand response is an approach undertaken by utilities to transfer power consumption from peak hours to the moderate or lower power consumption intervals. The utility or customers may shed the predefined noncritical load at peak hours.

Demand response technologies include end-user interfaces and load control devices. These technologies deployed in the smart grid are used to balance demand and supply of electricity.[11] End-user interfaces include various channels such as e-mail, mobile phones, web portals, and video conferencing devices. The energy usage and electricity pricing data of the smart grid can be communicated to the consumers using these technologies. We may also connect load control devices and smart appliances using a home area network to form a smart metering system. The IEEE standard P1588 prescribes a network protocol for precise synchronization of real-time clocks of load control devices in a distributed network.[12] Electrical utilities deploy load control devices to regulate heating and cooling from the control center during demand response events. The load control switches perform direct remote control switching action on both cooling and heating units during peak loads. The smart thermostats adjust the temperature of these units remotely.

By initiating a demand response event, the utilities can detect a load increment and take a control action in advance to avoid brownouts. Real-time energy usage and electricity pricing data can be communicated to the consumers using demand response, which increases grid efficiency. Demand response also initiates the real-time verification of critical operations such as consumer's load shedding. Smart grid innovations are in progress to develop certain technologies that would automatically sense the requirement of load shedding. This would help the efficient management of all the smart appliances in homes or offices.

2.3 Distributed Generations Technologies and Their Benefits in the Smart Grid

The energy generation and distribution nearer to its actual consumer employing small-scale technological infrastructure is

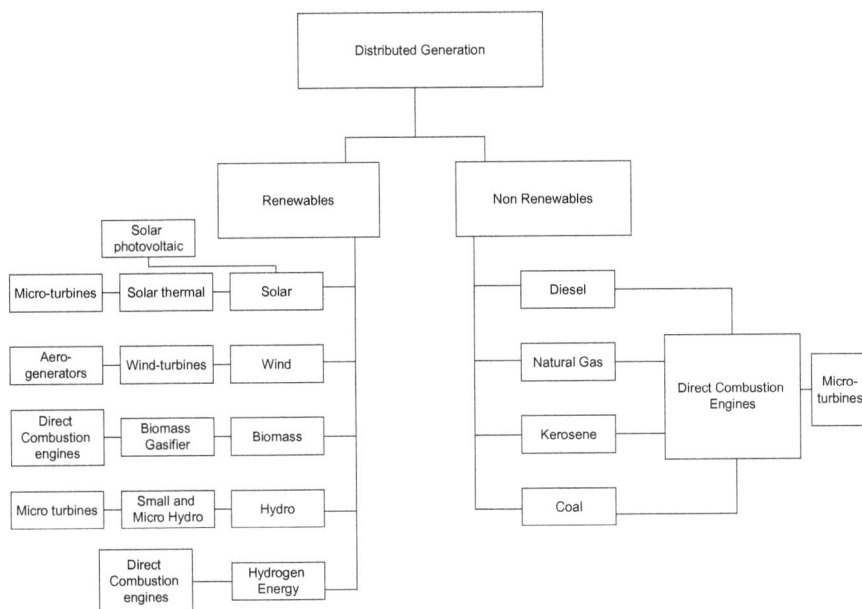

Fig. 1. Various distributed generation technologies in the smart grid.

known as distributed generation (DG). In the smart grid, DG units should be employed in grid's operational framework. The installed DGs can improve system reliability by acting as a backup generation in an isolated mode. They can also provide voltage support and reactive power control in the grid-connected mode.[13] Various distributed generation technologies employed in the existing smart grid are shown in the Fig. 1.

1. Microturbines: Microturbines are becoming popular in distributed generation, combined heat and power applications, and for powering hybrid electric vehicles because they have enough reserve margins to improve the power system reliability.[14] For microgrid applications, their capacity ranges from 40 to 450 kilowatts.
2. Fuel cells: Various types of fuel cells such as proton exchange membrane, phosphoric acid, molten carbonate, and direct methanol are being employed for standby power application mode. They have high conversion efficiency and ranges from 15 to 1000+ kW.[15]
3. Wind turbines: Wind turbines produce clean electricity using wind energy and do not require any new transmission infrastructure. They can be easily utilized to provide electricity to remote areas.
4. Photovoltaic: Photovoltaic (PV) in the form of solar panels whether on rooftops or freestanding consists of an array of solar cells producing DC (direct-current) power. DC power is then converted into AC power by using inverters. The grid-connected PV system for permanent energy injection into the grid is more popular due to preferential PV tariffs.[16]

IEEE standard P1547a specifies the requirements pertaining to the maintenance, operation, safety measures, and testing of the interconnection between distributed energy sources and the utility grid.[12]

Distributed generation improves the system reliability and energy management in cases where the microgrid operates in isolation to the utility grid (islanded mode). In the islanded mode of microgrid, its operations such as management of power as well as voltage and frequency control become quite easy due to the integration of distributed generation. Congestion management also becomes easier due to the power backup of the energy storage devices. Locally generated electric supply through distributed generation during peak hours can mitigate the problem of power shortage. Also, rural areas can get electrification in an economical manner by implementing distributed generation. Due to the nearness of the power generation source and the load, the implementation of distributed generation rules out the need for transmission and large distribution networks. This, in turn, reduces transmission and distribution losses.

2.4 Communication Technologies in the Smart Grid

The combination of intelligent electronic devices and telecommunication technologies in the smart grid setup provides better bidirectional information exchange between smart meters, local control units, intelligent relays, and the centralized control station. Integration of communication technologies in the smart grid will not only provide a more stable energy management setup but also a transparent information exchange between operators and consumers.

The term smart grid refers to the intelligent integration of the behavior of all users connected to an electricity network.[17] The modern power grid is made smart by the interconnection of power equipment in order to optimize bidirectional flow of

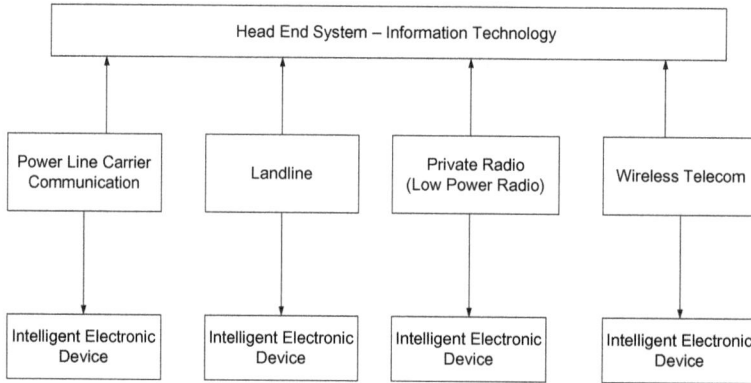

Fig. 2. Communication technologies in the smart grid.

electricity and information. This will enhance many intelligent power management applications such as accurate relay protection, quick demand response, and so forth.[18] Along with distributed intelligence, the communication technologies make the real-time reporting and control quite functional. Also, the intelligent electronic devices connected to the head end system through these communication channels make it easier for the control center operator to monitor the smart grid in an efficient and secure manner.

Some of the most important communication technologies that are implemented in the smart grid are low power radio (LPR), wireless, power line carrier communication, and landline[19] as shown in Fig. 2. LPR is formed by trunk mobile dispatching channels and meshed meter networks. Mobile phone communication includes GSM/GPRS/EDGE technology or the CDMA technology. Wi-Fi systems consist of various wireless communication technologies used in smart grid communication. Power line carrier communication is implemented by the superimposition of control signals over power signals using the technology of broadband-over-power line at the distribution voltage level.[20] Land line is a combination of analog subscriber lines, digital subscriber lines, coaxial cable, and fiber optics.

The congestion management in the microgrid can be handled by transmitting the microgrid data during off-load periods through the mobile cellular network from the data collector boxes to the control stations. The smart grid communication infrastructure is becoming more reliable and scalable in supporting other smart grid technologies such as AMI and advanced wide area network. The IEEE standard 1379-2000 recommends best practices for communication between substation installed remote terminal units and intelligent electronic devices connected to either distributed generators or load centers.[12]

2.5 Electric Energy Storage (EES) Technologies in the Smart Grid

EES compensates the variations in the output of the distributed generators by acting as a backup power source. This also minimizes the dependency of microgrids on utility grid power.[21] Keeping in mind the intermittency of renewable energy sources, EES development is of utmost importance.

Flywheels, batteries, double layer capacitors, thermal storage, and compressed air energy storage are various EES technologies being used in the smart grid. A flywheel is a rotating mechanical device with huge moment of inertia, which can store energy proportional to the square of its constant rotational speed. A connected transmission device increases or decreases its acceleration by injecting or taking out electricity.

Applications of the battery in the smart grid include powering electric vehicles, transient power backup, and grid integration. Various types of batteries used in the smart grid projects have different characteristics. Sodium sulfur (NaS) batteries have a high transient power rating. Lithium bromide batteries are rechargeable and can be used to power electric vehicles. Lead acid batteries, although used for renewable and distributed power systems, require huge space and maintenance. Vanadium redox flow batteries have a net efficiency up to 85% and are quite compatible with photovoltaic based microgrid systems.[22]

Electric double layer capacitors are the ultra-capacitors that are used for large storage applications and have a long life. Thermal storage technology is based on the concept of heat storage using hot water. It is used for systems involving large-scale renewable energy integration, mainly solar. In compressed air energy storage systems, we compress air in an underground storage cavern by using electricity, and later the combustion of the mixture of compressed air and natural gas drives a modified gas turbine with a round trip efficiency of 70%.

2.6 Electric Vehicles (EVs) Technologies and Their Importance in the Smart Grid

Electric vehicles are growing in popularity because they are environment friendly, energy efficient, and have a low maintenance cost.[23] In vehicle to grid (V2G) mode, EVs can be connected to the grid when not running. In this mode, EVs will provide electrical energy to the grid during peak hours. This is achieved by discharging the batteries during the day and charging them back during the night when there is abundant power generation.

The Indian government has formed a council named the National Council for Electric Mobility that will involve the participation of various ministries to encourage the development of EVs infrastructure in the country. The Ministry of Heavy Industry in India proposed a monetary incentive of Rs 6,000 crore along with full relaxation from custom duty and a waiver in the excise duty on imported batteries to boost the production of electric and hybrid cars in India.[24] At present, EV production in India is relatively naïve with Mahindra Reva being the most prominent product in the electric car market. The national utilities have to work hand-in-hand with the industry to improve the battery technology and charging infrastructure. This would make EVs more affordable and efficient for the customers. The importance of EVs could be assessed from the fact that the inclusion of about 3 lakh electric vehicles on Indian roads by 2020 could reduce the carbon dioxide, nitrous oxide, and hydrocarbon emissions by 16 lakh metric tons. This can also lead to a saving of more than Rs 3700 crores foreign exchange because India is dependent on foreign countries for 70 to 80% of its fossil fuel needs.[25]

2.7 Cyber Security

The smart grid involves communication technologies for bidirectional information exchange between utility and consumers. This information may include energy usage, status of grid availability, and electricity prices in real time for trading purposes. With an increase in cybercrime, the protection of such valuable information being exchanged is of utmost importance. The cyber security development should be based on the seven important guidelines given by the National Institute of Standards and Technology (NIST). As per NISTIR 7628 report, any smart grid should follow certain guidelines[26]:

1. The regulatory bodies should form policy and regulatory framework that provides a supportive environment for cyber security objectives.
2. Risk assessment techniques should be developed for threat assessment and its impact.
3. Four aspects of user privacy, i.e., personal information, personal privacy, behavioral privacy, and personal communication privacy should be protected.
4. Security architecture based on a smart grid conceptual reference model should be developed.
5. Certification schemes for smart grid devices, networks, systems, and processes should be developed enabling the stakeholders in benchmarking their infrastructures.
6. Cultivate research and development (R&D) program targeting four R&D issues, namely device level security, cryptographic management, networking related security, and system level security.
7. Security awareness and training programs that adhere to the regulatory framework and vital policies of the organization should be developed that back up the overall smart grid security.

In addition to NISTIR 7628 report, IEEE standard P2030.102.1 (a standard for interoperability of Internet protocol security) establishes interoperable configuration profiles for the information exchange in the smart grid energy sector.[12]

3. Development of Transmission Tariff Design for Renewable Energy Integration in the Smart Grid

A transmission tariff design, which is favorable for renewable energy in the smart grid sector, is an essential factor for the promotion of green energy. Various aspects of such a tariff structure are discussed in this section.

3.1 Transmission Tariff Structure for Renewable Energy Promotion in the Smart Grid

The renewable-based electricity transmission charges largely depend on the transmission pricing of the electricity market. The Public Utility Regulatory Policies Act (PURPA) has identified certain power generation facilities as the qualifying facilities (QFs) that can sell their energy to the local (host) utility. However, an open access electricity market will benefit the renewable energy generating companies in selling electricity to a better competitive market than a host utility. On the other hand, open access in transmission will also increase the number of customers and sellers of green energy. This will promote the participation of green energy retailers. An increased competition in the electricity market due to cheaper conventional power sources may put costly versions of renewable energy sources in danger. In such circumstances, one of the best options to promote renewable energy generation is through the inclusion of renewable portfolio standards (RPS) in the state restructuring legislation.[27] The inclusion of RPS would ensure that a specified portion of total electricity being injected in the grid is coming from renewable energy generation.

The consumers and producers of renewable energy are affected by the structure of its transmission tariff. This tariff structure will further influence the technological and economic growth of RESs in the smart grid sector. The structure of the transmission tariff is so designed that the marginal cost and fixed cost of the transmission system is recovered from the consumers and the utility. The marginal cost in general includes ancillary services cost, transmission losses, and congestion cost. The fixed cost is recovered by specifying the transmission tariff more than the marginal cost. This design of transmission tariff affects the electricity cost, which in turn influences the competition in the electricity trade market. Further, renewable energy generation facilities are also competing in the same trade market.[28] Thus, in order to nurture the renewable energy integration, certain preferential tariff scheme should be deployed to safeguard the interests of this newly born generation technology. The national utilities are formulating transmission tariffs that give a better risk to return ratio. Fig. 3 shows various components of renewable energy tariff design.

One of the popular methods to determine solar and wind tariffs in the smart grid sector is by calculating levelized electricity generation cost (LEC). It is defined as the net cost of installing the renewable energy system divided by its expected lifetime energy output. LEC of a RE system includes all the costs over its lifetime such as initial investment, operations and maintenance cost, cost of fuel, and cost of capital. It is expressed as[29]

$$LEC = \frac{\sum_{t=1}^{n} \frac{I_t + M_t + F_t}{(1+r)^t}}{\sum_{t=1}^{n} \frac{E_t}{(1+r)^t}} \qquad (1)$$

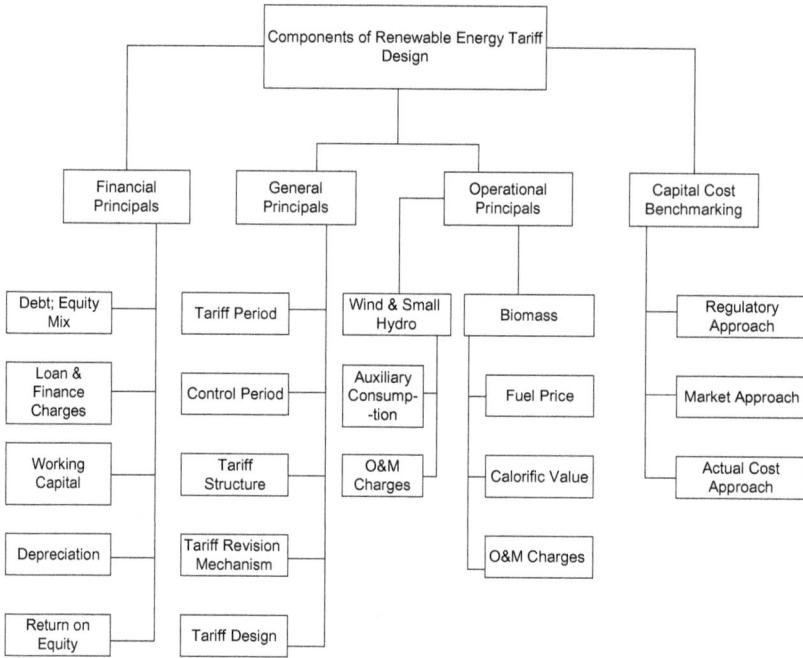

Fig. 3. Components of renewable energy tariff design.

where

LEC = Average lifetime levelized electricity generation cost
I_t = Investment expenditures in the year t
M_t = Operations and maintenance expenditures in the year t
F_t = Fuel expenditures in the year t
E_t = Electricity generation in the year t
r = Discount rate
n = Life of the system

A case study on preferential tariff design for promoting RES in India is presented here. Many Indian state utilities are now formulating transmission tariffs that give a better risk to return ratio for increasing the flow of investments in the renewable energy generation sector. Gujarat, which used to trade wind power at Rs 3.37 per unit, now provides an attractive tariff of Rs 3.50 per unit. Tamil Nadu has planned to increase the transmission tariff for purchasing wind energy from Rs 2.90 per unit to Rs 3.40 per unit. Accelerated depreciation (AD) was initially introduced for wind energy promotion in India, but later it was withdrawn in April 2012 owing to certain misuse by wind generation companies. By the efforts of the Indian Wind Power Association, the Indian government restored AD benefits in its union budget of 2014. Generation based incentives (GBI) are kept at Rs 0.5/unit of wind power. Table 1 shows the levelized tariff (with and without AD benefit) for different wind energy zones in India. The factors such as renewable facility project cost, interest on the equipment, interest on the land, and plant utilization greatly affect the structure of the transmission tariff.[30]

Table 1. Levelized tariff (LV) for different wind energy zones in India.

Indian wind energy zones	Wind power density (W/m^2)	Levelized tariff with AD (Rs/kWh)	Levelized tariff without AD (Rs/kWh)
Wind Energy Zone 1	Up to 200	6.08	6.37
Wind Energy Zone 2	201–250	5.49	5.78
Wind Energy Zone 3	251–300	4.80	5.07
Wind Energy Zone 4	301–400	4.10	4.25
Wind Energy Zone 5	More than 400	3.75	3.98

3.2 Preferential Tariff for Solar Energy in Modern Power Sector

The solar PV systems are more economical than diesel generator-based setups when installed on a large scale.[31] In computing the capital cost of a solar PV system, the unit cost is the major component, labor cost is the medium, and transportation cost is the smallest one.[32] To encourage grid connected solar PV rooftops with battery energy storage system (BESS) in the domestic setup, a preferential feed in tariff (FIT) with the optimal sizing of PV-BESS system is formulated as follows[33]:

$$Maximize\ f\ (PV,\ BESS,\ operation)$$

where

$$f\ (PV,\ BESS,\ operation) = Savings + Subsidy - Costs \quad (2)$$

$$Saving = Electr.bill\ without\ PV\&BESS$$
$$- Electr.bill\ with\ PV\&BESS \quad (3)$$

$$Subsidy = FIT * (E_{pv} - E_d) \quad (4)$$

$$Costs = Installment\ (PV) + Installment\ (BESS) \quad (5)$$

E_{pv} = Energy output from rooftop PV panel
E_d = Energy dispatched to the consumer load
$Installment\ (PV)$ = Total cost of installing PV panels
$Installment\ (BESS)$ = Total cost of installing battery energy storage system

A case study on preferential tariff structure of solar energy in the Indian power sector is discussed here. Many state utilities of India have set the solar tariff at attractive rates in order to promote the growth of solar photovoltaic and solar thermal power generation in their respective states. The Electricity Regulatory Commission (ERC) of Rajasthan has decided to promote the growth of solar energy generation by fixing its solar photovoltaic (PV) generation tariff at Rs 15.78 per kWh and solar thermal power generation tariff at Rs 13.78 per kWh. This includes the incentives given by the Ministry of New and Renewable Energy (MNRE). Tamil Nadu, on the other hand, has fixed its solar PV and solar thermal generation tariff at Rs 3.15 per kWh on a provisional basis for the distribution utility. The capped price of power from grid connected photovoltaic units (not receiving any incentive from MNRE) will be Rs 11.00 per kWh for trading with licensed distribution agencies. Such a tariff is applicable to solar power plants, which started operating by 2010 and is valid for 10 years. For the projects, which do not receive MNRE incentives and are commissioned between 2010 and 2012, the solar power tariff would be Rs 10/kWh. Table 2 shows solar PV tariffs for different states in India. Levelized energy cost (LEC) of solar energy in various states can be analyzed in order to identify the areas where cost savings would be the maximum. This would help the utilities to focus on solar power development in the economically profitable areas.[30]

4. Regulatory Policies for Promoting Renewable Energy Integration in the Smart Grid Sector

Regulatory policies and associated directives that safeguard the interests of the smart grid technologies and that of renewable energy are discussed here.

Table 2. Solar PV tariff rates for different states in India.

Name of the state	Tariff application period, financial year (FY)	Solar PV feed-in-tariff (Rs/kWh)	Solar PV (SPV) capital cost (lakh/MW)
Tamil Nadu	FY 2014–2016	LEC (with AD)-15.38 LEC (without AD)-18.78	1715
Karnataka	Up to March 31 2015, for 2013 Commissioning	14.5 (includes rooftop and small solar PV plants)	1550
Maharashtra	FY 2010–2015	LEC (with AD)-9.51 LEC (without AD)-11.16	1000
Haryana	FY 2013–2015	9.18 (SPV crystalline) 8.90 (SPV thin film)	894 (SPV crystalline) 863 (SPV thin film)
Orissa	Commissioning FY (2012–13) onward	LEC (with AD)-14.77 LEC (without AD)-17.80	1690
Kerala	for plants commissioned before 31 Dec. 2009	15.18 (incentives included)	2250
West Bengal	FY 2013–17 (5 Years)	10 (for capacity from 100kW to 2 MW)	Not yet defined
Gujarat	Jan. 2012 to March 31, 2015	For MW Scale Plants: FY 2014–15: 8.97 (without AD) 8.03 (with AD) For KW scale Plants FY 2014–15: 10.76 (without AD) 9.63 (with AD)	1000 (for MW capacity size plants) 1200 (for kW capacity sized plants)
Punjab	FY 2012–16	LEC (with AD)-9.35 LEC (without AD)-10.39	1000

4.1 Economic and Technical Aspects of the Smart Grid Policy Framework

Regulatory policies for the promotion of renewable energy (RE) in the modern power sector are characterized mainly by government directives, which enforce utilities to buy renewable energy. The government provides fiscal incentives to private companies for setting up RE projects. Incentives are also given to RE equipment manufacturers for manufacturing allied equipment. These policies also aim at encouraging private and government agencies to undertake research and development (R&D) projects linked to RE. Central and state electricity regulatory commissions have designed certain policies to promote renewable sources in the competitive electricity market. Two such policies are renewable portfolios standards (RPS) and renewable purchase obligation (RPO), which mandates a region to derive a fixed percentage of its electricity from renewable energy sources. Under these policies, each region can choose a combination of various available renewable energy sources to fulfill this mandate. Also, there are feed-in laws designed in order to designate where and in what quantum the renewable energy would be injected. Remunerative pricing of green power, easy banking, tariff support, and tax breaks are few additional economic aspects of such policies that encourage private players to take up the task of green energy generation. A vibrant domestic capital market and increased private equity investment have benefited the green energy development in the smart grid. A cost-benefit course to grid integration of renewables would support the main objective of the smart grid concept, which aims at better utilization of existing and future electricity assets.[34] Any cost-benefit regulatory norms that give tax incentives or cross-subsidy surcharge relaxation to the green power producers would definitely help them meet their debt service obligation. Designing of market-based energy policies would provide a competitive market framework that may enhance energy security, environmental protection, and economic efficiency for accelerated promotion of renewable energy in the smart grid (SG) setup.

In addition to this, liberalization of energy markets and minimization of direct government interventions in the energy sector would lure more private players to enter into renewable energy business. A coordinated public–private role in developing green energy infrastructure and market would promote the renewable energy generation by the enhancement of energy services. The main objectives of such policies include meeting of minimum energy needs through green energy integration and providing decentralized power to domestic and commercial sectors in rural or urban areas. The economic objective of these policies is encouraging private investments by providing fiscal and financial incentives such as tax holidays, rebates in excise duty, and so forth. The *IEEE Global Consumer Socialization of the Smart Grid* report describes consumer behavior toward adoption of smart grid technologies. This report divides consumers into various grades on the basis of their priorities of the smart grid technologies and SG services. Different methods of promoting smart grid technologies such as preferential tariff, buy back, wheeling, and tax benefits are described in the following section.[35]

The regulatory commissions in the whole world now stand united to address economic as well as technical aspects of the smart grid policy framework. The Federal Energy Regulatory Commission has focused on a few major technical aspects of the smart grid policy framework such as system security, intersystem communication, wide-area situational awareness, demand response, electric storage, electric vehicles, advanced metering, and distribution system automation.[36] The US Department of Energy in collaboration with technical institutions is working on technical aspects such as telecommunications, the Internet, and power industry dynamics that are integrated with smart grids. The Illinois Power Agency Act, Senate Bill No. 1592, ordered the utilities in Illinois to reduce peak load by 0.1% every year for 10 years.[37] Such policies have encouraged the state utilities to adopt smart grid technologies including the deployment of huge numbers of smart meters. The Illinois Institute of Technology has planned to integrate power controller systems, demand response controllers, and advanced meters to make the distribution grid more reliable. The Empower Maryland Energy Efficiency Act of 2008, Senate Bill No. 205, has ordered the utilities to reduce peak demand by 15% by 2015. This policy act has empowered the Maryland Public Service Commission to encourage power utilities to adopt smart grid technologies.

A European network and information security agency has identified cyber security components in the information and communication infrastructure of the smart grid. The cyber security components such as human factors, information sharing protocols, and physical security must be addressed by regulatory commissions in order to make smart grid communication setups more secure.[38] NIST cyber security framework discussed in Section 2 is also a vital contribution in the formation of smart grid security policies.

4.2 Case Study on the Regulatory Policies Implemented in the Indian Power Sector to Promote Renewable Energy Generation

Various policies were introduced by the government of India in the past few years to promote renewable energy integration in the Indian power sector. Most vital among them were the National Electricity Policy introduced in 2005 and the Electricity Act (2003), which delicensed distributed standalone generation and distribution system in rural areas.[39] Another important policy named the New Tariff Policy (2006) fixed a minimum percentage of green energy to be purchased by each state utility.[40] The various directives being deployed by such national policies for the promotion of renewable energy development in the smart grid in India are discussed here. These include renewable portfolio standard, renewable purchase obligation, industrial policy, foreign investment policy, buy back, and wheeling.

A renewable portfolio standard is a regulatory tool in the hands of State Electricity Regulatory Commissions (SERCs) to ensure that a specified portion of total electricity being injected in the grid is coming from renewable energy generation. This would ensure a safe position for these newly born green energy technologies in the competitive electricity market. The eligible renewable energy credit certificates given to green energy producers certify that each unit (kWh) of the electricity injected was being contributed by a genuine renewable energy producer.

The renewable purchase obligation (RPO) is a similar regulatory instrument, which is defined on the basis of the fixed percentage of total inserted power in the grid being purchased

Table 3. Growth of solar and wind RPO target of Gujarat state from 2011 to 2016.

	RPO target in percentage of energy (kWh) for Gujarat state		
Financial year	For wind generation	For solar generation	Wind + solar RPO target
2011–2012	5.0	0.5	5.5
2012–2013	5.5	1.0	6.5
2013–2014	5.5	1.0	6.5
2014–2015	6.25	1.25	7.50
2015–2016	7.0	1.5	8.5
2016–2017	7.75	1.75	9.5

from green energy producers. This would create favorable conditions for the growth of renewable energy in their respective areas. Karnataka, Gujarat, Maharashtra, Tamil Nadu, and Rajasthan were the leading states to incorporate RPS quota in their electricity regulatory norms. Table 3 gives growth of the fixed percentage of solar and wind RPO target for Gujarat state from 2011 to 2016.[40]

Various small-scale industries that produce renewable energy technological equipment and systems are being encouraged by MNRES under its industrial policy. No clearance is needed from the Central Electricity Authority to set up renewable energy generating plants costing up to Rs 1,000 million. For green

energy producers and RE device manufacturing firms, a tax exemption for five years is offered by the government of India. The fiscal incentives on RE instruments provided by the Indian government are (a) exemption/reduction in excise duty, (b) exemption from central sales tax and custom duty, and (c) full rebate on direct taxes in the first year of setting up of the RE venture.[30]

The Foreign Investment Promotion Board (FIPB) permits foreign investors to set up renewable energy (RE) generation plants or RE equipment manufacturing units, in a joint venture with Indian industrialists, where foreign equity investment up to even 100% can be utilized. These RE generation projects could be on a Build, Own, and Operate (BOO) basis. Reserve Bank of India exempts Indian enterprises from taking permission on using foreign investment under "automatic route," to set up RE generation plants or related projects.[41]

The Buy Back Policy allows customers to sell the surplus electricity generated though rooftop solar panel, wind turbines, and so forth, installed in domestic areas to the utility grid. The customers under this scheme can also avail many fiscal incentives such as interest and capital subsidy along with the income tax benefits. Wheeling involves the transmission facility provided by a third party between producer and consumer of renewable energy. For the utilities that are rescheduling its generation for the wheeling of RE, the government of India is giving subsidies on the transmission charges. In Section 42 (clause 1 and 2) of Electricity Act 2003, it is specified that SERC shall specify the wheeling charges and define the related surcharge. Table 4 gives

Table 4. Regulatory policy framework in different Indian states for grid connected solar PV rooftops.

Name of the state	Solar rooftop capacity target	Buy back metering mechanism	Capping of PV capacity/generation	Incentives on SPV rooftops
Chhattisgarh	500–1000 MW by March 2017	Net metering (excess units are billed at 50% of the solar tariff)	Injection not more than 49% of the yearly net generation	Wheeling, banking, and cross-subsidy, VAT exempted on equipment
Tamil Nadu	350 MW by 2015	Net metering	Capped commercially at 90% of the power consumption. Excess energy over 90% cap would be treated as lapsed	Subsidy of Rs. 20,000 per kW for 1 kWp system is provided. Electricity tax is exempted
Gujarat	30 MW in 6 cities	Feed-in-tariff	Not mentioned	Exemption from wheeling, banking, and cross-subsidy surcharge
Karnataka	400 MW by 2018	Net metering (excess units to be billed as per solar tariff)	Not mentioned	Wheeling, banking, and cross-subsidy surcharge exempted for 10 years
Haryana	50 MW till 2017	Net metering	Not defined	10% subsidy in addition to 30% MNRE subsidy, exemption from scrutiny, application fee, external development charges
Uttarakhand	5 MW per year (2013–2015)	Net metering	Not defined	No transmission and wheeling charges
West Bengal	34 MW by 2018	Net metering	Injection not more than 90% of the consumption from the yearly licensee's supply	Wheeling, cross subsidy charges exempted

the details of the regulatory policy framework in different Indian states for grid connected solar PV rooftops.

All the mentioned tariffs and regulatory policies have a deep impact on the rapid growth of RE generation. It is also evident from the preceding discussion that there is an accelerated increment of RE generation with each benchmark achieved in the development of tariff designs and regulatory policy framework. Consequently, these policies have encouraged various stakeholders for investing in RE generation, which will promote smart grid development.[30]

5. Conclusion

The smart grid deployment is an optimum solution to the present-day power sector problems such as environmental pollution caused by conventional power generation, grid losses, as well as poor reliability and accessibility of power in rural areas. For the sustainable growth of the smart grid, the allied technologies, tariff design, and regulatory framework must be developed in a customized manner. To accomplish this mission, all the national agencies and private players must work in close association to develop the smart grid technologies and policies. This requires the creation of dedicated departments for technological R&D and regulatory policy formation. The government tariff and regulatory policies must encourage more private industry participation in manufacturing areas of the smart grid technologies (SGTs) for its accelerated commercialization. The service and financial commitments in the market design of the SGTs must be driven by consumer satisfaction for rapid adoption of the SGTs.

References

1. Garg, P. Energy Scenario and Vision 2020 in India. *J. Sustain. Energ. Environ.* **2012**, *3*, 7–17.
2. Government of India, Central Electricity Authority. Revised Power Supply Position Report. http://www.cea.nic.in/reports/monthly/gm_div_rep/power_supply_position_rep/energy/Energy_2013_03.pdf (accessed August 7, 2014).
3. Government of India, Central Electricity Authority. Peak Demand Power Supply Position Report. http://www.cea.nic.in/reports/monthly/gm_div_rep/power_supply_position_rep/peak/Peak_2013_03.pdf (accessed August 9, 2014).
4. Goyal, M.; Jha, R. Introduction of Renewable Energy Certificate in the Indian Scenario. *Renew. Sustain. Energ. Rev.* **2009**, *13*, 1395–1405.
5. Upadhyay, A. Review of Regulatory Promotional Initiatives for Development of Renewable Energy Projects in India. *Int. J. Elec. Electron. Eng. Res. (IJEEER)* **2014**, *4*, 9–24.
6. Power Sector Reports: India. http://www.cea.nic.in (accessed August 14, 2014).
7. Tariff Order for Solar Power Generation—FY14–18. http://mnre.gov.in/file-manager/UserFiles/Grid-Connected-Solar-Rooftoppolicy/Karnataka_tariff_order-2014_2018.pdf (accessed August 25, 2014).
8. Singh, R.; Sood, Y. R.; Padhy, N. P.; Venkatesh, B. Analysis of Renewable Promotional Policies and Their Current Status in Indian Restructured Power Sector. Proc. of IEEE Transmission and Distribution Conference and Exposition (IEEE PES-2010), New Orleans, 19–22 April 2010; 1–8.
9. Smart Grid Vision and Roadmap for India. http://indiasmartgrid.org/en/Pages/Index.aspx (accessed August 28, 2014).
10. Staff Report of Federal Energy Regulatory Commission on Demand Response & Advanced Metering. http://www.ferc.gov/legal/staff-reports/12-08-demand-response.pdf (accessed September 2, 2014).
11. Rahiman, F. A.; Zeineldin, H. H.; Khadkikar, V.; Kennedy, S. W.; Pandi, V. R. Demand Response Mismatch (DRM): Concept, Impact Analysis, and Solution. *IEEE Trans. Smart Grid* **2014**, *5*, 1734–1743.
12. IEEE Approved Proposed Standards Related to Smart Grid. http://smartgrid.ieee.org/standards/ieee-approved-proposed-standards-related-to-smart-grid (accessed September 4, 2014).
13. Kahrobaeian, A.; Mohamed, Y. A. R. I. Interactive Distributed Generation Interface for Flexible Micro-Grid Operation in Smart Distribution Systems. *IEEE Trans. Sustain. Energ.* **2012**, *3*, 295–305.
14. Davis, M. W.; Gifford, A. H.; Krupa, T. J. Microturbines—An Economic and Reliability Evaluation for Commercial, Residential, and Remote Load Applications. *IEEE Trans. Power Sys.* **1999**, *14*, 1556–1562.
15. Li, W.; Li, W.; Deng, Y.; He, X. Single-Stage Single-Phase High-Step-Up ZVT Boost Converter for Fuel-Cell Microgrid System. *IEEE Trans. Power Electron.* **2010**, *25*, 3057–3065.
16. Sechilariu, M.; Wang, B.; Locment, F. Building Integrated Photovoltaic System With Energy Storage and Smart Grid Communication. *IEEE Trans. Ind. Elec.* **2013**, *60*, 1607–1618.
17. Yu, X.; Cecati, C.; Dillon, T.; Simões, M. G. The New Frontier of Smart Grids. *IEEE Ind. Elec. Mag.* **2011**, *5*, 49–63.
18. Lu, X.; Wang, W.; Ma, J. An Empirical Study of Communication Infrastructures Towards the Smart Grid: Design, Implementation, and Evaluation. *IEEE Trans. Smart Grid* **2013**, *4*, 170–183.
19. Gungor, V. C.; Sahin, D.; Kocak, T.; Ergut, S.; Buccella, C.; Cecati, C.; Hancke, G. P. Smart Grid Technologies: Communication Technologies and Standards. *IEEE Trans. Ind. Info.* **2011**, *7*, 529–539.
20. Bumiller, G.; Lampe, L.; Hrasnica, H. Power Line Communication Networks for Large-Scale Control and Automation Systems. *IEEE Commun. Mag.* **2010**, *48*, 106–113.
21. Srivastava, A. K.; Kumar, A. A.; Schulz, N. N. Impact of Distributed Generations with Energy Storage Devices on the Electric Grid. *IEEE Sys. J.* **2012**, *6*, 110–117.
22. Xin, Q.; Nguyen, T. A.; Guggenberger, J. D.; Crow, M. L.; Elmore, A. C. A Field Validated Model of a Vanadium Redox Flow Battery for Microgrids. *IEEE Trans. Smart Grid* **2014**, *5*, 1592–1601.
23. Byung-Gook, K.; Ren, S.; van der Schaar, M.; Lee, J. W. Bidirectional Energy Trading and Residential Load Scheduling with Electric Vehicles in the Smart Grid. *IEEE J. Selected Areas Comm.* **2013**, *31*, 1219–1234.
24. National Electric Mobility Mission Plan 2020 published in 2012. http://dhi.nic.in/NEMMP2020.pdf (accessed September 5, 2014).
25. Industry Analysis Report of Automobile Industry. http://www.scribd.com/doc/24945418/Industry-Analysis-Report-of-Automobile-Industry-Two (accessed September 12, 2014).
26. Chan, A. C.; Zhou, J. On Smart Grid Cybersecurity Standardization: Issues of Designing with NISTIR 7628. *IEEE Comm. Mag.* **2013**, *51*, 58–65.
27. Singh, R.; Sood, Y. R. Transmission Tariff for Restructured Indian Power Sector With Special Consideration to Promotion of Renewable Energy Sources. Proc. of IEEE Region 10 Conference (TENCON 2009), Singapore, 23–26 Jan. 2009; 1–7.
28. A Report on Overview of Renewable Potential of India: Government Commitment to Renewable Energy. http://www.geni.org/globalenergy/library/energytrends/currentusage/renewable/Renewable-Energy-Potential-for-India.pdf (accessed September 15, 2014).
29. Short, W.; Packey, D.; Holt, T. A Manual for Economic Evaluation of Energy Efficiency and Renewable Energy Technologies, National Renewable Energy Laboratory, March 1995; 1–120. http://www.nrel.gov/csp/troughnet/pdfs/5173.pdf.
30. Ministry of New and Renewable Energy Sources. Review of Government of India Policies for Promotion of Renewable and Biomass Energy Utilization in India. mnre.gov.in/file-manager/UserFiles/strategic_plan_mnre_2011_17.pdf (accessed September 17, 2014).

31. Kolhe, S. K. M.; Joshi, J. C. Economic Viability of Stand-alone Solar Photovoltaic System in Comparison with Diesel-powered System in India. *Energ. Econ.* **2002**, *24*, 155–165.

32. Das, I.; Bhattacharya, K.; Canizares, C. Optimal Incentive Design to Facilitate Solar PV Investments in Ontario. IEEE Power and Energy Society General Meeting, San Diego, 22–26 July 2012; 1–6.

33. Belli, G.; Brusco, G.; Burgio, A.; Menniti, D.; Pinnarelli, A.; Sorrentino, N. A Feed-in Tariff to Favorite Photovoltaic and Batteries Energy Storage Systems for Grid-connected Consumers. Proc. of 4th IEEE/PES Conference on Innovative Smart Grid Technologies Europe (ISGT EUROPE), Lyngby, Denmark, 6–9 Oct. 2013, pp. 1–5.

34. Strbac, G. Technical and Regulatory Framework for Smart Grid Infrastructure: A Review of Current UK Initiatives. Proc. of IEEE Power and Energy Society General Meeting, Minneapolis, 25–29 July 2010; 1–3.

35. Bhat, K.; Sundarraj, V.; Pandita, S.; Sinha, S.; Kaul, A. IEEE Global Consumer Socialization of Smart Grid. *IEEE Standard 2013, Smart Grid Research: Consumer Socialization*, pp. 1–64, Sept. 3, 2013.

36. Mason, W. E. Testimony on Energy Policy and Innovation. US Federal Energy Regulatory Commission. http://www.ferc.gov/EventCalendar/Files/20100701105022-Emnett-Testimony-07-01-10.pdf (accessed September 19, 2014).

37. Department of Energy, Washington, DC. Enhancing the Smart Grid: Integrating Clean Distributed and Renewable Generation. Office of Electricity Delivery and Energy Reliability. http://www.oe.energy.gov/DocumentsandMedia/RDSI_fact_sheet-090209.pdf (accessed October 2, 2014).

38. Enisa Security Aspects of the Smart Grid. https://www.enisa.europa.eu/activities/Resilience-and-CIIP (accessed October 5, 2014).

39. Government of India. The Electricity Act, 2003. Gazette of India, June 2003, New Delhi.

40. Ghosh, D.; Shukla, P. R.; Garg, A.; Ramana, P. V. Renewable Energy Strategies for Indian Power Sector. http://www.decisioncraft.com (accessed October 9, 2014).

41. Behera, L. K. India Needs a Liberal FDI Policy for Its Defense Industry. www.idsa.in/system/file/Book_CoreConcernsinIndianDefence.pdf (accessed October 17, 2014).

2

Effects of Pressure and Friction Parameters on a Concrete Bed Energy Storage System

A. A. ADEYANJU*

Mechanical Engineering Department, Ekiti State University, Ado-Ekiti, Ekiti, Nigeria

Abstract: This study carried out the effect of pressure drop and friction factor as performance parameters on a concrete bed energy storage system. Spherical-shaped concrete of diameters 0.11 m, 0.08 m, and 0.065 m were used in this analysis, and 10 air flow rates were studied between 0.0094 and 0.045 m^3/s. It was discovered that the pressure drop in all the cases increases with airflow rate with spherical-shaped concrete of diameter 0.065 m possessed the highest pressure drop, and this may be attributed to its low porosity, while spherical-shaped concrete of diameter 0.11 m had the lowest pressure drop. In the laminar range, the friction factor is a function of the Reynolds number. It is independent of the surface roughness of the bed. The transition from laminar flow to turbulent flow, the friction factor, increases from laminar to turbulent flow across the bed while it was laminar across the copper tube. The Reynolds number is a function of both the fluid properties and the concrete size. The Reynolds number across the packed bed moves from laminar to turbulent condition as air flow rate and pressure drop increases, while that within the copper tube was laminar.

Keywords: Pressure drop, friction factor, Reynolds number, concrete, packed-bed, laminar, turbulent

1. Introduction

Energy storage is an extremely important part of today's society. In almost every facet of science and technology, energy storage plays a significant role whether the energy is needed in chemical, heat, mechanical, electrical, or potential form. Though the motivation for the recent technological advancements in various fields of energy storage varies, the overall impetus is the same; energy supply whether it comes from the earth or the sun is never a constant. Day turns to night, winds die down, oil fields eventually run dry, and the geothermal heat from the crust of the earth, although seemingly constant, will eventually diminish. There is then a need to store energy so that it can be extracted when it is not readily available. This is clearly evident in solar panels, which convert the sun's radiation into electricity for later use. In fact, the storage of thermal energy perhaps dates back as far as civilization itself; since the beginning of recorded history, people have harvested ice to keep things cool when warmer weather approaches. It is this type of thinking that has provided the desire to store many other types of energy from various sources both for economic and ecologic purposes.

For some decades, the world's energy supply has not been keeping up with the increasing demand. Burgeoning countries undergoing industrial reform are consuming an increasing amount of crude oil, coal, and electricity, which has contributed to increases of overall energy prices to an unprecedented level. As a result, energy conservation has been on the rise lately, and new sources to feed the human energy hunger are sought without relent. Moreover, the search for more efficient, ecologically friendly, and cost-effective ways to capture and store energy for later use is always a popular topic.

A thermal-storage unit in which particulate materials contained in an insulated vessel is known as a packed bed (pebble bed or rock pile) storage unit. It uses the heat capacity of loosely packed pebbles or rocks to store energy; fluid, usually air, is circulated through the bed to add or remove energy. The most commonly used solid storage materials are rocks, concrete, bricks, and walls. The materials are invariably in porous form, and heat is stored or extracted by the flow of gas or liquid through the pores or voids.

For the subject of flow and heat transfer through porous media, there have been extensive investigations covering broad ranges of applications since the early work of Darcy in the nineteenth century.[1,2] Darcy correlated the pressure drop and flow speed experimentally by defining a special constant property of the medium called permeability. However, it is applicable only to low-speed (creeping) flow and low-porosity saturated medium. It is well known that in flow through porous media, the pressure drop caused by the frictional drag is proportional to the speed at

*Address correspondence to: A. A. Adeyanju, Mechanical Engineering Department, Ekiti State University, P.M.B. 5363, Ado-Ekiti, Ekiti State, Nigeria. Email: anthonyademolaadeyanju@yahoo.co.uk

the low Reynolds number range. In addition, this famous Darcy's law also neglects the effects of solid boundary and the inertial forces on fluid flow and heat transfer.

There are two interesting aspects that arise in the research of porous media. They are hydrodynamic and thermal effects. The dynamics of fluid flow through a porous medium is a relatively old topic. Since the nineteenth century, Darcy's law has traditionally been used to get quantitative information on flow in a porous medium. This law is reliable when the representative Reynolds number is low while the viscous and pressure forces are dominant. As Reynolds number increases, deviation from Darcy's law grows due to the contribution of inertial terms to the momentum balance.[1,2] It was shown that for all investigated media, the axial pressure drop was represented by the sum of two terms, one being linear in the speed (viscous contribution) and the other being quadratic in speed (inertial contribution). The inertial contribution is known as Forchheimer's modification of Darcy's law.[3] Beavers and Sparrow[4] proposed a similar model for fibrous porous media. A general expression can be obtained from Bear.[1] Lage et al.[5] suggested that an extra cubic term of fluid speed be included in the above equation in the regime of higher speed.

Another significant work for predicting momentum transport in porous media is by Brinkman.[6] Brinkman first introduced a term that superimposed the bulk and boundary effects together for flows with bounding walls. In Brinkman's model, an effective viscosity was postulated from experiments performed on beds of spheres to replace the viscosity of fluid by taking into account the porosity effect. There have been changes on the above function to describe different types of porous media. Computational modeling has also been used to give detailed flow fields. There are also results obtained by the asymptotic solutions.[7]

While study of porous media flow is an old topic in fluid mechanics, the convective heat transfer of flows through porous medium has emerged as a new interest due to new technological developments. Forced convection in porous media arises wherever the energy is delivered, controlled, used, converted, or produced. The widely used cellular microstructure materials have found implementation in the technologies of thermal dissipation media, impact absorbers, and compact heat exchangers. Their thermal attributes enable applications as heat dissipation media and as recuperation elements. Consequently, these enable high heat transfer rates and can be effectively used for either cooling or efficient heat exchange.

Hence, it has become important to understand the interaction between mass and thermal transports and the resulting effects on the thermo-mechanical characteristics of porous media. There has been much analytical, numerical, and experimental work done in the past to measure or estimate the overall heat transfer rate of convective heat transfer in porous media. Some early experimental studies have investigated the flow of fluids through packs of spheres.[8]

Resurging interests in applying complex microstructures to chemical industries and transpiration cooling have prompted studies in experiments of highly porous media.[9] In general, the experimental results usually are described statistically as empirical relationships in terms of dimensionless parameters of permeability, Reynolds, Nusselt, Prandtl, and Peclet numbers. However, those correlations fail to discuss general mass and

thermal transfer phenomena other than specific configurations, such as packed beds of particles or highly porous open-cellular foams. Non-intrusive experimental techniques have been applied to allow some detailed visualizations of microscopic flow and mass transfer occurring within porous media. Among these are laser Doppler velocimetry (LDV), particle image velocimetry (PIV), and photo luminescent volumetric imaging (PVI).

Dybbs and Edwards used a laser anemometer for measurement inside a model porous media.[10] The measurement of local velocity fields above the porous medium was first attempted by Vignes-Adler using LDV techniques and by Stephenson and Steward using PIV techniques.[10] Since then, the experimental measurement techniques and devices have been advanced from differential pressure transducers and hotwires in packed beds of glass beads or cylinders[10] to LDV and 100 μm thermocouples or magnetic resonance imaging (MRI). They used PIV to measure velocity close to porous media and pointed out the deficiency of the classical semi-empirical Brinkman equation.[6]

Concerning analytical studies, Darcy's law ignores the effects of solid boundary or the inertial forces on fluid flow and heat transfer. While these effects become significant near the boundary and in highly porous materials, relatively little attention had been directed to study these effects. Brinkman[6] proposed a model to account for the presence of the solid boundary by adding a viscous term to Darcy's law. He took inertial effect into account by adding a speed squared term to Darcy's law. However, both works do not consider boundary and inertia effects simultaneously.

Vafai and Tien[11] used a volume averaged momentum and energy equations to numerically study both effects simultaneously. The volume averaging method is widely used for investigating inertial and boundary effects on flow and heat transfer in porous media. It uses local volume averaging method and supplementary empirical relations to set up the macroscopic governing equations of momentum and energy for porous media.[11,12] These empirical parameters include the inertial coefficient C_f in the momentum equation and the effective thermal conductivity K_{eff} in the energy equation, which consists of stagnant K_o and dispersion thermal conductivity K_d.

Conventionally, the stagnant thermal conductivity is determined from porosity and thermal conductivity of solid and fluid phase. Moreover, some models consider extra terms of interfacial heat transfer to account for diffusion between two phases. An empirical model for its determination was presented by Vafai and Tien.[11]

This research work studies the effects of performance parameters on a packed bed in porous media where spherical-shaped concrete embedded with a copper tube was used as the storage medium. Thermal storage in concrete relies on sensible heat storage where the stored thermal energy is defined by the heat capacity of the concrete and the temperature difference between the charged and the discharged states.

2. Solid Storage Media

Energy can be stored in rocks or pebbles packed in insulated vessels. This type of storage is used very often for temperatures up to 100°C with solar air heaters. It is simple in design and relatively inexpensive. Typically, the characteristic size of the pieces of

Table 1. Solid media properties for sensible heat storage.

Medium	Density (kg/m³)	Specific heat (J/kgK)	Heat capacity $\rho c \times 10^{-6}$ (J/m³K)	Thermal conductivity (W/mK)	Thermal diffusivity $\alpha = k/\rho c\ 10^6$ (m²/s)
Aluminum	2707	896	2.4255	204 at 20°C	84.100
Aluminum oxide	3900	840	3.2760	–	–
Aluminum sulfate	2710	750	2.0325	–	–
Brick	1698	840	1.4263	0.69 at 29°C	0.484
Brick magnesia	3000	1130	3.3900	5.07	1.496
Concrete	2240	1130	2.5310	0.9–1.3	0.356–0.514
Cast iron	7900	837	6.6123	29.3	4.431
Pure iron	7897	452	3.5694	73.0 at 20°C	20.450
Calcium chloride	2510	670	1.6817	–	–
Copper	8954	383	3.4294	385 at 20°C	112.300
Earth (wet)	1700	2093	3.5581	2.51	0.705
Earth (dry)	1260	795	1.0017	0.25	0.250
Potassium chloride	1980	670	1.3266	–	–
Potassium sulfate	2660	920	2.4472	–	–
Sodium carbonate	2510	1090	2.7359	–	–
Stone, granite	2640	820	2.1648	1.73 to 3.98	0.799–1.840
Stone, limestone	2500	900	2.2500	1.26 to 1.33	0.560–0.591
Stone, marble	2600	800	2.0800	2.07 to 2.94	0.995–1.413
Stone, sandstone	2200	710	1.5620	1.83	1.172

rock used varies from 1 to 5 cm. An approximate rule of thumb for sizing is to use 300 to 500 kg of rock per square meter of collector area for space heating applications. Rock or pebble bed storage can also be used for higher temperatures up to 1000°C. At present, most thermal storage devices use sensible heat storage, and good technology is developed for the design of such systems. However, above 100°C, the storage tank must contain water as its vapor pressure, and the storage tank cost rises sharply for temperatures above this point.

Organic oils, molten salts, and liquid metals do not show the same pressure problems, but their use is limited because of their handling, containment, storage capacities, and cost. Among liquid materials, water appears to be the most convenient because it is not expensive and has a high specific heat. The difficulties and limitations on liquids can be avoided by using solid materials for storing thermal energy as sensible heat. But larger amounts of solids are needed than for using water, due to the fact that solids, in general, show a lower storing capacity than water. The cost of the storage media per unit energy stored is, however, still acceptable for rocks. Direct contact between the solid storage media and a heat transfer fluid is necessary to decrease the cost of heat exchange in a solid storage medium.

The use of rocks for thermal storage provides the following advantages:

1. Rocks are not toxic and non-flammable;
2. Rocks are inexpensive;
3. Rocks act both as heat transfer surface and storage medium;
4. The heat transfer between air and rock bed is good due to very large heat transfer area.

Refractory materials such as magnesium oxide, aluminum oxide, and silicone oxide are also suitable for high-temperature sensible

heat storage. Bricks made of magnesia have been used in many countries for many years for storing heat. They are available in the form of devices with electric heater elements embedded in bricks. The heat is stored at night (when electricity rates are low) by switching on the electric heaters, and it is supplied during the day for space heating purposes by allowing air to pass through the devices. The properties of solid media storage are listed in Table 1.

3. Methodology

3.1 Test and Equipment

The schematic diagram of the experimental setup is shown in Figure 1. For indoor experimentation, an air duct with an electric heater was used.[13] The packed bed storage system consists of an inlet plenum chamber, outlet plenum chamber, and packed spherical-shaped concrete embedded with 4 copper tubes of two

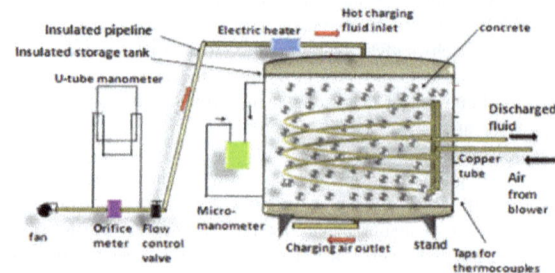

Fig. 1. A diagrammatic sketch of simultaneous charging and discharging packed bed energy storage system components.

passes and a radius 0.115 m. The copper tube (L type) was 0.00635 m standard size; 0.02223 m outside diameter, 0.01994 m inside diameter, 0.01143 m wall thickness, and 1.32 m length. The spherical-shaped concrete had a ratio 1:1.2:1.1 of cement, sand, and gravel.

The dimension of the heating section was $2 \times 0.5 \times 0.0254$ m, and the size of the duct was $3 \times 0.5 \times 0.0254$ m. A electric heater having a size of 2×0.5 m was fabricated by combining series and parallel loops of heating wire wound on an asbestos sheet. In order to decrease the heat losses, the backside of the heater was insulated with fiberglass. The heater was fixed on the top of the duct between entry and exit lengths. Electric supply to the heater was controlled by a variac.

A centrifugal blower was used to force hot air from an air duct (insulated) to a storage tank through a 0.051-m diameter orifice. The blower motor was 5 horsepower and 3800 rpm. Flow was varied by controlling the blower speed using the variable transformer, which could supply any voltage from 0 to the rated voltage of 120 V. The blower produced a flow of 0.047 m^3/s at a pressure of 13699.9 N/m^2 at standard conditions.

A 3-m long plastic pipe of 0.15 m diameter connected to a 0.53 m long barrel galvanized steel pipe was used between the blower and the electric heater. The length of the pipe was important because it gives a fully developed air velocity distribution inside the pipe to accurately measure the air flow rates with an orifice meter.

A cylindrical storage tank of diameter 0.70 m, height 1.07 m, made of 3.00-mm thick MS sheet was insulated with fiberglass to decrease heat loss. Silicone rubber was used for sealing the joint connections to avoid air leakage.

Holes were drilled at different cross-sections of the tank to insert the thermocouple wires. An air velocity meter was used for the measuring of air flow rate in the pipeline. This meter simultaneously measures and data logs the air parameters using a single probe with multiple sensors. The model measures velocity and temperature and calculates flow. It has a telescopic articulated probe. An orifice meter with a U-tube manometer was installed on the pipeline for pressure drop measurement. A control valve was provided in the pipeline for adjusting the flow rate of air. A micro-manometer was attached with the taps at the top and bottom of the bed for measuring pressure drop through the bed. Temperatures of air and solid at different points along different cross-sections in the bed were measured with thermocouples.

The temperatures of the flowing air through the packed bed were measured at intervals of 10 minutes at different locations. These temperatures were measured by thermocouples connected to three data loggers each with 8 channels. Each of the channels are independent of each other and can be independently enabled or disabled. Type J thermocouples were used for all tests performed. This instrument was capable of 0.1°C resolution with readings displayed in °C and can continuously record and export results to a remote computer. A centrifugal fan with a control valve was installed to provide air at varied flow rates through the copper tube in the storage tank.

3.2 Test Procedures

The experimentation involved studying the pressure drop, energy loss, air passage through the packed bed, and the fan energy used for optimization of the packed bed storage system. The spherical-shaped concrete of diameters 0.065 m, 0.08 m, and 0.11 m were casted, and its physical properties such as weight, density, and compressive strength were determined. The void fraction (ε) was calculated using the relationship between porosity and concrete bulk density (ρ_b), which is given by the following equation:

$$\varepsilon = 1 - \frac{\rho_b}{\rho_c} \qquad (1)$$

The bulk density of the concrete was determined from the volume of the storage section of the packed bed and the weight of the concrete filling the volume.

Testing was carried out with spherical-shaped concrete of diameters 0.11 m, 0.08 m, and 0.065 m, respectively. The concrete balls were dropped and arranged in the storage tank for proper void fraction.

Before the packing of storage materials, thermocouples were fixed in small grooves in material particles and placed at different points in different cross-sections of the bed along with thermocouples for measurement of air temperature.

Holes were drilled at different locations across the height of the storage tank where 22 thermocouples were inserted. Four thermocouples were inserted to measure the air temperature within the void of the packed bed at different height of the tank. Four thermocouples each were also inserted to measure the surface and internal temperature of the spherical-shaped concrete and another four each for the copper tube surface and internal air temperature at different heights. One thermocouple each was also inserted at the entry and exit of the copper tube and at entry and exit of the storage tank.

Three runs of air flow rates were conducted for spherical-shaped concrete of diameters 0.11 m, 0.08 m, and 0.065 m at the normal drop. The designed air flow rates were 0.0094 m^3/s, 0.013 m^3/s, and 0.019 m^3/s per square meter of total cross-sectional area of the storage tank. The corresponding superficial velocities were about 0.1 m/s, 0.15 m/s, and 0.20 m/s.

Temperature measurements were carried out for air, concrete surfaces, copper tube surfaces, concrete core, copper tube internal part, storage tank inlet and outlet, and copper tube inlet and outlet and recorded at four levels via a data logger connected with the computer. These four levels were located at different heights above the base, 117.5 cm, 235 cm, 352.5 cm, and 470 cm.

The measurements were taken automatically at an interval of 10 minutes for 12 hours. In order to test the storage capacity of the spherical-shaped concrete and the copper tube, the measurements were also taken during the night period when the simulated heat was no longer in supply to the packed bed. Upon analysis of all measuring equipment, the error calculated for these experiments was found to be ±5%.

The second phase of the experimentation involved studying the air resistance through the packed bed, the blower, and the copper tube. The pressure drop measurements were taken at varying air flow rates of 0.0094 m^3/s, 0.012 m^3/s, 0.014 m^3/s, 0.017 m^3/s, 0.019 m^3/s, 0.021 m^3/s, 0.024 m^3/s, 0.026 m^3/s, 0.028 m^3/s, and 0.031 m^3/s.

The pressure drops were taken for spherical-shaped concrete of diameters 0.11 m, 0.08 m, and 0.065 m. The following measurements were taken:

i. Pressure drops of air across the pipe (barrel) leading to storage tank inlet;
ii. Pressure drops of air across the pipe entry of the copper tube inlet;
iii. Pressure drops of air across the packed bed.

From this experimentation, blower characteristic performance and the power used for the operation were established. The volume flow rate handled by the blower expressed the inlet conditions. The blower total pressure is expressed as follows:

$$P_T = P_{TO} - P_{TI} \tag{2}$$

where P_T is blower total pressure, P_{TO} is outlet blower total pressure, and P_{TI} is inlet blower total pressure.

The blower total pressure calculated from Equation 2 represents the pressure drop across the blower. The blower efficiency can also be expressed as follows:

$$\eta_{blower} = \frac{\text{power output}}{\text{power input}} \tag{3}$$

3.3 Experimental Uncertainties

It is critical to assess the accuracy of the measurement process at the start of the study. Inaccurate measurements may lead to false signals on control charts. Due to significant error in the measurement process, a capable process may be confused with an incapable process. Several factors affect the reliability of measurements, including

i. differences in measurement procedures;
ii. differences among operators;
iii. instrument repeatability and reproducibility;
iv. instrument calibration and resolution.

Repeatability of a measuring instrument refers to how well the instrument is repeatedly able to measure the same characteristic under the same conditions. Reproducibility is the variation due to different operators using the same measuring instrument at different time periods.

In this study, the linear heat source method was used to measure simultaneously the coefficients of thermal conductivity and thermal diffusivity of the 21 concrete samples. The storage capacity was determined according to the relationship with conductivity and diffusivity. The 21 concrete samples were measured 3 times (trials) each. The study was conducted so that each operator (one at a time) was presented, one of the samples (selected randomly from the 21 samples), and a KD2-thermal properties analyzer was used for the measurement of thermal conductivity, diffusivity, and resistivity of the concrete.[14] The measurement process was repeated for the other 20 samples. The same 21 samples were presented (in random order) for the second trial, then again for the third trial. This same study procedure was used for each operator.

The material and equipment measurement error test was analyzed using Equations 5 to 8.

$$\delta_{Tins} = \delta_P + \delta_m \tag{4}$$

Table 2. K_1 and K_2 values.[15]

No. of trials	2	3	4	5
K_1	4.56	3.05	2.50	2.21
K_2	3.65	2.70	2.30	2.08

where δ_{Tins} is the total variance in instruments, δ_P is variance in process, and δ_m is the variance in measurements.

Repeatability is the variation in equipment (EV).

Reproducibility is the appraisal variation (AV).

$$EV = K_1 \bar{R} \tag{5}$$

$$AV = K_2 \bar{X}_D \tag{6}$$

where K_1 and K_2 are constants as shown in Table 2, $\bar{R}$ is the overall average of ranges and $\bar{X}_D$ is the difference between the largest and smallest average.

$$R \& R = \sqrt{EV^2 + AV^2} \tag{7}$$

$$\%R \& R = \frac{R \& R}{\text{Tolerance range}} \times 100 \tag{8}$$

The statistic known as R&R was then calculated, representing the amount of variation associated with repeatability and reproducibility. The result of repeatability was 8.56%; reproducibility, 7.51%; and % R&R was 10%, which was satisfactory according to guidelines for acceptable results according to American Society of Quality.[15]

4. Results and Discussion

Pressure drop in packed beds is influenced by certain parameters such as the particle Reynolds number; mean void fraction, particle geometry, and particle size distribution. The Reynolds number defines the through regime, whether it is laminar, transitional, or turbulent flow. Dimensionless pressure drop is inversely proportional to the pressure drop in streamline flow[16] and becomes independent of the Reynolds number approaching 6×10^4. The dimensionless pressure drop is very much dependent on the particle Reynolds number as it changes gradually with increasing Reynolds number, indicating a smooth transition from laminar to turbulent flow. The second value is the mean void fraction. The dimensionless pressure drop is highly sensitive with the changes in the mean void fraction. This influence is described empirically by using dimensional analysis[17] and theoretically by employing hydraulic radius concept.[18]

The third parameter is the particle geometry. Proper choice of packing form will result in reduced pressure drop in packed beds and thus improve the overall heat transfer performance as shown in Figure 1. The fourth parameter is the particle size distribution. It can be treated like different particle geometries by introducing an equal particle diameter. The variables on which the pressure drop seems to depend are summarized by Leva[19] as variables of the fluid: rate of flow, fluid density, and fluid viscosity and variables of the storage system: rock bed diameter, packing diameter, voids of fraction, particle shape, particle surface roughness, and possibly bed configuration. The pressure drops occurring across the porous medium are attributed to several factors such as viscous drag from bounding walls and inertia force.[20, 19]

In this study, to find the pressure drop of the packed bed energy storage system for spherical-shaped concrete of diameters 0.11 m, 0.08 m, and 0.0 65m, 10 different experiment were run at an air flow rate (m³/s) of 0.0094, 0.012, 0.014, 0.017, 0.019, 0.021, 0.024, 0.026, 0.028, 0.031, 0.033, 0.035, 0.038, 0.040, 0.042, and 0.045. It is known that pressure drop through a packed bed is proportional to the fluid velocity at low flow rates and to the square of the velocity at high rates. So the pressure required to force air through the bed is a function of airflow.

The pressure drop of air across the pipe, copper tube, packed bed, barrel, and blower, respectively, at an air flow rate (m³/s) of 0.0094, 0.012, 0.014, 0.017, 0.019, 0.021, 0.024, 0.026, 0.028, 0.031, 0.033, 0.035, 0.038, 0.040, 0.042, and 0.045 for spherical-shaped concrete of size 0.065 m, 0.08 m, 0.11 min diameters are shown in Figures 2 to 6. The pressure drop in all these cases increases with air flow rate with spherical-shaped concrete of diameter 0.065 m possessed the highest pressure drop of 196.1 N/m² across the packed bed, and this may be attributed to its low porosity, while spherical-shaped concrete of size 0.11 m diameter had the lowest pressure drop of 93.17 N/m² due to higher void fraction as shown in Figure 4. Pressure drop across the copper tube is shown in Figure 3. Spherical-shaped concrete of diameter 0.065 m also possessed the highest pressure drop of 68 N/m², while spherical-shaped concrete of size 0.11 m diameter had the lowest pressure drop of 65.39 N/m².

The pressure drop decreases down the length of the bed, and thus the volumetric flow rate increases. This is line with Ergun,[18] where he used an extensive set of experimental data covering a range of particle size and shapes and presented a general equation

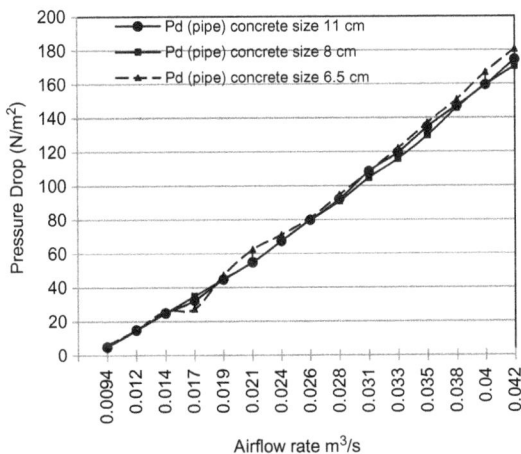

Fig. 3. Pressure drop across the copper tube for spherical-shaped concrete of size 0.065 m, 0.08 m, and 0.11 m diameter.

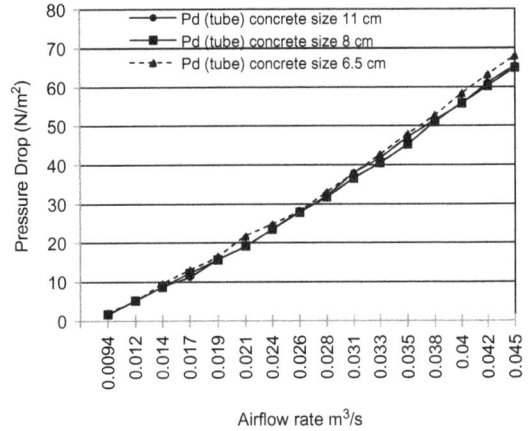

Fig. 4. Pressure drop across the packed bed for spherical-shaped concrete of size 0.065 m, 0.08 m, and 0.11 m diameter.

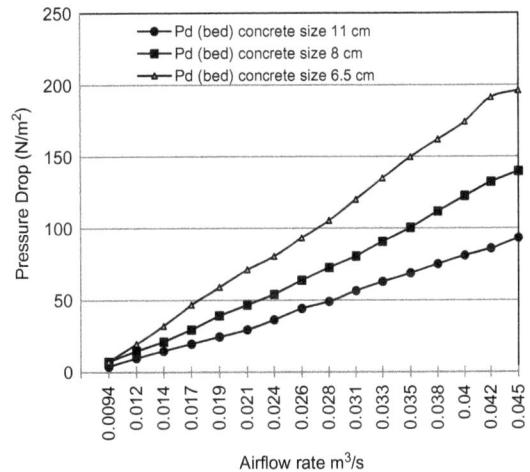

Fig. 2. Pressure drop across the pipe for spherical-shaped concrete of size 0.065 m, 0.08 m, and 0.11 m diameter.

to calculate the pressure drop across a packed bed for all flow conditions (laminar and turbulent). The equation is commonly referred to as the Ergun equation for flow through a randomly packed bed of spheres. This equation combines both the laminar and turbulent components of the pressure loss across a packed bed. In laminar flow conditions, the first components of the equation dominate with the Ergun equation essentially reducing to the Carman-Koreny equation, although with a slight variation in the constants used due to variations in the experimental data with which the correlations was developed. In the laminar region, the pressure drop through the packed bed is independent of fluid density and has a linear relationship with superficial velocity. Under turbulent conditions, the second part of the Ergun equation dominates. Here the pressure drop increases with the square of the

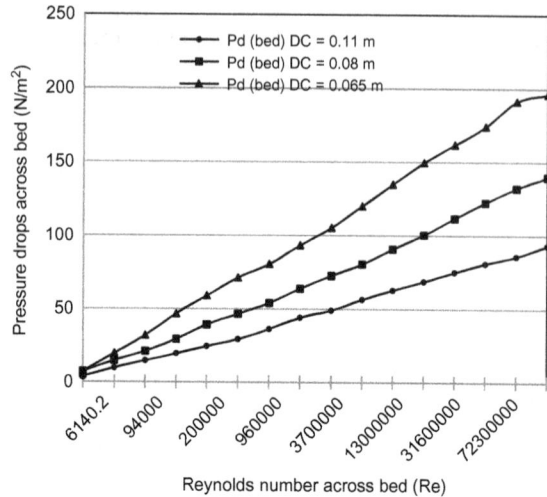

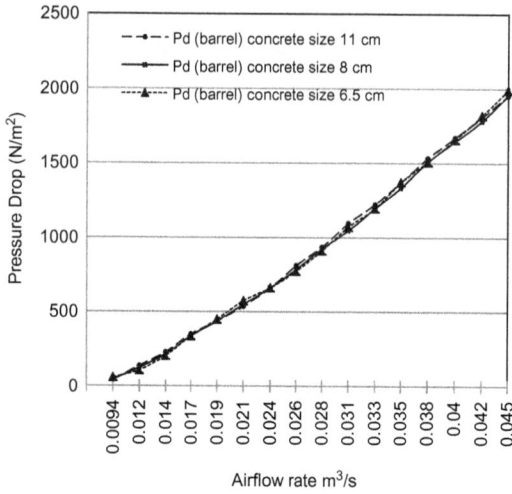

Fig. 5. Pressure drop across the barrel for spherical-shaped concrete of size 0.065 m, 0.08 m, and 0.11 m diameter.

Fig. 7. Reynolds number versus pressure drop across the packed bed for spherical-shaped concrete of size 0.065, 0.08, and 0.11 m diameter.

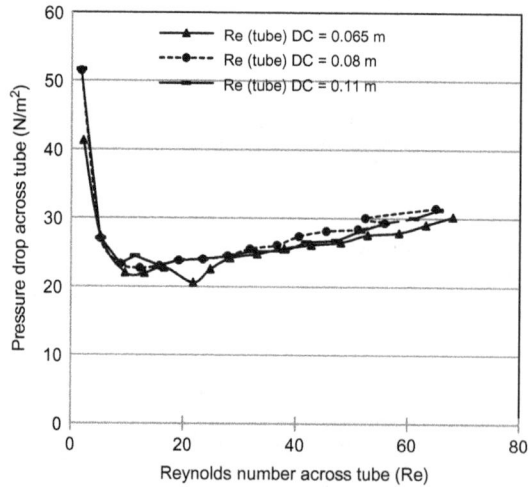

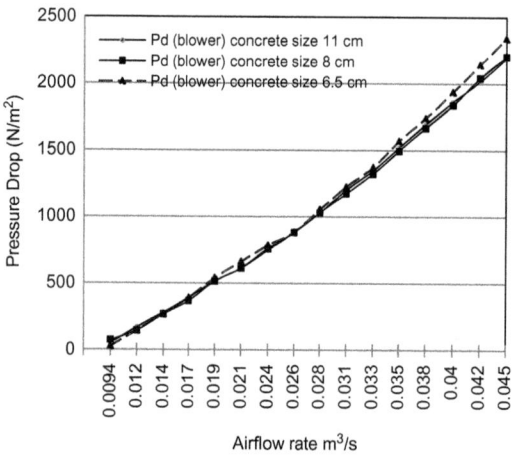

Fig. 6. Pressure drop across the blower for spherical-shaped concrete of size 0.065 m, 0.08 m, and 0.11 m diameter.

Fig. 8. Reynolds number versus pressure drop across the copper tube for spherical-shaped concrete of size 0.065, 0.08, and 0.11 m diameter.

superficial velocity and has a linear dependence on the density of the fluid passing through the bed.

Figures 7 and 8 illustrate the variability in pressure drop that might be expected with packed beds and copper tubes against the Reynolds number. The packed bed Reynolds number describes the ratio of inertial to viscous forces for fluid flow through a packed bed, which can be defined as

$$\text{Re} = \frac{\rho_a V_{average} D_c}{\mu_a} \qquad (9)$$

where D_c is the concrete spherical diameter; μ_a is the air viscosity, and ρ_a is the air density.

The Reynolds number is a function of both the fluid properties and the concrete size. The Reynolds number across the packed bed move from laminar to turbulent condition as air flow rate and pressure drop increases as shown in Figure 7 while that of the copper tube was of laminar flow as shown in Figure 8.

The flow behavior in packed beds for a wide range of friction factors in both laminar and turbulent regions as shown in Figure 9 is very important. The friction factor is expressed as the pressure drop due to friction when the packed bed is properly described. It relates the shear stress to the average kinetic energy of the fluid.

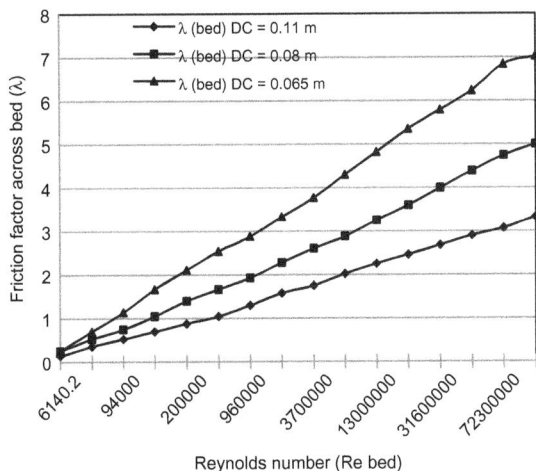

Fig. 9. Reynolds number versus friction factor across the packed bed for spherical-shaped concrete of size 0.065, 0.08, and 0.11 m diameter.

In the laminar range, the friction factor is a function of the Reynolds number. It is independent of the surface roughness of the bed. The transition from laminar flow to turbulent flow, the friction factor, increases from laminar to turbulent flow across the bed for spherical-shaped concrete bed of size 0.08 m and 0.065 m diameter. It was laminar up to friction factor of 1.58 before entering the turbulent range as shown in Figure 9. The turbulence was more pronounced with 0.065 m diameter concrete size due to low porosity. The flow was approximately laminar for concrete size 0.11 diameter due to high fraction of voids in the packed bed. The flow was laminar across the copper tube as shown in Figure 10.

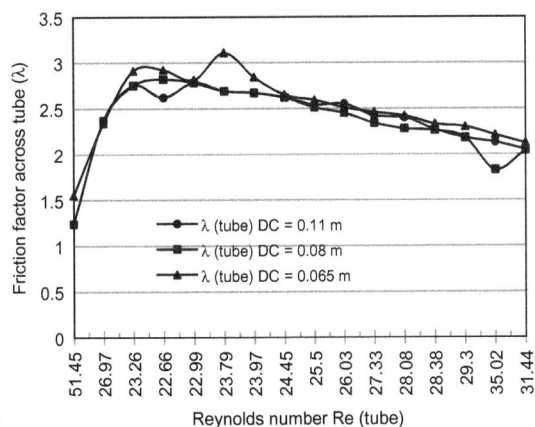

Fig. 10. Reynolds number versus friction factor across the copper tube for spherical-shaped concrete of size 0.065, 0.08, and 0.11 m diameter.

The friction factor is a complex function of surface roughness, fraction of voids, concrete size, and the bed configuration. Rose[17] has shown the edge effect to be negligible if the ratio of container diameter to particle diameter is greater than 20. Also, if the ratio of the bed diameter to particle diameter is six or more, the wall effect in pressure drop is neglected.

Overall, the test was a success, but there were many areas that contributed to error in the results. Most important was maintaining a full seal throughout the device. Since the apparatus used for the experimentation has been previously used, the seals were not as strong and had the potential to leak or rupture due to degradation over time. During the initial testing/calibration, the seal around the plenum ruptured due to a large internal pressure that rapidly accumulated from too much inlet pressure. Once the integrity of the seal was compromised, the entire apparatus was resealed to make sure no other seals burst during actual testing. The seals were properly checked to make sure accuracy in the results.

A second error was pipe leakage from the inlet. This discovery was made when a large variance was observed between the experimental and theoretical results. After further inspection, it was discovered that leakage was the cause. Instead of submerging the apparatus in water or using a soapy solution to observe bubble generation, data variance was used as a check. Overall leakage was reduced by using Teflon tape on the connection threading and ensured that the fittings were tight. Even with these precautions, there were still minor leaks that could not be avoided.

A third significant source of error was found in the flow meter. Experiments conducted by other groups determined that surface roughness does, in fact, alter the pressure, but not by a significant amount at the macro level.[8]

5. Conclusions

The effect of pressure drop and friction factor as performance parameters on a concrete bed energy storage system was analyzed, and from this analysis it was discovered that the pressure drop in all the cases increases with airflow rate with spherical-shaped concrete of diameter 0.065 m possessed the highest pressure drop, and this may be attributed to its low porosity, while spherical-shaped concrete of size 0.11 m diameter had the lowest pressure drop. In the laminar range, the friction factor is a function of the Reynolds number. It is independent of the surface roughness of the bed. The transition from laminar flow to turbulent flow, the friction factor, increases from laminar to turbulent flow across the bed while it was laminar across the copper tube. It was concluded that the pressure drop characteristics for the fluid flow through the concrete bed are controlled by many variables. Some of these variables are related to the thermo-physical properties of the fluid passing through the packed bed, such as density, kinematic viscosity, mass velocity, and gravity of fluid. Laminar, transient, and turbulent flow also affect the value of the friction factor. Some variables are related to the geometrical characteristics of the packing materials such as particle diameter, shape factor, surface roughness, surface area, fractional effective void, and orientation. Other variables are related to the configuration of the unit such as packing bed dimensions such as diameter and length.

References

1. Bear, J. *Dynamics of Fluids in Porous Media*. Elsevier: New York, 1972.
2. Kaviany, M. *Principles of Heat Transfer in Porous Media*. Springer-Verlag: New York, 1991.
3. Reynolds, O. *Papers on Mechanical and Physical Subjects*. Cambridge University Press: Cambridge, 1900.
4. Beavers, G. S.; Sparrow, E. M. Non-Darcy Flow through Fibrous Porous Media. *ASME J. Appl. Mech.* **1969**, *36*, 711–714.
5. Lage, J. L; Antohe, B. V.; Nield, D.A. Two Types of Nonlinear Pressure-drop versus Flow-rate Relation Observed for Saturated Porous Media. *J. Fluids Eng.* **1997**, *119*, 700–706.
6. Brinkman, H. C. A Calculation of the Viscous Force Extended by a Flowing Fluid on Dense Swarm Particles. *Appl. Sci. Res.* **1947**, *1*, 27–34.
7. Chapman, A. M.; Higdon, J. J. L. Oscillatory Stokes Flow in Periodic Porous Media. *J. Phys. Fluids* **1992**, *4*, 2099–2116.
8. Kocecioglu, I.; Jiang, Y. Flow Through Porous Media of Packed Spheres Saturated With Water. *Trans. ASME* **1994**, *116*, 164–170.
9. Hunt, M. L.; Tien, C. L. Non-Darcian Convection in Cylindrical Packed Beds. *ASME J. Heat Trans.* **1998**, *110*, 378–384.
10. Vafai, K.; Alkire, R. L.; Tien, C.L. An Experimental Investigation of Heat Transfer in Variable Porosity Media. *ASME J. Heat Trans.* **1985**, *107*, 642–647.
11. Vafai, K.; Tien, C. L. Boundary and Inertia Effects on Flow and Heat Transfer in Porous Media. *Int. J. Heat Mass Trans.* **1981**, *24*, 195–203.
12. Calmidi, V. V.; Mahajan, R. L. The effective thermal conductivity of high porosity fibrous metal foams. *ASME J. Heat Trans.* **1999**, *121*, 466–471.
13. Singh, R.; Saini, R. P.; Saini, J. S. Nusselt Number and Friction Factor Correlations for Packed Bed Solar Energy Storage System Having Large Sized Elements of Different Shapes. *Solar Energy* **2005**, *80*, 760–771.
14. Bristow, K.L.; White, R.D.; Kluitenberg, G.J. Comparison of Single and Dual Probes for Measuring Soil Thermal Properties with Transient Heating. *Australian J. Soil Res.* **1994**, *32*, 447–464.
15. ANSI/ISO/ASQC Q9000 Series, *Quality Management and Quality Assurance Standards*; ASQC Quality Press: Milwaukee, 1994.
16. Carman, P.C. *Fluid Flow through Granular Beds*. Institution of Chemical Engineering Transactions: London, pp 150–166, 1937.
17. Rose, H.E. *Fluid Flow through Beds of Granular Materials*. Institute of Physics Conference, Leamington, Spain. October 25–28, 1950. Edward Arnold and Company: London, pp 136–163, 1951.
18. Ergun, S. Fluid Flow through Packed Columns. *J. Chem. Eng.* **1952**, *48*, 89–94.
19. Leva, M. A Report Covering Bureau of Mines Research during the Calendar Year 1947 on the Packed Tubes. *Industrial and Chemical Engineering Fundamental. ACS Publication* **1947**, *39*, 857–859.
20. Adeyanju, A. A. Effect of Fluid Flow on Pressure Drop in a Porous Medium of a Packed Bed. *J. Eng. Appl. Sci.* **2009**, *4*, 83–86.

3

Integration Potentials of Insular Energy Systems to Smart Energy Regions

PARIS A. FOKAIDES[1]* , ANGELIKI KYLILI[1], ANDRI PYRGOU[1], and CHRISTOPHER J. KORONEOS[2]

School of Engineering and Applied Sciences, Frederick University, Nicosia, Cyprus
[2]*Laboratory of Heat Transfer and Environmental Engineering, Aristotle University of Thessaloniki, Thessaloniki, Greece*

Abstract: Smart energy regions are defined as regions that offer maximal quality of living to their inhabitants with a minimal consumption of energy by intelligently joining of infrastructure (energy, mobility, transport, communication, etc.) on different hierarchical levels (building, district, city). The development of insular energy systems into smart energy regions, due to their special character, is presented with some challenges. The focus of this article is on presenting the potential of insular energy systems transforming into smart energy regions. Insular energy systems are defined based on data retrieved from Energy Information Administration (EIA) and classified according to their size and the nature of their isolation. In terms of this study, two novel indexes are introduced: the necessity index, which quantifies the need, and the ability index, which represents the capability of an insular energy system to develop into a smart one. These indexes are defined for those insular systems that are considered potentially upgradable. The analysis revealed that the main prerequisites to achieve the development of insular energy systems into smart ones are the reduction of GHG emissions, the introduction of political obligation toward promoting environmentally friendly policies, and the increase of RES utilization for energy production.

Keywords: Smart energy regions, necessity index, ability index, energy insular systems

1. Introduction

The emerging environmental concerns arising from the increase of carbon dioxide (CO_2) and greenhouse gases (GHG) emissions, the uncertainty of the international and regional energy supply, and the ever-increasing cost of energy have induced the necessity of promoting measures toward low-carbon societies. Smart energy regions are expected to play a crucial role in achieving the decarbonization of Europe's economy, in line with the EU 2020 energy and climate goals.

IEA's World Energy Outlook (WEO)[1] presents an assessment of the prospects for global energy markets in the period to 2035 and draws out the implications for energy security, environmental protection, and economic development. The objective is to provide policymakers, industry, and the general public in countries all over the world with the data, analysis, and insights needed to make judgments about our energy future, as a basis for sound energy decision-making. Demographic factors will continue to drive changes in the energy mix. Energy technologies that are already in use or that are approaching commercialization are expected to achieve ongoing cost reductions as a result of increased learning and deployment. Although there are exceptions that create some basis for optimism, recent progress in deploying clean energy technologies has not matched policy expectations, and, in many cases, their future uptake hinges on dedicated policy support or subsidies.

The European Strategic Energy Technology Plan (SET Plan), the strategic plan to accelerate the development and deployment of cost-effective low-carbon technologies in Europe, defines "smart cities" as systems of people interacting with and using flows of energy, materials, and services and financing to catalyze sustainable economic development, resilience, and high quality of life.[2] According to the SET Plan, these flows and interactions become smart through making strategic use of information and communication infrastructure and services in a process of transparent urban planning and management that is responsive to the social and economic needs of society.[3] However, a broad set of challenges, including the lack of skills related to implementing low-carbon technologies in the built environment, misunderstanding of capital and operational costs, where technologies can be implemented, the impact on quality of life and policy, and planning for the future have yet to be overcome in order to enable technologies to be widely applicable and transferable

*Address correspondence to: Paris A. Fokaides, School of Engineering and Applied Sciences, Frederick University, 7 Y. Friderickou Str., 1036 Nicosia, Cyprus. Email: p.fokaides@frederick.ac.cy

within and between regions within the context of a low-carbon built environment.[4]

The distinctiveness of insular energy systems, defined by the country's inability to interconnect with other energy generators and/or consumers through a wider transmission grid outside its national borders, is what drives their need to develop into smart ones.[5] This interconnection inability could be attributed to the smallness of the system, its remoteness, or political aspects. The great dependency of insular systems on imported conventional fossil fuels typically makes their power generation costly and insecure. Furthermore, the potential of these energy systems for renewable energy generation extends only up to the point allowed by the locally available environmental conditions.

This study aims to introduce quantitative parameters for the definition and classification of energy insular systems as well as their capability of being integrated to energy smart ones. In terms of this study, the main features of energy insular systems, as well as the existing gap between the current status and the vision of energy intelligence, will be analyzed and discussed. This work has been introduced by providing definitions for smart energy regions and insular energy systems. The following section will present the necessary provisions for the upgrading of an energy system into an intelligent one and explain the special nature of an insular energy system, while Section 3 will describe the methodology for the classification of energy insular systems. The development of two novel indices, the necessity and ability indices that are introduced in this article, will also be presented in Section 3. The results of the analysis and the relevant discussion are given in Section 4, and the last section provides the final concluding remarks.

2. Theoretical Background

2.1 Energy Smart Regions Development Prerequisites

Global environmental problems such as global warming caused by human activities are expected to be alleviated once the current society of heavy energy consumption transforms into a recycling-oriented society with systematic energy- and resource-saving measures.[6] The prerequisites for the development of an intelligent energy region are the following:

- The existence or the potential of developing the necessary energy distribution and storage infrastructure in terms of smart grids
- The existence of advanced renewable energy sources (RES) applications as well as the required renewable energy potential for the diversification of the energy mix
- The existence of indigenous energy sources and their contribution toward reducing the cost of energy
- The existence of political obligations such as those imposed under European directives or other policy initiatives such as the Kyoto Protocol

2.1.1 Gross Domestic Product (GDP) and Economic Growth Rate

The development of smart energy regions requires the existence or the potential to develop the required infrastructure. This potential can be defined through financial vectors such as the country's gross domestic product (GDP) and its economic growth rate.

Energy consumption and GDP exert a positive impact on each other.[7] Countries that increase investment in energy infrastructure and promote the development of energy efficiency measures avoid as well the negative effect on the country's economic growth by reducing energy dependency. Kjaer elaborates even further, stating that the development of infrastructure solutions not only reduces the transfer of a country's wealth to oil exporters but also increases the input money in its own economy and, furthermore, gains control of its own energy future.[8]

2.1.2 Renewable Energy Technologies (RET)

Renewable energy technologies (RET) consist of a prerequisite for the establishment of smart energy regions. RET exploit the potential of solar, wind, geothermal, and tidal energy, as well as the biomass and the hydropower potential and thus contribute to the diversification of the energy mix. Czisch[9] has suggested a "supergrid," a high-voltage direct-current grid across the EU that would be able transmit power from RES and deal with the variability of RET, while Elliott[10] confirms that a country's energy security requires a wide mix of renewable sources, scales, and locations. Moreover, Roiniotia et al.[11] explain that the development of a more efficient, flexible, and environmentally acceptable energy system will arise through its energy diversification and efficiency, the co-operation of the energy market participants, and the orientation of the economy toward RES investments. The available solar applications in the built sector include solar photovoltaics (PV) and concentrating solar power (CSP) for electricity generation, as well as solar thermal collectors for heat production. Kylili and Fokaides defined the potentials of achieving zero energy buildings using building integrated photovoltaics.[12] Wind turbines are used for electricity generation, while biomass can be exploited for all three energy sectors, namely electricity, heat, and transportation. The available geothermal applications include high enthalpy systems for electricity generation and mild enthalpy systems for heat production. Hydropower and tidal power also generate electricity by employing hydro turbine and wave generators, respectively.

2.1.3 Indigenous and Imported Energy Sources

The potential of an energy system to be upgraded into a smart one greatly depends on the availability of indigenous energy sources. The indigenous resources define the nature of the energy mix and the energy technologies, as well as the potential of the system's further development. Michalena and Angeon[13] reported that before the adoption of RETs' promotion strategy, the special characteristics of territories, such as the geo-morphologic particularities, the energy, socioeconomic, and demographic characteristics, the local natural resources, and the level of green awareness should be studied. Also, Kaldellis et al.[14] stated that the introduction of a variety of RET to the energy mix of an insular system by employing the locally available resources can provide a cheaper alternative to the high fossil-fuel electricity generation and also provide additional security to the power system. Other works, including Xydis et al.[15] and Koroneos[16] also encourage the utilization of indigenous energy sources to satisfy the local energy demands toward the development of an intelligent community. Additionally, the low cost of energy provided by the indigenous energy sources gives the possibility of investing in other development areas of intelligent energy systems such as in infrastructures for smart grids as well as in the promotion of renewable energy technologies.

2.1.4 Motivation Imposed by Political Obligations

Another significant parameter that defines the potential of achieving smart energy systems is the political or other obligations, which commit the countries to the promotion of relevant policies. Bojnec and Papler[17] concluded that management strategies and policies are improving the country's economic efficiency performance and, consequently, contributing to energy saving sustainable economic development. Examples include the Directive 2009/28/EC on the promotion of the use of energy from renewable sources and the Kyoto Protocol. The Kyoto Protocol is considered the first attempt of humanity to mitigate climate change, by committing countries internationally to achieve at least an 80% reduction of their GHG emissions from their 1990 baseline levels by 2050. Encouraged by the Kyoto Protocol, the EU has set 2020 as a milestone year, targeting for the 20% contribution of the renewable energy sources (RES) to its overall energy consumption and a 10% contribution of RES to the transport sector, under the Directive 2009/28/EC. The Member States were obligated to implement this directive into their legislations and develop their National Action Plans (NAP) to fulfil these targets.

In Europe, the motivation imposed by political obligations is driven mainly by the climate and energy package, a set of binding legislation that aims to ensure the EU meets its ambitious climate and energy targets for 2020.[18] These targets, known as the "20-20-20" targets, set three key objectives for 2020:

- A 20% reduction in EU greenhouse gas emissions from 1990 levels;
- Raising the share of EU energy consumption produced from renewable resources to 20%;
- A 20% improvement in the EU's energy efficiency.

The climate and energy package comprises four pieces of complementary legislation that are intended to deliver on the 20-20-20 targets. Apart from the promotion of the RES target, analyzed in the previous paragraph, the other three targets are the following:

- Reform of the EU Emissions Trading System (EU ETS)
- Set national targets for non-EU ETS emissions
- Promote carbon capture and storage

The EU ETS is the key tool for cutting industrial greenhouse gas emissions most cost-effectively. The climate and energy package includes a comprehensive revision and strengthening of the legislation that underpins the EU ETS. Also under the so-called effort sharing decision, Member States have taken on binding annual targets for reducing their greenhouse gas emissions from the sectors not covered by the EU ETS, such as housing, agriculture, waste, and transport. The last element of the climate and energy package is a directive creating a legal framework for the environmentally safe use of carbon capture and storage technologies. Carbon capture and storage involves capturing the carbon dioxide emitted by industrial processes and storing it in underground geological formations, where it does not contribute to global warming.

2.2 Energy Mix in Insular Energy Systems

The current energy mix of most insular energy systems is dominated by diesel and heavy fuel oil (HFO).[19] The imports of energy fossil fuels to cover the demands of the insular system create a potential instability in the economy and the security of the communities.[20] The absence of indigenous energy resources, the limited infrastructure of energy delivery, the lack of storage, and the flexibility of the power generators to meet seasonal needs promote further the domination of fossil fuels. The high cost of fuel transportation and the risk of spills undermine their utilization.[21]

Insular energy systems typically have only a few independent power producers, and there is not a variety of power generation technologies. The reduction of GHG emissions is one of the greatest challenges for insular energy systems as most of their electricity production is based on fossil fuels.[22] Typically, there are not adequate support schemes and incentives in insular systems to promote the transition to alternative low-emission energy sources, and their current state makes it difficult and costly to comply with international or European legislations, as also expressed in Sanseverino et al.[23]

Existing external networks are based on a centralized supply model similar to what dominates today's national and regional energy systems, the nature of which already requires significant transmission capacities due to the long distance between centers of supply and demand and significant built-in redundancy in capacity.[24] Establishing flexible demand response to real-time price signals for all end-users is a central component of local energy grids. Local smart energy grids may include all the elements of a distributed micro-grid, such as the integration of different renewable energy technologies, the integration of micro heat and power cogeneration units (CHP) and heat pumps in distributed generation, and the development of real-time markets that allows for both households and distributed operators to provide balancing and ancillary services.[25–27]

Archipelagos represent a special case of energy insular systems. The islands' interconnection, where possible, may gradually replace the local autonomous power stations' operation and, at the same time, allow the absorption of large amounts of renewable energy without causing the instability effects noticed in autonomous grids. On the other hand, such an electricity production strategy has to face the significant technological problems related to the undersea electricity transportation, the rather high first installation cost (approx. €3 million per km of transportation grid), and the strong opposition of local societies claiming important environmental impacts.[28]

The energy transition of insular systems to energy smart ones can be achieved through various measures and policies such as the promotion of energy efficiency measures, the establishment of smart grids, the utilization of RET, and the installation of large storage systems. All the applications should preliminarily consider the local conditions of the energy system and prior consideration, and a feasibility study of the cost and investigation of the latest technologies should be conducted, as Kaldellis et al.[29] have done for the case of Agathonisi in Greece. Exploiting RES, as indigenous energy sources, as well as mixing RET and/or hybrid RET models, can play an important role in reducing the level of energy imports of insular energy systems with positive impacts for the balance of trade and security of supply.[30,31]

3. Classification of Energy Insular Systems

In this study, the main assumption for defining an insular energy system was zero electricity imports and exports from adjacent countries. An energy-isolated system can easily be identified in the case of an island as the surrounding sea is a barrier in the power interchange. However, other mainland nations may be isolated due to political issues or lack of energy infrastructure and power grids.

In terms of this study, insular systems were defined based on data retrieved from the Energy Information Administration.[32] It should be noted that only oil- producing countries and island nations that were defined as energy insular systems were included in the analysis. Islands and non-self-governing territories whose legislation is regulated by a mainland central government were excluded. Energy insular systems were divided into three categories according to the following parameters:

- Installed power capacity (up to 100 MW and from 100 MW to 15 GW)
- Location (island or mainland nation)

Based on the above criteria, three categories of insular energy systems were defined (Table 1).

The first category includes islands having installed generation capacity of up to 100 MW. The countries in this category have limited energy demand and, in many cases, large distance from the mainland. The small energy requirements aggrandize the impact of establishing renewable energy power stations as the small economy of these regions is vulnerable to large-scale investments. Only a few of these countries utilize renewable energy sources, mainly hydroelectric power stations, and these contribute up to 5% of the energy mix. This category includes a total of 17 countries: four in Africa, six in the United States, one in Asia, and six in Oceania.

The second category includes islands with installed electricity capacity of over 100 MW and up to 15 GW, which are shown in Figure 1. It is noted that most of the islands in this category exploit, to a small or large extent, renewable energy technologies due to their larger consumption demand and their higher GDPs. Australia and Japan could also have been included in this category; however, the grid capacity of these two countries exceeds significantly the generation potential of the rest of the included island nations.

Table 1. Classification criteria of insular energy systems.

Category	Installed Capacity	Location
A	Up to 100 MW	Island nations
B	From 100 MW up to 15 GW	Island nations
C	No limit	Mainland nations with no grid interconnection

The third category of insular energy systems includes the regions that are situated in the mainland but are isolated due to lack of electric grid connection. Most of the countries of this category are situated in Africa, and their GDP is lower than $5,000 (Figure 2). The political situation in these countries does not allow investing in electricity infrastructure as health and military issues usually are more alarming and urgent to resolve. However, several of these countries successfully managed to integrate renewable energy sources into their power grid system, mainly hydroelectric energy. There are cases where the isolation results from political issues (e.g., South and North Korea). There are also some mainland nations that do not export or import electricity to their neighbors and are as well major petroleum-exporting countries (Oman, Qatar, Kuwait, UAE, Saudi Arabia, Equatorial Guinea). It is noticeable that this adequacy of fossil fuels has not alarmed these countries into turning to renewable energy resources.

4. Evaluation Method of the Integration Potentials of Insular Energy Systems to Smart Energy Regions

According to the classification of energy insular systems introduced in this study, the countries classified in groups 1 and 3 face more difficulties in promoting measures toward the development of energy smart infrastructures, mainly due to their limited energy system size and their low GDPs. Countries in group 2 with GDPs higher than $15,000 as well as the rich oil exporters of group 3 fulfill the necessary conditions, due to the larger size of their energy systems as well as their higher GDPs.

The required prerequisites for a region to become energy smart are the availability of the necessary funds and infrastructures that allows the investment in opportune steps that enable this transition. These requirements may be better fulfilled in countries with relatively high GDP. The size of the energy system is also another parameter that affects the ability of a country to develop energy smart infrastructures. The presence of political obligations as well as the potential of RES and indigenous energy source utilization also contribute to integrating insular systems into a system of energy smart techniques.

In terms of this study, two indexes for the characterization of the energy insular systems were developed, one that could quantitatively indicate the necessity in developing a smart energy region (necessity index) and the other that could quantitatively reveal the ability of a country to become smart (ability index). Each index resulted as the weighted sum of several parameters affecting the index under investigation. The value of the parameters was normalized based on a unitary linear approach, by means of the following expression:

$$I_i = \frac{P - MinP}{(MaxP - MinP)} \quad (1)$$

where P is the parameter under investigation, min and max represent the minimum and maximum value of the parameter in the examined sample, and I is the resulting index value.

Furthermore, in order to take into consideration the importance of each of the above parameters, a weighting system was

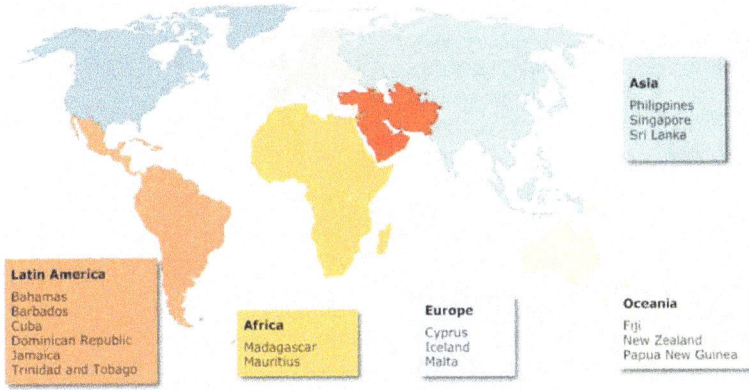

Fig. 1. Island nations with insular energy systems (Category B).

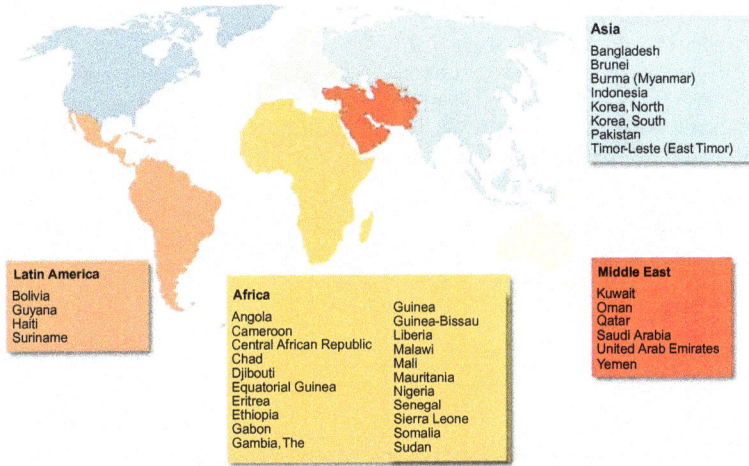

Fig. 2. Mainland nations with insular energy systems (Category C).

introduced. These weighting factors help establish the importance of the parameters in determining the overall rating.

$$NI_i = \frac{1}{n} \sum_1^{i=n} W_i NP_i \; ; \qquad AI_i = \frac{1}{n} \sum_1^{i=n} W_i AP_i \qquad (2)$$

The parameters used to define the necessity index and the ability index were based on the literature analysis findings regarding the prerequisites of smart energy regions conducted in Section 2.1.

4.1 Necessity Index

The necessity index shows how necessary is for a country to become energetically smart in order to reduce its dependence on energy imports, reduce the cost of energy, mitigate GHG emissions, and have financial and environmental sustainability. The quantifiable parameters of this index are

1. the specific electricity cost per country's GDP (NI_1)
2. the independence of a country to imported energy sources (NI_2)
3. the specific GHG emissions per country's capita (NI_3)
4. the political obligations of a country toward promoting environmentally friendly policies (Kyoto or EU 2020 framework) (NI_4).

The political obligation index was quantified based on the commitment of the investigated countries to the Kyoto Protocol annexes as well as to the EU 2020 roadmap. The value of the index for each commitment was increased by 0.5. The above-mentioned parameters fully cover all the aspects of the need of an energy system to be upgraded into an intelligent one.

4.2 Ability Index

The ability index shows the capability of a country to be upgraded into an energy smart region. The quantifiable parameters of this index that fully cover all the aspects of the ability of this upgrading are

1. the GDP per capita (AI_1)
2. the growth rate (AI_2)
3. the electricity produced from renewable energy sources (AI_3).

An alternative would be to use the renewable potential rather than the installed renewable energy capacity. However, apart from the solar and the wind potential, for which remote sensing techniques (satellites) are employed, the rest of the renewable resources (hydro, geothermal, and biomass) are still not documented for all countries. To this end, the authors were restricted to use the installed RES instead of the potential resources.

With regard to the role of the distance of the insular systems from the mainland energy networks, the investigation of this aspect would have been of interest if the interconnection of insular energy systems to mainland energy networks would mean, at the same time, the integration of these systems to smart ones. However, the kickoff point of this analysis is that the development of smart energy regions in insular systems will not be based on their interconnection with advanced systems but rather with the promotion of specific measures that would improve the quality of the examined energy systems per se. To this end, this aspect was not considered in the ability index.

All these values were normalized, and any negative values of the growth rate were considered 0.

5. Results and Discussion

Three scenarios were investigated in this work, which were defined by the weighting factors of the parameters shown in Table 2:

- Basic scenario that assumes equal weighting among all parameters
- Economy-dominant scenario, where the weighting favors the cost-driven parameters
- Technology/environment-dominant scenario, where the weighting favors the technological and/or environmentally driven parameters

The investigated scenarios illustrate the necessity and ability of each of the countries' energy system to develop into a smart one.

Table 2. Weighting factors of the three investigated scenarios.

Scenario	Index						
	NI_1	NI_2	NI_3	NI_4	AI_1	AI_2	AI_3
Basic	0.25	0.25	0.25	0.25	0.33	0.33	0.33
Economy-driven	0.35	0.35	0.15	0.15	0.40	0.40	0.20
Technology/ Environment- driven	0.15	0.15	0.35	0.35	0.20	0.20	0.40

The horizontal axis indicates the necessity index and the vertical the ability index of each country, while the size and the color of the bubble represent its installed power capacity and the continent of the country, respectively. The higher the ability index—in other words, the further to the right on the chart a country is located—the more feasible is its development. Countries that are located in the bottom part of the chart have the greater necessity to be energetically upgraded, due to either high electricity prices, high dependency on imported energy sources, high GHG emissions or political obligations, or a combination of these.

5.1 Basic Scenario

The basic scenario assumed that all necessity and ability parameters were equally important. The chart, shown in Figure 3, indicates seven different groups of countries:

1. The rich oil-producing countries of the Middle East, even though they can easily develop into intelligent energy regions (high ability indices), do not have strong motivations indicated by the low necessity indexes. It is also remarkable that these countries present the largest insular energy systems; therefore, measures toward supporting their upgrade are considered crucial.
2. The Central American countries are presented with the lowest ability indices and average necessity indices.
3. The African island nations (Fiji as well) are among the smallest insular energy systems and are found on the medium scale of both the necessity and ability indices.
4. The southern European insular island states, including Cyprus and Malta, have very high necessities to be energetically upgraded, driven mainly by political obligations. At the same time, the impact of the great financial recession on these countries as well as the low contribution of RET to their energy mix is made obvious by their (low) ability indices.
5. The Eastern Asian countries have high necessity indices, due to their high dependency on imported energy, while their ability to develop into smart energy regions is on the average scale
6. The good ability indices of Oceania countries, including New Zealand and Papua New Guinea, can be attributed to the high levels of RES utilization and high growth rate, respectively, while their necessity is found to be average
7. Iceland, as a northern European country with high percentages of RES in its energy mix, presents a special case, as it has the highest ability and necessity indices. It is the most capable insular energy system to develop into a smart one, primarily due to its high GDP, high percentage of renewable electricity generation, and appropriate motivation and need from political obligations.

5.2 Economy-Driven Scenario

The economy-driven scenario assumed that an insular country's cost of electricity, independence to imported energy sources, GDP, and growth rate were the dominant factors for the development of its energy system into an intelligent one. The results, given in Figure 4, illustrate a greater dispersion among the countries in the chart. The necessity index of the Middle East

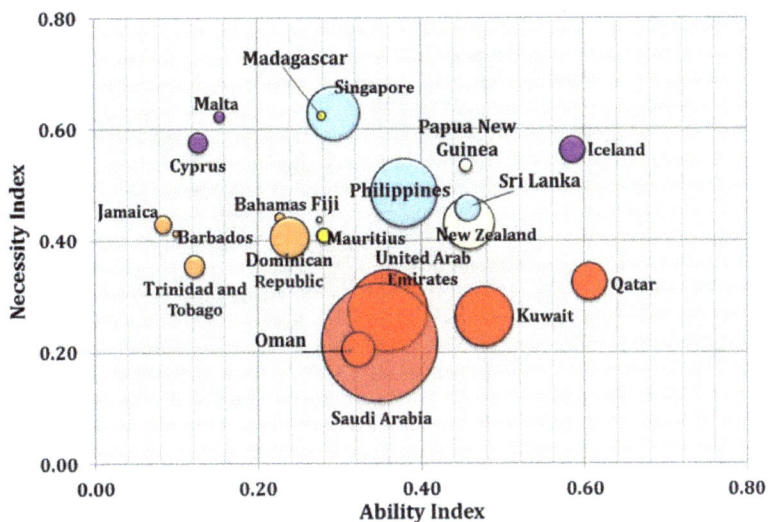

Fig. 3. Necessity Index and Ability Index of the integration potentials of insular energy systems to smart energy regions for basic scenario.

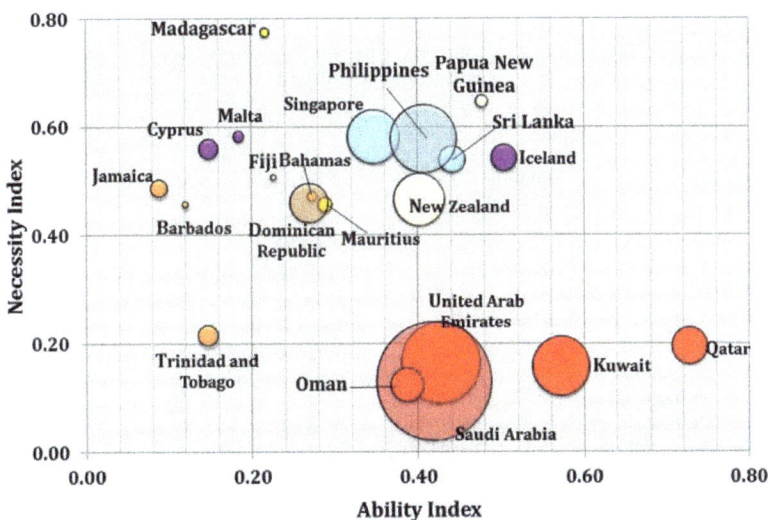

Fig. 4. Necessity Index and Ability Index of the integration potentials of insular energy systems to smart energy regions for economy-dominant scenario.

oil producer drops even further. The necessity index of the Central American countries slightly increases with the exception of Trinidad and Tobago, whose index drops significantly due to its low electricity prices and dependency on imported energy sources. The European and Oceania countries stay relatively unaffected by the weighting variation, whereas the African island countries and Fiji show a slight decrease in their ability index. In this scenario, Madagascar stands out with its necessity index rising almost to 0.8. On the other hand, the ability and necessity indices of the Eastern Asia countries slightly decrease and increase, respectively, with the exception of Singapore.

5.3 Technology/Environment-Driven Scenario

In this scenario, it was assumed that the technological and environmental upgrade of the energy systems were the main parameters for its development into a smart energy country. Figure 5 illustrates a more concentrated distribution of the countries. The majority indicate a movement toward the center of the chart, meaning that the necessity index of the oil producers has significantly increased. The Central American countries and the European countries with political obligation toward promoting environmentally friendly policies, including Iceland, Cyprus, and Malta, have remained relatively constant, while the ability index

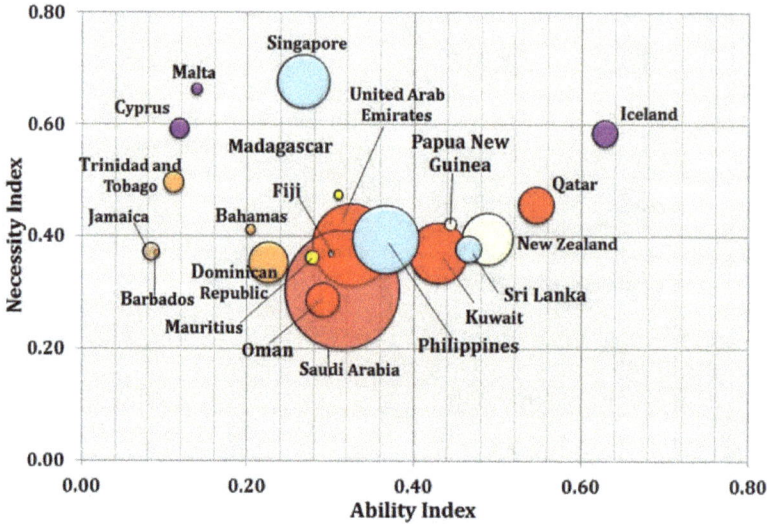

Fig. 5. Necessity Index and Ability Index of the integration potentials of insular energy systems to smart energy regions for technology and environment-dominant scenario.

of the African countries is increased. The Eastern Asia countries, namely Philippines and Sri Lanka, and the Oceania countries have also moved toward the center of the chart by decreasing their necessity index. The exception is Singapore, whose high emissions place it high in the necessity index. The results of the technology/environment-driven scenario highlight the need for motivation to encourage the development of insular energy systems into smart ones.

6. Model Validation Based on the Current Situation: The Cases of Iceland, Cyprus, and United Arab Emirates (UAE)

In this section, the validity of the derived model is performed by means of three case studies: Iceland, Cyprus, and United Arab Emirates (UAE). The choice of these systems was based on their placing on the necessity–ability chart: Iceland possesses the highest ability and necessity potentials, Cyprus has high necessity but low ability dynamics, and UAE stand in the middle of the chart, thus with average necessity and ability capacities.

6.1 Iceland

According to this study's analysis, Iceland presents the optimum performance in terms of its necessity and ability potential to upgrade its energy system into a smart one. Natural hydro- and geothermal resources have turned Iceland into the world's largest green energy producer per capita. Presently, the hydro- (20%) and geothermal (65%) resources cover almost the entire demand of Iceland's consumption of electricity and approximately 85% of Iceland's total primary energy consumption. Iceland possesses the world's highest share of renewable energy in any national total energy budget. Iceland's current power generation per capita totals about 55 MWh annually, which makes Iceland

the world's largest electricity generator per capita. In comparison, the EU average is less than 6 MWh. Iceland's renewable energy sources are not only abundant, relative to the size of the nation as a whole, but they are also available at a comparatively low cost. For this reason, electricity prices in Iceland are much lower than in most other OECD countries.[33]

6.2 Cyprus

The accession of Cyprus into the European Union in 2004 was a milestone for the energy system of the island. Cyprus was obliged to follow European directives in the field of energy, resulting in the mandatory upgrade of its energy system.[31] The main consequences were the following:

- The liberalization of the energy market
- The establishment of national policies and action plans for the promotion of renewable energy
- The adoption of an energy-saving strategy.

From 2010 onward, four wind farms with a cumulative power of 140 MW were licensed and operated.[34] The biomass sector followed with the licensing and operation of 8 MW power plants, using biogas derived from livestock waste. With regard to medium- and large-scale PV power plants, so far 25 MW PV systems were installed and operated, whereas the outlook for the upcoming months includes the licensing of at least another 80 MW of PV installations.[5]

Having achieved so much in the energy sector in such a short time, Cypriot society experienced a major setback in 2011. A large accident that occurred near the main power station of the island incapacitated about 60% of the installed capacity. Due to the need for a quick response, in order to recover the energy deficit in Cyprus, unsurprisingly, all attention was focused on the renewable energy sector. The interested parties came, though, across a small surprise: the state was not able to determine precisely the number of renewable energy projects that were already licensed as well as the number of projects that were still in the process of getting all their licenses. The answer to this question was also not possible to be given, in view of the very tangled way of licensing RET projects in Cyprus. This fact led to considerable concerns and initiated a discussion on licensing procedures of RET projects.[35] Progress was also decelerated due to the great 2013 Cypriot financial crisis, which resulted to a €10 billion international bailout by the Eurogroup, the European Commission (EC), the European Central Bank (ECB), and the International Monetary Fund (IMF). The findings of our analysis are hence in good agreement with the told story, namely the high necessity and the low ability of Cyprus to improve the technological status of its energy system.

6.3 UAE

According to the necessity–ability model, the UAE government currently presents moderate trends to further upgrade its energy system. Realizing the risks of being largely exposed to fuel prices, the government of the UAE effectively sought diversification of its energy mix away from fossil fuels and toward clean energy. The UAE enjoys reasonable renewable energy resources, with the average vertical solar irradiance being 2120 kWh/m^2/year, and the average monthly wind speed in coastal areas lying between 4.2–5.3 m/s in coastal areas. Having these facts as drivers, the UAE aims to attract significant investment in alternative and sustainable energy projects by 2020. Both Abu Dhabi and Dubai are targeting the generation of 7% and 5%, respectively, of total power demand from renewable sources by 2030. In a strong reaffirmation of the country's commitment to the wider carbon and sustainable living agenda, Abu Dhabi launched its Masdar Sustainable City initiative, which will house 50,000 people and which will be completely reliant on renewable sources for its power needs. The UAE's track record of delivering against plans and growing evidence of a project pipeline is gradually contributing to an increased trust by the international community in the UAE's renewables program. However, the program still remains highly dependent on government investments in large utility-scale projects, which presents a clear sustainability risk.[36]

7. Conclusion

The focus of this article was on insular energy systems and their potential to develop into smart energy regions. Due to the unique characteristics of insular energy systems, a special approach was required toward defining the dominant parameters that can drive this development. In terms of this work, the main features of the smart energy regions, as well as of the insular energy systems, were defined. An overview of the insular energy systems worldwide was performed, and the systems were classified based on specific criteria, resulting in three categories. This work also examined the necessity and the ability of a number of energy insular countries to develop into intelligent ones. Two factors, namely the ability index and the necessity index, were defined and calculated for all insular energy systems taken into account. Each of the indices has resulted as the weighted sum of a variety of parameters, namely the cost of energy, the dependency on imported energy sources, GHG emissions, political obligations, GDP, growth rate, and level of RET utilization. The main outcome of this analysis was that the development of insular energy systems into smart ones will be driven mainly by technological and environmental motivations including the reduction of GHG emissions, the introduction of political obligation toward promoting environmentally friendly policies, and the increase of RES utilization for the diversification of their energy mix.

Appendix 1: Insular energy systems of island nations with installed capacity up to 100 MW (Category A) (2013).[37]

Country	Installed capacity [MW]	Electricity from RET [%]	CO$_2$ per capita [tons/capita]	GDP growth rate [%]	GDP per capita [Intl. $]	Continent
Cape Verde	90	2.44	0.77	4.8	4.095	Africa
Comoros	6	12.50	0.18	2.5	1.300	Africa
Sao Tome and Principe	14	33.33	0.72	4.5	2.077	Africa
Seychelles	89	0.00	10.79	3.0	25.788	Africa
Antigua and Barbuda	27	0.00	7.72	1.0	18.492	America
Dominica	97	30.35	1.82	0.4	13.288	America
Grenada	50	0.00	2.61	0.5	10.837	America
Saint Kitts and Nevis	55	0.00	5.00	0.0	13.144	America
Saint Lucia	76	0.00	2.48	0.7	11.597	America
Saint Vincent/Grenadines	47	19.12	1.79	1.2	10.715	America
Maldives	62	0.03	3.35	3.4	8.871	Asia
Kiribati	5	0.00	0.60	−0.4	5.900	Oceania
Nauru	1	0.00	18.25	0.0	5.000	Oceania
Samoa	41	45.92	0.90	1.5	4.475	Oceania
Solomon Islands	36	0.00	0.38	7.4	1.517	Oceania
Tonga	12	0.00	1.80	1.4	4.886	Oceania
Vanuatu	28	0.00	0.48	2.6	4.379	Oceania

Appendix 2: Insular energy systems of island nations with installed capacity from 100 MW up to 15 GW (Category B) (2013).[37]

Country	Installed capacity [MW]	Electricity from RET [%]	CO_2 per capita [tons/capita]	GDP growth rate [%]	GDP per capita [Intl. $]	Continent
Madagascar	430	58.71	0.09	1.9	966.000	Africa
Mauritius	900	24.57	3.67	3.4	14.420	Africa
Bahamas	493	0.00	10.25	2.5	31.978	America
Barbados	239	0.00	5.14	0.7	19.320	America
Cuba	5855	3.54	2.54	3.1	10.200	America
Dominican Republic	5701	9.92	1.98	4.0	9.796	America
Jamaica	1175	6.75	3.39	0.9	7.083	America
Trinidad and Tobago	1605	0.25	39.38	0.7	25.074	America
Philippines	16320	27.41	0.80	6.6	4.119	Asia
Singapore	10250	2.68	43.03	1.3	60.688	Asia
Sri Lanka	2685	53.82	0.58	6.0	5.582	Asia
Cyprus	1527	2.41	10.29	−2.3	32.254	Europe
Iceland	2579	99.99	10.12	2.7	36.483	Europe
Malta	573	0.10	19.28	1.2	27.504	Europe
Fiji	245	52.25	1.55	2.0	4.757	Oceania
New Zealand	9679	74.42	8.28	2.2	31.082	Oceania
Papua New Guinea	700	37.31	0.47	7.7	2.676	Oceania

Appendix 3: Insular energy systems of mainland nations with no grid interconnection (Category C) (2013).[37]

Country	Installed capacity [MW]	Electricity from RET [%]	CO_2 per capita [tons/capita]	GDP growth rate [%]	GDP per capita [Intl. $]	Continent
Angola	1155	68.39	1.29	6.8	5.920	Africa
Cameroon	1115	74.22	0.44	4.7	2.359	Africa
Central African Republic	44	84.38	0.07	4.1	810.000	Africa
Chad	31	0.00	0.02	7.3	1.498	Africa
Djibouti	130	0.00	1.34	4.8	2.296	Africa
Equatorial Guinea	38	7.22	3.38	5.7	36.202	Africa
Eritrea	141	0.68	0.08	7.5	585.000	Africa
Ethiopia	2061	99.41	0.07	7.0	1.109	Africa
Gabon	415	45.47	2.71	6.1	15.852	Africa
Gambia	62	0.00	0.26	−1.6	1.809	Africa
Guinea	395	53.56	0.13	4.8	1.124	Africa
Guinea-Bissau	26	0.00	0.24	−2.8	1.270	Africa
Liberia	197	0.00	0.12	9.0	700.000	Africa
Malawi	287	87.84	0.07	4.3	893.000	Africa
Mali	304	57.69	0.05	−4.5	1.091	Africa
Mauritania	263	16.55	0.55	5.3	2.532	Africa
Nigeria	5900	25.37	0.42	7.1	2.533	Africa
Senegal	638	10.91	0.44	3.7	1.967	Africa
Sierra Leone	102	68.97	0.23	21.3	1.131	Africa
Somalia	80	0.00	0.08	2.6	600.000	Africa
Sudan	2338	53.43	1.64	−55.0	2.325	Africa

Appendix 4: Insular energy systems of mainland nations with no grid interconnection (Category C) (2013).[37]

Country	Installed capacity [MW]	Electricity from RET [%]	CO_2 per capita [tons/capita]	GDP growth rate [%]	GDP per capita [Intl. $]	Continent
Bolivia	1655	35.21	1.28	5.2	5.099	America
Guyana	363	0.00	2.08	3.7	3.438	America
Haiti	261	31.23	0.20	2.8	1.171	America
Suriname	410	55.10	3.76	4.0	8.355	America
Bangladesh	6673	3.85	0.37	6.1	1.777	Asia
Brunei	759	0.00	21.49	2.7	51.760	Asia
Burma (Myanmar)	1713	68.80	0.22	6.2	1.400	Asia
Indonesia	34074	16.74	1.74	6.0	4.636	Asia
Korea, North	9500	63.08	2.60	0.8	1.800	Asia
Korea, South	84662	1.36	11.62	2.0	29.834	Asia
Pakistan	22269	35.16	0.77	3.7	2.745	Asia
Timor-Leste (East Timor)	0	0.00	0.48	10.0	1.578	Asia
Kuwait	12679	0.00	24.15	6.3	54.283	Middle East
Oman	4265	0.00	13.77	5.0	28.684	Middle East
Qatar	4893	0.00	34.76	6.3	88.314	Middle East
Saudi Arabia	49050	0.00	16.05	6.0	24.268	Middle East
United Arab Emirates	23248	0.02	25.95	4.0	47.893	Middle East
Yemen	1334	0.00	0.86	−1.9	2.333	Middle East

Appendix 5: Necessity index and ability index of the integration potentials of insular energy systems to smart energy regions for basic scenario.

Country	NI1	NI 2	Ni 3	Ni 4	AI1	AI2	AI3	Capacity (normalized)
Kuwait	0.00	0	0.56	0.5	0.61	0.82	0.00	0.26
Qatar	0.00	0	0.81	0.5	1.00	0.82	0.00	0.10
United Arab Emirates	0.00	0	0.60	0.5	0.54	0.52	0.02	0.47
Saudi Arabia	0.00	0	0.37	0.5	0.27	0.78	0.00	1.00
Oman	0.01	0	0.32	0.5	0.32	0.65	0.00	0.09
Trinidad and Tobago	0.01	0	0.91	0.5	0.28	0.09	0.00	0.03
Singapore	0.02	1	1.00	0.5	0.68	0.17	0.03	0.21
Iceland	0.02	1	0.23	1.0	0.41	0.35	1.00	0.05
Bahamas	0.03	1	0.24	0.5	0.36	0.32	0.00	0.01
Barbados	0.03	1	0.12	0.5	0.21	0.09	0.00	0.00
New Zealand	0.03	1	0.19	0.5	0.34	0.29	0.74	0.20
Malta	0.05	1	0.45	1.0	0.30	0.16	0.00	0.01
Mauritius	0.06	1	0.08	0.5	0.15	0.44	0.25	0.02
Cyprus	0.07	1	0.24	1.0	0.36	0.00	0.02	0.03
Dominican Republic	0.08	1	0.04	0.5	0.10	0.52	0.10	0.11
Jamaica	0.14	1	0.08	0.5	0.07	0.12	0.07	0.02
Fiji	0.22	1	0.03	0.5	0.04	0.26	0.52	0.00
Sri Lanka	0.32	1	0.01	0.5	0.05	0.78	0.54	0.05
Philippines	0.43	1	0.02	0.5	0.04	0.86	0.24	0.33
Papua New Guinea	0.63	1	0.01	0.5	0.02	1.00	0.34	0.01
Madagascar	1.00	1	0.00	0.5	0.00	0.25	0.59	0.01

Appendix 6: Necessity index and ability index of the integration potentials of insular energy systems to smart energy regions for economy-driven scenario.

Country	NI1	NI 2	Ni 3	Ni 4	AI1	AI2	AI3	Capacity (normalized)
Kuwait	0.00	0	0.56	0.5	0.61	0.82	0.00	0.26
Qatar	0.00	0	0.81	0.5	1.00	0.82	0.00	0.10
United Arab Emirates	0.00	0	0.60	0.5	0.54	0.52	0.02	0.47
Saudi Arabia	0.00	0	0.37	0.5	0.27	0.78	0.00	1.00
Oman	0.01	0	0.32	0.5	0.32	0.65	0.00	0.09
Trinidad and Tobago	0.01	0	0.91	0.5	0.28	0.09	0.00	0.03
Singapore	0.02	1	1.00	0.5	0.68	0.17	0.03	0.21
Iceland	0.02	1	0.23	1.0	0.41	0.35	1.00	0.05
Bahamas	0.03	1	0.24	0.5	0.36	0.32	0.00	0.01
Barbados	0.03	1	0.12	0.5	0.21	0.09	0.00	0.00
New Zealand	0.03	1	0.19	0.5	0.34	0.29	0.74	0.20
Malta	0.05	1	0.45	1.0	0.30	0.16	0.00	0.01
Mauritius	0.06	1	0.08	0.5	0.15	0.44	0.25	0.02
Cyprus	0.07	1	0.24	1.0	0.36	0.00	0.02	0.03
Dominican Republic	0.08	1	0.04	0.5	0.10	0.52	0.10	0.11
Jamaica	0.14	1	0.08	0.5	0.07	0.12	0.07	0.02
Fiji	0.22	1	0.03	0.5	0.04	0.26	0.52	0.00
Sri Lanka	0.32	1	0.01	0.5	0.05	0.78	0.54	0.05
Philippines	0.43	1	0.02	0.5	0.04	0.86	0.24	0.33
Papua New Guinea	0.63	1	0.01	0.5	0.02	1.00	0.34	0.01
Madagascar	1.00	1	0.00	0.5	0.00	0.25	0.59	0.01

Appendix 7: Necessity index and ability index of the integration potentials of insular energy systems to smart energy regions for technology/environment-driven scenario.

Country	NI1	NI 2	Ni 3	Ni 4	AI1	AI2	AI3	Capacity (normalized)
Kuwait	0.00	0	0.56	0.5	0.61	0.82	0.00	0.26
Qatar	0.00	0	0.81	0.5	1.00	0.82	0.00	0.10
United Arab Emirates	0.00	0	0.60	0.5	0.54	0.52	0.02	0.47
Saudi Arabia	0.00	0	0.37	0.5	0.27	0.78	0.00	1.00
Oman	0.01	0	0.32	0.5	0.32	0.65	0.00	0.09
Trinidad and Tobago	0.01	0	0.91	0.5	0.28	0.09	0.00	0.03
Singapore	0.02	1	1.00	0.5	0.68	0.17	0.03	0.21
Iceland	0.02	1	0.23	1.0	0.41	0.35	1.00	0.05
Bahamas	0.03	1	0.24	0.5	0.36	0.32	0.00	0.01
Barbados	0.03	1	0.12	0.5	0.21	0.09	0.00	0.00
New Zealand	0.03	1	0.19	0.5	0.34	0.29	0.74	0.20
Malta	0.05	1	0.45	1.0	0.30	0.16	0.00	0.01
Mauritius	0.06	1	0.08	0.5	0.15	0.44	0.25	0.02
Cyprus	0.07	1	0.24	1.0	0.36	0.00	0.02	0.03
Dominican Republic	0.08	1	0.04	0.5	0.10	0.52	0.10	0.11
Jamaica	0.14	1	0.08	0.5	0.07	0.12	0.07	0.02
Fiji	0.22	1	0.03	0.5	0.04	0.26	0.52	0.00
Sri Lanka	0.32	1	0.01	0.5	0.05	0.78	0.54	0.05
Philippines	0.43	1	0.02	0.5	0.04	0.86	0.24	0.33
Papua New Guinea	0.63	1	0.01	0.5	0.02	1.00	0.34	0.01
Madagascar	1.00	1	0.00	0.5	0.00	0.25	0.59	0.01

Appendix 8: Nomenclature

Abbreviations

CHP	Cogeneration of heat and power
CSP	Concentrating Solar Power
EC	European Commission
ECB	European Central Bank
EIA	Energy Information Administration
ETS	Emissions Trading System
EU	European Union
GDP	Gross Domestic Product
GHG	Greenhouse Gases
HFO	Heavy Fuel Oil
HVDC	High Voltage Direct Current
IEA	International Energy Agency
IMF	International Monetary Fund
NAP	National Action Plan
PV	Photovoltaic
RES	Renewable Energy Sources
RET	Renewable Energy Technologies
SET	Strategic Energy Technology
UAE	United Arab Emirates

Symbols

AI_i	Ability Index
I_i	Index
max	Maximum
min	Minimum
NI_i	Necessity Index
P_i	Parameter
W_i	Weighting factor

Acknowledgment

The authors are indebted to COST Action TU 1104 "Smart Energy Regions" for supporting this work.

References

1. International Energy Agency. World Energy Outlook 2013. http://www. worldenergyoutlook.org/publications/weo-2013/ (accessed November 2013).
2. European Energy Research Alliance (EERA).Smart Cities: JP Review [Online], 2013, http://www.eera-set.eu/lw_resource/datapool/_items/item_755/reviewsmartcites_web.pdf (accessed Nov 21, 2013).
3. European Commission. European Innovation Partnership on Smart Cities and Communities: Strategic Implementation Plan. http://ec.europa.eu/eip/smartcities/files/sip_final_en.pdf (accessed Nov 21, 2013).
4. European Cooperation in Science and Technology (COST). Memorandum of Understanding for the Implementation of a European Concerted Research Action Designated as COST Action TU1104: Smart Energy Regions, 2011.
5. Fokaides, P.A.; Kylili, A. Towards Grid Parity in Insular Energy Systems: The Case of Photovoltaics (PV) in Cyprus. *Energ. Policy* **2014**, *65*, 223–228.
6. Yantovski, E.; Gorski, J. Zero Emissions Future City. In C*lean Energy Systems and Experiences;* Eguchi, K., Ed.; InTech [Online] **2010**, 165–178.http://www.intechopen.com/books/clean-energy-systems-and-experiences/zero-emissions-future-city (accessed Nov 27, 2013).
7. Mishra, V.; Smyth, R.; Sharma, S. The Energy-GDP Nexus: Evidence from a Panel of Pacific Island Countries. *Resour. Energy Econ.* **2009**, *31*, 210–220.
8. Kjaer, C. Energy Transition Can Bring Lasting Peace and Generate Wealth. *Energy Strategy Rev.* **2013**, *1*, 140–142.
9. Czisch, G. *Scenarios for a Future Electricity Supply*; IET: London, 2011.
10. Elliott, D. Emergence of European Supergrids—Essay on Strategy Issues. *Energy Strategy Rev.* **2013**, *1*, 171–173.
11. Roinioti, A.; Koroneos, C.; Wangensteen, I. Modeling the Greek Energy System: Scenarios of Clean Energy Use and Their Implications. *Energ. Policy* **2012**, *50*, 711–722.
12. Kylili, A., Fokaides, P.A. Investigation of Building Integrated Photovoltaics Potential in Achieving the Zero Energy Building Target. *Indoor Built Environ.* **2014**, *23*, 92–106.
13. Michalena, E.; Angeon, V. Local Challenges in the Promotion of Renewable Energy Sources: The Case of Crete. *Energ. Policy* **2009**, *37*, 2018–2026.
14. Kaldellis, J.K.; Kavadias, K.; Christinakis, E. Evaluation of the Wind-Hydro Energy Solution for Remote Islands. *Energ. Convers. Manage.* **2001**, *42*, 1105–1120.
15. Xydis, G.A.; Nanaki, E.A.; Koroneos, C.J. Low-enthalpy Geothermal Resources for Electricity Production: A Demand-Side Management Study for Intelligent Communities. *Energ. Policy* **2013**, *62*, 118–123.
16. Koroneos, C.; Zairis, N.; Charaklias, P.; Moussiopoulos, N. Optimization of Energy Production System in the Dodecanese Islands. *Renew. Energ.* **2005**, *30*, 195–210.
17. Bojnec, S.; Papler, D. Economic Efficiency, Energy Consumption and Sustainable Development. *J. Bus. Econ. Manag.* **2011**, *12*, 353–374.
18. European Commission. The 2020 Climate and Energy Package. http://ec.europa.eu/clima/policies/package/index_en.htm (accessed Aug 8, 2014).
19. Poullikkas, A.; Hadjipaschalis, I.; Kourtis, G. The Cost of Integration of Parabolic Trough CSP Plants in Isolated Mediterranean Power Systems. *Renew. Sustain. Energ. Rev,* **2010**, *14*, 1469–1476.
20. Shupe, J.W.; Weingart, J.M. Emerging Energy Technologies in an island environment: Hawaii. *Annu. Rev. Energy* **1980**, *5*, 293–333.
21. Ibrahim, H.; Younes, R.; Ilinca, A.; Dimitrova, M.; Perron, J. Study and Design of a Hybrid Wind-Diesel–Compressed Air Energy Storage System for Remote Areas. *Appl. Energ.* **2010**, *87*, 1749–1762.
22. Oikonomou, E.K.; Kilias, V.; Goumas, A.; Rigopoulos, A.; Karakatsani, E., Damasiotis, M.; Papastefanakis, D.; Marini, N. Renewable Energy Sources (RES) Projects and Their Barriers on a Regional Scale: The Case Study of Wind Parks in the Dodecanese Islands, Greece. *Energ. Policy* **2009**, *37*, 4874–4883.
23. Sanseverino, E.R.; Sanseverino, R.R.; Favuzza, S.; Vaccaro, V. Near Zero Energy Islands in the Mediterranean: Supporting Policies and Local Obstacles. *Energ. Policy* **2014**, *66*, 592–602.
24. Blarke M.B.; Jenkins B.M. Super Grid or Smart Grid: Competing Strategies for Large-Scale Integration of Intermittent Renewables? *Energ. Policy* **2013**, *58*, 381–390.
25. Faruqui A.; Harris D.; Hledik R. Unlocking the €53 Billion Savings from Smart Meters in the EU: How Increasing the Adoption of Dynamic Tariffs Could Make or Break the EU's Smart Grid Investment. *Energ. Policy* **2010**, *38*, 6222–6231.
26. Hargreaves, T.; Nye, M.; Burgess, J. Making Energy Visible: A Qualitative Field Study of How Householders Interact with Feedback from Smart Energy Monitors. *Energ. Policy* **2010**, *38*, 6111–6119.

27. Jackson, J. Improving Energy Efficiency and Smart Grid Program Analysis with Agent-Based End-Use Forecasting Models. *Energ. Policy* **2010**, *38*, 3771–3780.

28. Kaldelis, J.K.; Zafirakis, D. Present Situation and Future Prospects of Electricity Generation in Aegean Archipelago Islands. *Energ. Policy* **2007**, *35*, 4623–4639.

29. Kaldellis, J.K.; Gkikaki, A.; Kaldelli, E.; Kapsali, M. Investigating the Energy Autonomy of Very Small Non-Interconnected Islands: A Case Study: Agathonisi, Greece. *Energ. Sustain. Dev.* **2012**, *16*, 476–485.

30. Ribeiro, L.A.; Saavedra, O.R.; Lima, S.L.; Matos, J.G.; Bonan, G. Making Isolated Renewable Energy Systems More Reliable. *Renew. Energ.* **2012**, *45*, 221–231.

31. Koroneos, C.; Fokaides, P.A.; Moussiopoulos, N. Cyprus Energy System and the Use of Renewable Energy Sources. *Energy* **2005**, 30, 1889–1901.

32. Energy Information Administration (EIA). EIA's website. http://www.eia.gov/ (accessed Nov 27, 2013).

33. The Independent Icelandic Energy Portal. http://askjaenergy.org/iceland-introduction/iceland-energy-sector/ (accessed Aug 1, 2014).

34. Fokaides, P.A.; Miltiadous, C.I; Neophytou, M.K.A.; Spyridou, L.P. Promotion of Wind Energy in Isolated Energy Systems: The Case of the Orites Wind Farm. *Clean Techn. Environ. Policy* **2014**, *16*, 477–488.

35. Fokaides, P.A.; Poullikkas, A; Christofides, C. Lost in the National Labyrinths of Bureaucracy: The Case of Renewable Energy Governance in Cyprus. In E. Michalena and J. M. Hills (Eds.), Renewable Energy Governance, Lecture Notes in Energy 57; Springer-Verlag: London, 2013.

36. Nimer, A.A. Renewable Energy Review: United Arab Emirates. http://www.renewableenergyworld.com/rea/news/article/2012/12/renewable-energy-review-united-arab-emirates (accessed Aug 1, 2014).

37. Central Intelligence Agency (CIA). The World Factbook. http://www.cia.gov/index.html (accessed Nov 21, 2013).

4

Decision-Making Impacts on Energy Consumption Display Design

ANU K. GUPTA[1]*, DALE C. ROACH[2], SHELLEY M. RINEHART[1], and LISA A. BEST[3]

[1] Faculty of Business, University of New Brunswick, Saint John, New Brunswick, Canada
[2] Department of Engineering, University of New Brunswick, Saint John, New Brunswick, Canada
[3] Department of Psychology, University of New Brunswick, Saint John, New Brunswick, Canada

Abstract: Policy makers are considering smart grid technologies such as energy consumption displays (ECD) as a means of changing the behavior of energy consumers. This article provides a review of consumer energy management behavior, presents decision-making models, and considers consumer–technology interface (CTI) display guidelines in ECD design. Existing research finds the underlying drivers of energy management behavior to be existing motivators and sociodemographics. An analysis of ECD research demonstrates a preference for an energy consumption graphic, consumption rates in monetary units, the ability to provide detailed and historical information at the appliance level, and the avoidance of unknown parameters. Integrating CTI guidelines with ECD preferences and decision-making approaches yields two ECD models. The article presents two prototype ECD models that serve different decision-making approaches. The first ECD prototype is designed to serve routine or predisposition decision makers, presenting summary information and employing colors and graphics. The second ECD prototype is designed to serve rational decision makers, presenting detailed appliance-level information, goal-setting capabilities, and the ability to produce detailed history. Further research is recommended to establish decision-making preferences, predisposition elements, threshold parameters, and ECD prototype testing by region.

Keywords: Consumer energy management behavior, smart meter, smart grid, consumer-technology interface, cognitive psychology, engineering psychology, information processing, consumer design

1. Introduction

There are many benefits envisioned with the growth of smart grid infrastructures, including power grid optimization, integration of distributed generation, large-scale power storage, demand response capabilities, smart homes, and demand side management. As smart grid deployments grow, policy makers and power utilities are considering smart grid technologies such as energy consumption displays (ECD) as a means of changing the behavior of energy consumers assisting in demand side management. This is complex, requiring the understanding of the decision-making process and the interaction with consumer–technology interfaces (CTI)—specifically, displays. Smart grid proponents indicate that smart meter design will need to include specific display consideration for ECDs and are a key contributor to the success of the entire smart grid infrastructure.[1]

This article will provide a review of energy management behavior, present decision-making models, and CTI display design methodologies, as applicable to ECDs. It will integrate the three components; energy management, decision making, and CTI display design. Through this integration, it will present the energy management decision-making implications in ECD design and the development of associated ECD prototypes. The objective of this research is to demonstrate that by integrating the human decision-making process of energy management with CTI design and principles, better ECDs can be developed. These ECDs would have a greater potential to influence consumers and support policymakers in changing behaviors of energy management. The research conducted incorporates global data but is developed in the North American context.

Studies have been conducted on energy consumption behavior and the associated impacts of household interventions. Some are specific to ECDs; however, most are not. ECDs provide consumer-specific energy consumption information within the home with various levels of information and multiple design formats. A range of units have been designed by manufacturers and deployed by power utilities with highly inconsistent results.[2,3,4,5] The research to date has varied materially, from small-scale

*Address correspondence to: Anu K. Gupta, Faculty of Business, UNB Saint John, 100 Tucker Park Rd., Saint John, NB E2L 4L5, Canada. Email: agupta@unb.ca

experiments of a dozen homes with electronic in-home ECDs to large-scale studies with thousands of consumers, through paper billing information across varying periods of time from months to years.

Many of these studies demonstrated that feedback and, at times, interventions of consumer energy consumption information could impact the consumer to reduce energy consumption. However, reduction of energy consumption has had a wide range of results, from 0 to 20%.[6,7,8,9,10,11] While there is a greater awareness by consumers of the value and need for energy management, many consumers still fail to take steps toward this form of management and conservation.[12] There is a sizeable discrepancy between people's knowledge, values, attitudes, and intentions and their behavior around energy management and conservation.[12] Energy management behavior assessment is complex due to differing values and drivers that motivate behavior.

2. Energy Management Behavior Review

Many different forms of energy interventions can trigger a consumer's energy management action. Interventions can range from energy consumption information, through comparisons, to goal setting. Information presented may include rates, current and historical usage, averages, trending, comparisons year over year, neighborhood and regional information, and modeling through reports. There has been much research impacting consumer energy behavior but not necessarily through ECDs. Although the research on interventions may not have been directly through ECDs, it still provides valuable insight into how consumer behavior was impacted by intervention programs, being based on thousands of consumers, whereas ECD experiments have much smaller sample sizes.

Abrahamse et al.[7] reviewed research conducted on the intervention of household energy conservation in homes between 1977 and 2004. They analyzed 38 academic studies that had a measure against a consistent baseline of criteria and were targeted at the household environment. The meta-analysis represented an aggregated sample of 11,770. The 38 studies did not include ECD-based interventions. They identified that the underlying determinants of energy use were existing attitudes and sociodemographics.[7] In their meta-analysis they found interventions had varying degrees of success; information increased knowledge levels but did not necessarily bring about behavioral change.[7] They also found that rewards encouraged change but only in the short term or until the reward ended.[7]

Fischer[6] performed a review of 26 energy conservation behavior studies, some including ECDs. Her meta-analysis was grouped into three categories: written (often bill enhancements), electronic displays (some computerized or appliances), and surveys. The breakdown of the sample size by intervention method in the 26 studies is shown in Table 1; studies that did not include the sample size are not shown.

In the analysis, Fischer[6] defined success as enabling households to conserve energy or to improve customer satisfaction. She found that there was evidence the following five features provided the most successful feedback: it was given frequently and over a long period of time; it provided appliance specific breakdowns; it was presented in a clear and appealing way; and it employed computerized and interactive tools.[6] She also states

Table 1. Fischer meta-analysis sample sizes.[6]

	Studies	n
Written	13	14781
Electronic display	5	212
Surveys	3	1890

feedback is best when it successfully captures the consumer's attention, draws a close link between specific actions and their effects, and activates various motives that may be appealing to a specific consumer, such as cost savings, energy conservation, and competition.[6] Linking back to the psychological factors, Fischer[6] further states that feedback is best if it triggers an existing motivation. She found that the studies with the best results included computerized feedback, interactive elements, detailed appliance breakdown, and daily feedback.[6]

Ueno et al.[4] performed a study of 10 Japanese homes that were given a computerized information system providing personal energy consumption information. They tested the energy use for 28 days before and after the systems were deployed and found that energy consumption reduced by 12% on average. They attempted to control significant variables such as season and household members. However, many variables could not be controlled: the average number of days with sun, special functions that may require additional energy usage, and others, such as visitors. Uncontrollable variables may be found in many studies.

Hargreaves et al.[3] trialed three models of energy consumption displays with different functionality. They were trialed with 275 residential consumers over a 12-month period. It was found that over time the displays became "backgrounded" with household routines, the displays increased the knowledge of energy consumption, but the displays did not encourage energy conservation.[3] They believe the households, once aware of their consumption rates, may accept these as their regular level of usage and their limitations in saving energy. Initially, they found there was a reduction in energy usage; however, once the stabilized level was reached, it became consistent.[3] Faruqui et al.[10] reviewed a dozen power utility pilot programs that focused on the conservation impact of ECDs and found that ECDs encouraged electricity conservation by 7% when prepayment was not in place.

Murtagh et al.[13] identified the differences in behaviors between homes with ECDs and those without. They performed a qualitative study with 21 households that had ECDs for more than 6 months. They found that there was a wide variation between households' energy consumption behavior and suggested that attempts to reduce energy were tied to social and physical contexts. They also found that there were variations between the energy users within the households in terms of their behavior toward energy consumption.

Oltra et al.[11] found that studies demonstrate savings in electricity consumption with ECDs from the range of 0 to 20%. To understand the range, they followed a sample of 17 people; eight individuals had ECDS, and nine did not. Their results found that savings may be impacted by the level of the user's interest in the ECD, determined by the user's motivation to save energy, prior attitudes, and the level of engagement induced by the intervention.

Abrahamse et al.[7] and Fischer's[6] meta-analysis found underlying drivers to energy behavior; prior attitudes, existing motivators, and sociodemographics drove energy management behavior. Additionally, each of their best-in-class results supported frequent, long-term, and computerized information supporting the employment of ECDs. Both meta-analyses together comprised 54 studies and 28,653 participants. The research specifically conducted with ECDs, although with small sample sizes, also supports both of these findings.[3,10 11,13]

Several additional themes that assisted energy management and conservation were found and can be applied to the analysis and design of ECDs. Fischer's[6] meta-analysis demonstrates that energy information feedback with the following characteristics had impact. The feedback successfully captured the consumer's attention, drew close links between specific actions and their effects, and activated various motives that may appeal to different consumer groups, such as cost savings, resource conservation, and emissions reduction.[6] Abrahamse et al.'s[7] meta-analysis also found that, in some instances, providing frequent, specific, and relevant feedback had an impact on conservation.

The following were themes of interventions that were generally not very effective in energy conservation and can also be applied to the design of ECDs. These include information alone, intent compared to actions, nonspecific information and relating carbon dioxide emissions to electricity consumption.[7,14,15] Feedback on its own is a poor conservation tool as it does not change existing motivators; however, it can assist in triggering an existing motivation. A relevant precondition must exist for feedback to work; there must be an implicit or explicit motivation.

Energy management is a complex field with many variables. There are many conflicting findings; this may be due to many factors, such as predispositions, sample size, timing of studies (1980 vs. 2010), country of study (culture), and many other demographic factors. These factors appear to have led to differing results with the same interventions. There appears to be an underlying belief that all consumers use the same process to make decisions regarding energy management. The meta-analyses acknowledge the importance of attitudes and motivators but do not integrate the decision-making process into the research. This leads to an exploratory discussion on decision-making approaches developed through research in cognitive psychology.

3. Decision Making

Energy consumption displays can impact decisions. Understanding how decisions are made at the cognitive level and the greatest area of potential impact within decision-making approaches is valuable in the design of ECDs. Psychology offers a model to explain environmental behavior that has applicability to energy management decision-making behavior. The model distinguishes between two types of behavior: routinized and conscious decisions. Routinized behavior is automated, whereas conscious decisions require norm or more rational activation.[16] In norm activation, an individual must recognize there are options to choose from and that criteria are needed to evaluate those options. For the purposes of this article, the conscious decision-making approach will be further broken out into two specific decision-making approaches: irrational and

rational. Ariely[17] employs the term "irrational" to differentiate the decision-making process. However, it does not necessarily mean the decision is irrational to the individual, only different from the rational or normative approach as it does not maximize the utility function, but rather is driven by a predisposition. Maximizing motivation and commitment are the key drivers of irrational decision-making.[18]

The three decision-making approaches are presented in a gated sequence in the integrated decision-making model depicted in Figure 1—the three approaches being routine, irrational, and rational, which can be applied to energy-related decisions. Each of the three decision approaches has defining characteristics, decision-specific parameters, weighting, and scoring. The weighting and scoring apply to differing degrees in the three approaches. Additionally, the first two approaches are similar to gates and arrive before the rational approach. If neither the routine or irrational decision-making approaches are engaged, the individual moves to the rational approach.

The microeconomic theory of consumer choice, rational (normative) decision-making, assumes the individual desires to maximize utility, when given resource constraints. A decision resulting in higher utility would be consistently preferred to all other outcomes that yield a lower utility.[20] The rational decision-making approach presumes perfect and complete information is available and is closest to a scientific approach with a consistent methodology.[20] Given the same decision parameters, the same weighting of each parameter, the score should yield homogenous results by different individuals. The rational model attempts to be as objective as possible and is highly analytical. Once the decision criteria are established, a weighting is applied to each. In the rational model, the information found on ECDs would be considered relevant and integrated into the weighting system to yield a score. How this information is assessed through the cognitive process has the potential to set the direction of the decision. These rational decisions are highly conscious, where the ECD has the potential to impact the scoring process. The ECD would be required to detail specific information and employ terms the individual in the rational decision-making process would require such as kWh or Mt in the scoring or weighting process.

Behavioral economists seek to integrate a broader understanding, including psychological components of the decision-making process. There is a substantial amount of research and data demonstrating that individuals do not consistently make rational decisions.[20,21] Individual decisions violate one or more of rational decision-making foundations, to maximize the utility, and are thus considered to be irrational compared to the normative or rational approach.[20] For this article, the term "irrational" is employed to represent non-optimized and non-habitual behavior. It does not necessarily mean it is a bad decision; the irrational approach may yield the best results for a specific individual because, while different from the approach used by others, it is aligned to their personal motivators. In the irrational model, if the ECD's information is associated with a parameter that has been ranked by the individual with a 100% weighting, it has the potential to have a material impact; if not, it will be seen as superfluous information by the mind and discarded. In the case where the individual is motivated by cost savings, they may view only the values with dollar symbols and ignore all kWh values.

In the routinized decision-making approach or response behavior, consumers often have well-established criteria, may require small amounts of additional information, or simply choose what they know. Routinized behavior occurs often and may have initially been a rational or irrational decision. However, with frequency, the individual develops a consistent response, shifting the characteristics of complexity, knowledge level, and cognitive requirements. Graybiel's[16] research outlines five key characteristics of routine or habitual behavior. According to her, the routines are largely learned, become ingrained, are performed primarily subconsciously, have an ordered sequence triggered by specific stimulus, and comprise both cognitive and motor expressions of routine.[16] One may not realize one is actually making a decision. Many small home energy management decisions may fall into this category; they are highly automated and occur at the subconscious level. The ECD with motivating information could be considered Graybiel's stimulus—for instance, costs or carbon parameters for the individual triggered by cost savings or environmental protection.

The same decision by different individuals could be considered rational, irrational, or routine depending on their level of knowledge and the frequency with which the decision is made. Several factors of decision making would place an individual on one of the three decision-making paths; uncertainty, risk, time, frequency, familiarity, and expertise.[22] Each of these factors evolves with experience. The same or similar rational decisions made consistently can move to irrational or routinized decisions, depicted in Figure 1, as a feedback loop in the integrated decision-making model. If the decision is made frequently, yielding consistent results, it may move from the conscious to the subconscious level and take on the routine approach. The risk for the individual becomes that the criteria, weighting, and scoring may change over time, yielding suboptimal results.[23] The value is in the routine decision-making approach requiring less cognitive resources.[23] In the routine decision-making approach, the specific information on the ECDs may be disregarded. The ECD information, however, may still trigger an action.

Connecting information processing with decision making enables the optimization of ECD design. Engineering psychology presents an information processing model that can be applied to energy management decisions at the consumer level: the human as a limited information processor. The model consists of four stages plus cognitive memory functions and feedback: stage one—STSS (short-term sensory store), stage two—perception, stage three—response selections, and stage four—response executions,[22] as shown in Figure 2. STSS, stage one, takes real-time environmental input from our senses, sight,

smell, touch, and others. The ECD would be an input to the STSS. These sense inputs move to stage two, where they are perceived and where meaning, drawn from the long-term memory, is attached to the sensory signal. As the model depicts, memory functions are engaged between stages two and three.

Within the cognitive memory functions, two forms of memory are distinguished, working memory and long-term memory,[22] as shown in Figure 2. Working memory contains information under current consideration, from our senses and cognitive analysis capabilities. It is here that demographic factors, market segmentation, and other real-time information is incorporated in the decision making process. The working memory's key constraint is its limited capacity only to hold a few items at a time under consideration. Bettman et al.[24] identify the number of items to be held in the working memory to be seven items plus or minus two. This limitation drives two important consequences. First, people do not transform the information, but rather process it in the form presented as seen on displays. The non-transformation of the information explains why the same information presented in different formats can have different impacts on an individual's decision—known as the concreteness principle.[24] The second factor of the working memory limitation to seven items is that people employ heuristics to process information. Heuristics are mental shortcuts that ease the cognitive load of decision making; in this instance, they are procedures to simplify the search and the elimination of information, potentially that are seen on a display. Both the concreteness principle and the employment of heuristics can have profound impacts on ECDs in the decision-making process and thus materially impact how input from the ECD is processed.

Long-term memory holds the individual's mass of available knowledge, including facts, experiences, values, and beliefs.[22] It is considered infinite, though, due to the time and energy required to shift working memory to long-term memory, not all memory is shifted from working to long-term memory.[25] Further, experiences are converted into long-term memories. The ability to retrieve information from long-term memory impacts the decision-making process.[25] The information that is retrieved guides the decision-making process. It is here that knowledge considered relevant is shifted to the working memory—for instance, the individual's personal views on the environment, income, and other views on the criteria under consideration. The shift of information is still limited to the seven items. This is one of the primary drivers of organizations to create familiarity with brands or social movements such as environmental protection.[26]

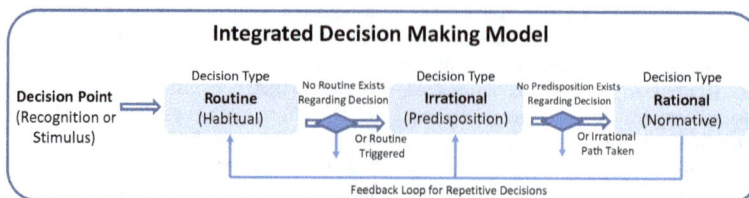

Fig 1. Integrated decision-making model.

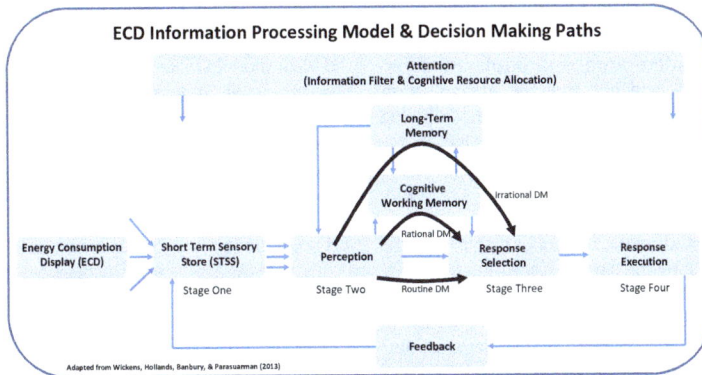

Fig 2. ECD information processing model and decision-making paths.[22]

An important implication of the limited capacity to process information is that humans struggle in the rational decision-making approach,[21] implying one must consider the cognitive processability over the availability of information in decision making. Further, the processability is, in part, a function of the way the information is presented. With the short-term inability to transform presented information and limitation to seven items, how ECDs are organized can have a major impact, again implying the ECD has the potential ability to make a material impact on energy management decisions. The integration of the two forms of memory with the STSS and perception stage will move the information processing that is occurring to stage three, the response selection, as shown in Figure 2. The decision-making approach determines which component of the information-processing model will play the prevalent role, as depicted in Figure 2. Once the individual traverses the path of the specific approach chosen, they move to stage four: response execution.[22] This can be instantaneous and at the subconscious level, as is often the case with routinized decision-making, or take a much a longer period of time if the rational approach is underway. The feedback loop implies the decision or action taken may change the STSS, setting the process in place with different sensory inputs. In addition, the stages may require cognitive resources, implying that available attention plays a role in the decision. The attention element can be broken apart into two components, first, as a filter of information and, second, the allocation of mental resources.[25]

An energy management feedback device or ECD would input directly into the STSS component of the process, where the display's sight, sound, and touch would start the information processing. For instance, an ECD may exhibit that the time-of-day rates have changed upward, depicting a shift in numeric values and potentially a change in colors from yellow to red. This information would move into the second stage of perception, where meaning of the sensory signal is applied, implying a high rate period has begun. In this instance, the perception would correlate that all electricity costs have increased in the home. This meaning is derived from past experiences from the long-term memory component; it is here that the individual is triggered or not and moves into the response selection mode, stage three.

The trigger may be an automated subconscious routine, as discussed earlier, with an action followed by a reward; it might be no action, maintaining the status quo; or it might be a third action at the conscious level. The role of the working memory cognitive functions is to tie the sensory information from the ECD with meaning and the long-term memory into response selection.

The information processing model becomes powerful when it is understood which of the three decision-making approaches an individual may take and what role is required of the ECD in each approach. In the routine or irrational approach, the ECD would be a trigger. In the rational approach, the specific information provided by the ECD would become part of the scoring methodology. In each of these scenarios, the ECD may be designed differently and present different information. In this manner, the ECD has an opportunity to influence and even drive energy management behavior. What, how, when, and where this information is presented and which of the three approaches the individual will employ to make the decision creates different levels of impact on each individual.

4. Consumer–Technology Interface Displays

Display technology has advanced materially over the recent past.[40] Modern technology has enabled consumer–technology interface displays to have great variety, including size, color, charts, numbers, sounds, thresholds, statistics, real-time, historical, modeling, reporting, mobility, and network connections, to name some. The combination and variability of design parameters has led to a plethora of design permutations with differing results. These vast options are certainly a component of the technical design. However, it is not clear how to integrate into ECDs the cognitive element of how information is processed, decisions made, and actions taken associated to energy management.

The relevance of the cognitive element of information processing in decision making being integrated into CTI displays is evident with the work of Johnson et al.[27] They perform two experiments testing preference reversal dependent on how information is displayed. Specifically, they test the frequency of preference reversal increases when probabilities are displayed in formats that are harder to process. Their experiments demonstrate that decision makers changed information processing strategies when the same information was displayed differently.[27] They further present evidence that the greater effort required to process the information, the greater the likelihood users will ignore or misuse the information presented in the display.[27] This applies to energy consumption displays as the decision-making process may also shift for the end users with different display formats.

Another instance demonstrating the impact of displays on the cognitive decision-making function is the well-known display illusion of Shepard's Tables.[28] The mind "sees" one of the tables as longer than the other. However, when measured they are found to be exactly the same length. The display illusion demonstrates an extreme case of the potential impact of images on ECDs and the inability to process information correctly, and the risk of misleading. It is possible to dismiss the display component as trivial within the larger smart grid infrastructure; Shepard's Tables[28] and Johnson et al.[27] reversal of decision is presented to highlight the importance of CTI design in decision making.

Egan[29] assessed the value of comparison-based energy information in improving the home energy decision environment. She tested four different graphic models with home energy information for 600 utility bill payers. The test included the consumer's interpretation of the graphical displays, display preferences, comparison group preferences, the likelihood of acting on a given display, and the willingness to pay for such a service. Participants commented positively on the graphical displays. She found distributional graphs to be most easily comprehended.[27]

Fischer[6] found that there were very few studies that had considered the relevance of graphic design or formulation of text on presentation. With the limited studies, she did find that, culturally, there was a significant difference in display preference. A design that ranked highest in the United States was not considered desirable in Norway and was thought of as childish.[6] She further found that in the UK and Sweden, citizens preferred comparisons on their own previous consumption, whereas in Finland and Japan, consumers were more interested in comparison among others than their own.[6] These opposing results demonstrate potential demographic impacts on displays.

Wood and Newborough's[14] model provides factors influencing the design of ECDs within a home. They identify six high-level categories: place of display, motivational factor, display units, display methods, timescale, and category.[14] These categories are a mix of energy components and display design. They provide several motivational factors for energy consumption that would work on an electronic display; monetary reward, self-competition/comparison, internal (in-home) competition, and consumer-specific goals (self-set or by others).[14] They indicate that little is known about how to best indicate energy consumption on displays but recommend employing home–computer interface guidelines.[14]

Anderson and White[5] define a core set of specifications for ECDs based on qualitative investigations of consumer experiences with existing ECDs in the UK market. They grouped 38 individuals into five different demographic groups and requested each group draw their ideal ECD. The participants then compared their ideal model to ECDs in market. They found

that the five groups' ideal display converged into a single model; however, the majority of the in-market ECDs did not align with the ideal model created by the participants.[5] The consensus display consisted of four key findings. The display included an analogue indicator of current rate of consumption, current rate of consumption as a monetary unit (pounds/dollars), cumulative daily spend, and a historical button showing spending over the last day, week, month, and quarter.[5] The researchers did acknowledge that not all decision making is rational and that this research did not include multiple decision-making approaches.

Fitzpatrick and Smith[2] assessed what data should be presented, how and in what format on ECDs. They drew on four sources of information in the UK: an online survey with 57 responses, two studies conducted with six households, and the comparison between the two. Their research was based on four existing ECDs. They found four key functional display design areas to integrate into ECDs: behavior and context, data, device and, engagement.[2] They also found that all the devices were designed by power utilities to service them.

Karjalainen[15] performed a qualitative study with 14 individuals in Finland, testing eight various ECD paper prototypes and gathering information on their preferential choices. They tested various factors such as charts, graphs, tables, text, units, comparison data, and goal-setting. They found consumers had two problems with displays; first, they were not familiar with scientific units such as W (watts) or kWh (kilowatt-hour).[15] Second, many individuals did not understand how carbon dioxide emissions related to electricity consumption.[15] They also found that there was a large preference that displayed a total overview of consumption as well as an appliance-specific breakdown, with consumption units in a currency format.[15] This display included the option to choose the period from minute, hour, day, week, month, and year. It did not provide comparison, goal-setting, or historical options. Further studies validated the consumer desire for appliance-specific breakdowns, however, with units associated to knowledge areas, specifically kWh as opposed to monetary values.[33]

Schultz et al. conducted a field experiment with 431 consumers in California for three months. They compared energy conservation results between ECDs that presented kWh, costs, and social norms. The social norms represented consumer information of similar homes. They found energy conservation occurred for those households that received the social norm framework with a reduction between 7% and 9%;[35] this occurred for both short-term (one week) and long-term periods (three months). They also found that residents reported a greater desire to reduce energy with the presence of an ECD; however, careful consideration must be given to framing of the feedback.[35]

An aggregated view of the CTI ECD findings yields preference for analogue presentation, consumption rates shown in monetary and energy units, ability to present historical information, appliance-specific breakdown, and the avoidance of uncommon parameters such as carbon production. The research to date on CTI and on energy management with electronic displays has been limited, and the existing analysis comes primarily from Europe. A pilot study in China by Xu et al. demonstrated that ECDs improved awareness, understanding, and attitudes toward energy conservation.[34] Fischer[6] demonstrated that region-based demographics are distinct and should not be extrapolated to apply to multiple regions, further limiting the research for the North American context.

Tufte[30] presents models to depict complex information through an optimal combination of words, numbers, and pictures, integrating the human element. Smith and Mosiers,[31] in their comprehensive work, provide guidelines for designing user interface software that can be applied to ECDs. The lessons from the ECD display research of Fischer[6] and the others plus the display guidelines for CTIs, from Tufte[30] to Smith and Mosiers,[31] can be brought together and applied through the product design process. The product design process by Ulrich and Eppinger[32] demonstrates how and where ECD design factors fit together and at what stage. Of the six stages, stages one, two, and three are most relevant to ECD analysis and design, as associated to energy management decision-making. In stage one—planning—there is recognition of the requirement for ECDs. In stage two—concept development—the needs of the consumer are identified, and concepts and alternatives developed. At this stage, an assessment may be performed through market research of the specific ECD consumer needs relative to objectives of the product. In the case of ECDs, stage two would incorporate the desired energy management outcomes and the concepts developed would integrate the three distinct decision-making approaches taken by consumers. Stage three, the system-level design, defines the product architecture, decomposes the product into subsystems, and develops a preliminary design. The preliminary design would include ECD specific display parameters, including character size, color, words, charts, and other display parameters identified.

5. Discussion

Energy management is complex, due to the many factors outlined earlier. The ideal ECD would reduce some of this complexity by understanding individuals' preconceptions, knowledge levels, and abilities to process information. The objective of the ECD is to present information that can be easily processed within the constraints of the information processing model, thus bringing value to the consumer and other stakeholders. The ECD would provide information in the decision-making form the consumer understands while still able to make decisions deemed correct by the consumer.

A view integrating the three components of energy management, decision making, and CTI display provides a holistic picture of ECDs. This picture presents the implications of energy management decision-making on ECD design. The energy management review yielded common themes; first, prior attitudes, existing motivators, and demographics play a material role in energy management decisions.[6,7] The integrated decision-making model revealed three specific decision-making approaches: routine, irrational, and rational. The attitudes, motivators, and demographics are drawn from long-term memory and play the largest role in irrational decision-making, as shown in Figure 2. Tying energy management decision-making knowledge to CTI experience, specific to ECDs, and display design guidelines would direct designers toward two potential, distinct display paths. The first ECD path would be an ECD presenting summary information in monetary terms that would trigger a routine or an existing predisposition, employing both colors and graphics. The

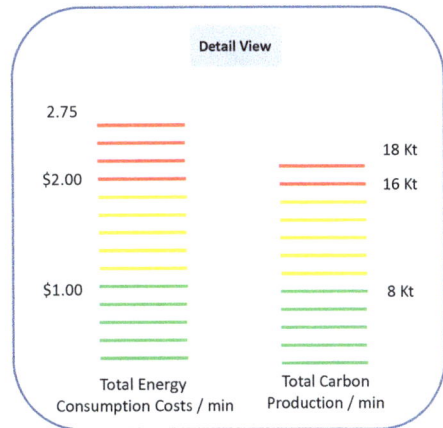

Fig 3. ECD One Prototype.

second ECD display path, for rational decision-makers, would show a detailed, appliance-specific breakdown in energy units with the ability to set goals and run detailed reports.

Figure 3 presents a prototype of the first path display, ECD One Prototype. It illustrates the integration of the energy management findings and display guidelines for routine and irrational decision makers. Two hypothetical predispositions and threshold values have been chosen for the prototype. Both of these need to be researched further and identified for specific regions. The prototype assumes that reducing costs or reducing carbon are the driving predispositions of the irrational decision. They could also be the trigger in the routine approach.

Figure 3 illustrates the prototype of a path one ECD unit designed for consumer energy management of routine and irrational decision makers, ECD One Prototype. The theoretical predisposition items are consumption costs and carbon production, in theory engaging consumers who have prior attitudes and existing motivators for either costs savings or environmental protection. Actual predisposition items would need to be researched and understood for each target region.

ECD One incorporates the CTI display guidelines determined to be optimal for routine and irrational decision-makers. These characteristics include familiar units; only necessary and immediate information is shown; graphics, colors, and complete words are also employed. ECD One design assumes those predisposed to environmental protection are familiar with the Kt units as a measure of carbon production. The values are based on approximate averages on a per capita basis; the top value represents the actual total energy consumption and total carbon production for that minute. Again, detailed research is required for each region to establish what these parameters would be. It also assumes a fixed energy rate, found in many regions. The ECDs are not, however, limited to fixed rates but rather highly suited to time-of-day rates; they are highly dynamic and feature-rich. They have real-time bidirectional network access able to communicate time-of-day rates as they change without delays.[36,37] As rates change from low to high, they can be color-coded to represent peak cost

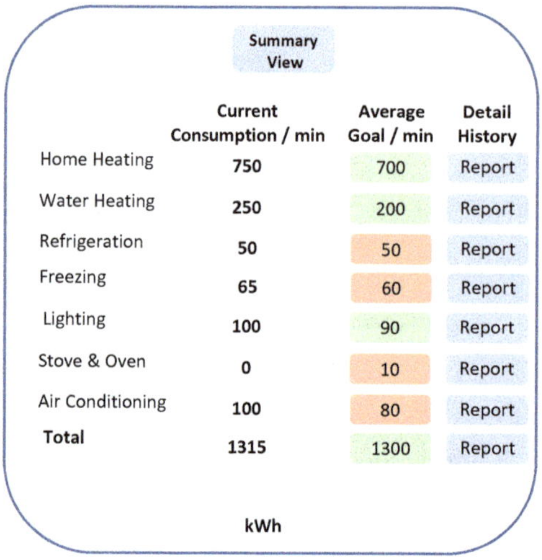

Summary View			
	Current Consumption / min	Average Goal / min	Detail History
Home Heating	750	700	Report
Water Heating	250	200	Report
Refrigeration	50	50	Report
Freezing	65	60	Report
Lighting	100	90	Report
Stove & Oven	0	10	Report
Air Conditioning	100	80	Report
Total	1315	1300	Report
	kWh		

Fig 4. ECD Two Prototype.

periods; additionally, they can provide detailed instantaneous rate impacts to the appliance level if desired by the consumer.

ECD One may also provide the opportunity in Graybiel's[16] routine process to be the stimulus, triggering action. Finally, there is a simple single button to toggle between the detail and summary view for other members in the home who may prefer the rational approach, requiring more detail. For additional simplicity, it is suggested that the summary view would be the default view upon start.

Figure 4 illustrates the prototype of a path two ECD unit designed for consumer energy management of rational decision-makers, ECD Two Prototype. Studies across the United States and Europe demonstrate that households consume the greatest electrical energy per kWh for home heating, heating of water, refrigeration, freezing, lighting, cooking, and air conditioning.[7] ECD Two incorporates appliance-specific energy consumption information, with the ability to set average target goals, and run detailed reports on each appliance. The report level of detail is not presented in this article but would employ the same CTI display guidelines. Again, the appliance values are arbitrary for illustrative purposes. ECD Two integrates the rational decision-making approach requiring greater amounts of information with the ability to obtain further details. The energy management review demonstrated that in certain instances, perhaps when taking the rational path, individuals would be motivated by setting and tracking goals. In ECD Two, the goal-setting button demonstrates the status by changing colors between green and red when the goal parameters are not being met. In addition, the history would provide greater detail, potentially through an output to an e-mail allowing the timeframe and level of detail to be set by the consumer. Again, a simple, prominently displayed toggle button allows routine or irrational decision-makers to change the view without complexity. ECD Two would not serve routine or irrational decision-makers if the information presented was not associated to an existing predisposition. The display information

would be filtered and considered surplus without cognitive processing.

There are still many unknowns; therefore, further research is recommended. A step-by-step approach is suggested for this research, beginning at a high level and becoming more specific. Researchers need to begin by targeting specific regions, as the results will vary by region and will not necessarily be pertinent to all areas. Demographics play an important role in energy management and are often associated with geographic regions.[6] Once the target region is established, a further breakdown of the decision-making approach employed in that region would set a focus, as the ECD design paths are distinct between routine/irrational and rational approaches. In addition, an understanding of existing predispositions in the region needs to be considered. Associated threshold parameters need to be established by region; for instance, per capita household energy consumption and carbon production show substantial global variations. Once the preferred decision-making approaches are established, predispositions found, and threshold parameters set by respective markets, policy makers and consumer stakeholders can choose or influence manufacturing designs in line with their market needs. Once this research is performed and integrated into the ECD design, a prototype should be developed and tested for desired results, similar to Figures 3–4 prototypes.

Demand side or load management opportunities of distributed micro-generation, such as renewable energy generation[38] and electric vehicles,[39] also exist. Individual predispositions may include household generation such as solar and wind generation or electric vehicle consumption and storage. An opportunity exists for the ECD to display the homeowner's respective feed-in generation or emergency storage in electric vehicles.[39] The flexibility of electrical in-home ECDs enables the future integration of emerging parameters, including tariff and carbon emissions data. In addition, the ECD can be a precursor to electronic billing with the ability to model invoice information and integrate multiple factors beyond consumptions such as distributed micro-generation[39] and carbon emissions, leading to a new model for billing and bill presentment.

Multiple methods exist for the dissemination of consumer energy consumption information. Information can be presented on displays targeting specific consumers—for instance, energy consumption information may be displayed on appliances[41] with high energy consumption such as dryers and freezers. Additionally, with real-time network connectivity[42] the ability to securely present household consumer energy information remotely exists in consumer work locations or during periods of travel. Additionally, the technology exists to present information in mobile forms through smartphones.[42] The findings supporting the prototypes ECD One and Two can be further extrapolated and researched to expand into these and other methods of consumer energy consumption information dissemination.

Much progress has taken place over the decades in CTI displays with a focus on industries such as avionics, medicine, control systems, and others. The majority of this work has been targeted at industry, where the user profile is homogenous and well defined through rigorous hiring, education, and training practices.[2] Through the decision-making analysis, it can be seen that the "one size fits all" approach to ECDs is not applicable. However, the literature would indicate that much of the research

6. Conclusion

Energy management behavior is complex, as many variables play a role, from demographics to cognitive abilities to process the information presented. The review of consumer energy management yielded a broad array of results, with energy management consumption varying from 0 to 20%.[6,7,8,9,10,11] Abrahamse et al.[7] and Fischer's[6] meta-analysis across 64 studies found the following underlying drivers to energy behavior to be prior attitudes, existing motivators, and sociodemographics. Through the information processing model, the prior attitudes and existing motivators of energy management were connected to distinct decision-making approaches. The routine, irrational, and rational approaches to decision making were considered to be applicable in consumer energy management. By understanding which decision-making path an individual employs, individual appropriate energy information can be presented. In the routine or irrational approach the ECD became trigger to action, whereas in the rational approach the ECD needed to provide detailed and specific information to assist in the weighting process of decision making. Further, connecting CTI display optimal design parameters and best-in-class results from Fischer's and Abrahamse's meta-analysis, the relevant energy information is sent into sensory input to be processed at the cognitive level.

The independent fields of consumer energy management, human information processing, and CTI display design have been brought together in this article. The ECD enabled the integration of the three fields by ensuring what is presented, how it is presented, and when it is presented are optimized to the specific individual and the respective decision-making approach chosen. The integration produced optimized ECDs to support both consumers and policy makers in energy management. Two ECD prototypes were developed. ECD One Prototype, designed for both routine and irrational decision-makers, and ECD Two Prototype, which served rational decision-makers. These ECD prototypes provide a starting point and templates for policy makers and power utilities toward supporting energy management behavior in their smart grid initiatives. Further research is recommended by target regions to establish regionally relevant parameters to include in ECD design.

and deployments of ECDs in smart grid infrastructure to date have followed this methodology, perhaps due to the influence of the industrial approach or physical (hardware) and economic constraints.

References

1. Depuru, S. S. S. R.; Wang, L. ; Devabhaktuni, V. Smart Meters for Power Grid: Challenges, Issues, Advantages and Status. *Renew. Sustain. Energ. Rev.* **2011**, *15*, 2736–2742.
2. Fitzpatrick, G.; Smith, G. Technology-Enabled Feedback on Domestic Energy Consumption: Articulating a Set of Design Concerns. *Pervas. Comp. IEEE* **2009**, *8*, 37–44.
3. Hargreaves, T.; Nye, M.; Burgess, J. Keeping Energy Visible? Exploring How Householders Interact With Feedback From Smart Energy Monitors in the Longer Term. *Energ. Pol.* **2013**, *52*, 126–134.
4. Ueno, T.; Inada, R.; Saeki, O.; Tsuji, K. Effectiveness of Displaying Energy Consumption Data in Residential Houses. Analysis on How the Residents Respond. *Proc. Eur. Council Energ. Efficient Econ.* **2005**, *6*, 19.
5. Anderson, W.; White, V. Exploring Consumer Preferences for Home Energy Display Functionality. Centre for Sustainable Energy, Technical Report, 2009.
6. Fischer, C. Feedback on Household Electricity Consumption: A Tool for Saving Energy? *Energ. Effic.* **2008**, *1*, 79–104.
7. Abrahamse, W.; Steg, L.; Vlek, C.; Rothengatter, T. A Review of Intervention Studies Aimed at Household Energy Conservation. *J. Environ. Psychol.* **2005**, *25*, 273–291.
8. Ehrhardt-Martinez, K. Changing Habits, Lifestyles, and Choices: The Behaviours That Drive Feedback-Induced Energy Savings. *Proceedings of the 2011 ECEEE Summer Study on Energy Efficiency in Buildings*, Toulon, France, 6–11. 2011.
9. Darby, S. *The Effectiveness of Feedback on Energy Consumption. A Review for DEFRA of the Literature on Metering, Billing and Direct Displays*; Environmental Change Institute, University of Oxford. **2006**, *486*.
10. Faruqui, A.; Sergici, S.; Sharif, A. The Impact of Informational Feedback on Energy Consumption—A Survey of the Experimental Evidence. *Energy* **2010**, *35*, 1598–1608.
11. Oltra, C.; Boso, A.; Espluga, J.; Prades, A. A Qualitative Study of Users' Engagement With Real-Time Feedback From In-House Energy Consumption Displays. *Energ. Pol.* **2013**, *61*, 788–792.
12. Frederiks, E. R.; Stenner, K.; Hobman, E. V. Household Energy Use: Applying Behavioural Economics to Understand Consumer Decision-Making and Behaviour. *Renew. Sustain. Energ. Rev.* **2015**, *41*, 1385–1394.
13. Murtagh, N.; Gatersleben, B.; Uzzell, D. 20: 60: 20: Differences in Energy Behaviour and Conservation Between and Within Households With Electricity Monitors. *PloS One* **2014**, *9*, e92019.
14. Wood, G.; Newborough, M. Energy-Use Information Transfer for Intelligent Homes: Enabling Energy Conservation With Central and Local Displays. *Energ. Build.* **2007**, *39*, 495–503.
15. Karjalainen, S. Consumer Preferences for Feedback on Household Electricity Consumption. *Energ. Build.* **2011**, *43*, 458–467.
16. Graybiel, A. M. Habits, Rituals, and the Evaluative Brain. *Annu. Rev. Neurosci.* **2008**, *31*, 359–387.
17. Ariely, D. *Predictably Irrational: The Hidden Forces That Shape Our Decisions*. New York: HarperCollins, 2008.
18. Brunsson, N. The Irrationality of Action and Action Rationality: Decisions, Ideologies and Organizational Actions. *J. Manag. Studies* **1982**, *19*, 29–44.
19. Edwards, W. The Theory of Decision Making. *Psychol. Bull.* **1954**, *51*, 380.
20. Wilson, C.; Dowlatabadi, H. Models of Decision Making and Residential Energy Use. *Annu. Rev. Environ. Resour.* **2007**, *32*, 169–203.
21. Gigerenzer, G.; Selten, R. Rethinking Rationality. In *Bounded Rationality: The Adaptive Toolbox*, Gigerenzer, G.; Selten, R., Eds.; The MIT Press: Cambridge, UK, 2001; pp 1–12.
22. Wickens, D.; Hollands, J.; Banbury, S.; Parasuarman, R. *Engineering Psychology and Human Performance* (4th ed.). New York: HarperCollins, 2013.
23. Lopes, M. A. R.; Antunes, C. H.; Martins, N. Energy behaviours as promoters of energy efficiency: A 21st century review. *Renew. Sustain. Energ. Rev.* **2012**, *16*, 4095–4104.
24. Bettman, J. R.; Johnson, E. J.; Payne, J. W. Consumer Decision Making. *Handbook of Consumer Behavior* **1991**, *44*, 50–84.
25. Weiten, W.; McCann, D. *Psychology: Themes and Variations*. Cengage Learning: Toronto, Ontario, Canada, 2013.
26. Schiffman, L.; Kanuk, L.; Wisenbilt, J. *Consumer Behavior* (19th ed.). Upper Saddle River, NJ: Prentice Hall, 2010.
27. Johnson, E. J.; Payne, J. W.; Bettman, J. R. Information Displays and Preference Reversals. *Org. Behav. Human Dec. Proc.* **1988**, *42*, 1–21.
28. Shepard, R. N. *Mind Sights: Original Visual Illusions, Ambiguities, and Other Anomalies, With a Commentary on the Play of Mind in Perception and Art*. W.H. Freeman/Times Books/Henry Holt: New York, 1990.

29. Egan, C. Graphical Displayers and Comparative Energy Information: What Do People Understand and Prefer. In *Summer Study of the European Council for an Energy Efficient Economy*, Proceedings of the Summer Study of the European Council for an Energy Efficient Economy; European Council for an Energy Efficient Economy: Stockholm, 1999; 2–12.

30. Tufte, E. R.; Graves-Morris, P. R. *The Visual Display of Quantitative Information* (vol. 2). Cheshire, CT: Graphics Press, 1983.

31. Smith, S. L.; Mosier, J. N. *Guidelines for Designing User Interface Software*. Bedford, MA: Mitre Corporation, 1986.

32. Ulrich, K.; Eppinger, S. *Product Design and Development* (5th ed.). New York: McGraw-Hill, 2012.

33. Krishnamurti, T.; Davis, A. L.; Wong-Parodi, G.; Wang, J.; Canfield, C. Creating an In-Home Display: Experimental Evidence and Guidelines for Design. *Appl. Energ.* **2013**, *108*, 448–458.

34. Xu, P.; Shen, J.; Zhang, X.; Zhao, X.; Qian, Y. Case Study of Smart Meter and In-home Display for Residential Behavior Change in Shanghai, China. *Energ. Procedia* **2015**, *75*, 2694–2699.

35. Schultz, P. W.; Estrada, M.; Schmitt, J.; Sokoloski, R.; Silva-Send, N. Using In-Home Displays to Provide Smart Meter Feedback About Household Electricity Consumption: A Randomized Control Trial Comparing Kilowatts, Cost, and Social Norms. *Energy*, **2015**, *90*, 351–358.

36. Depuru, S. S. S. R.; Wang, L.; Devabhaktuni, V. Smart Meters for Power Grid: Challenges, Issues, Advantages and Status. *Renew. Sustain. Energ. Rev.* **2011**, *15*, 2736–2742.

37. Donohoe, M.; Jennings, B.; Balasubramaniam, S. Context-Awareness and the Smart Grid: Requirements and Challenges. *Comp. Networks* **2015**, *79*, 263–282.

38. Zaheeruddin; Manas, M. Analysis of Design of Technologies, Tariff Structures, and Regulatory Policies for Sustainable Growth of the Smart Grid. *Energy Tech. Pol.* **2015**, *2*, 28–38.

39. Niesten, E.; Alkemade, F. How Is Value Created and Captured in Smart Grids? A Review of the Literature and an Analysis of Pilot Projects. *Renew. Sustain. Energy Rev.* **2015**, *53*, 629–638.

40. Masia, B.; Wetzstein, G.; Didyk, P.; Gutierrez, D. A Survey on Computational Displays: Pushing the Boundaries of Optics, Computation, and Perception. *Comp. Graphics* **2013**, *37*, 1012–1038.

41. Mahmood, A.; Khan, I.; Razzaq, S.; Najam, Z.; Khan, N. A.; Rehman, M. A.; Javaid, N. Home Appliances Coordination Scheme for Energy Management (HACS4EM) Using Wireless Sensor Networks in Smart Grids. *Procedia Comp. Sci.* **2014**, *32*, 469–476.

42. Fan, Z.; Kulkarni, P.; Gormus, S.; Efthymiou, C.; Kalogridis, G.; Sooriyabandara, M.; Chin, W. H. Smart Grid Communications: Overview of Research Challenges, Solutions, and Standardization Activities. *Comm. Surv. Tutorial IEEE* **2013**, *15*, 21–38.

5

Review of Design, Operating, and Financial Considerations in Flue Gas Desulfurization Systems

ANDREAS POULLIKKAS *

Department of Electrical Engineering, Cyprus University of Technology, Limassol, Cyprus

Abstract: In this work, measures available for the reduction of sulfur dioxide stack emissions are discussed. The various flue gas desulfurization (FGD) technologies available in the market, for the reduction of sulfur dioxide emissions, are presented. The process descriptions are discussed and the capital and operating costs of the various methods are presented. Also, possible sources of raw materials required in each process and the viable means of disposal of the end products are discussed. In the FGD market throughout the world, limestone wet scrubbers take the lead, the byproduct of which is a marketable gypsum. Most of the second (other wet scrubbing) covers the similar process but produces a disposal product. This sector also includes the seawater scrubbing process. Other significant sectors include spray dry scrubbers, regenerable processes, and sorbent injection systems. Combined SO_2/NO_x processes have a small share, and the trend is not expected to change.

Keywords: flue gas desulfurization, sulfur dioxide emissions, wet scrubbers, spray dry scrubbers

1. Introduction

Carbon monoxide, nitrogen oxides, volatile organic compounds, sulfur dioxide, and particulates are commonly referred as "criteria pollutants" because of their contribution to the formation of urban smog. These also have an impact on global climate, although their impact is limited because their radiative effects are indirect, since they do not directly act as greenhouse gases but react with other chemical compounds in the atmosphere. The combustion of fossil fuels, such as coal and heavy fuel oil (HFO), liberate three of the major air pollutants, such as sulfur dioxide (SO_2), nitrogen oxides (NO_X), and particulates.

Particulates can be removed satisfactorily by electrostatic precipitators or cyclones, whereas the nitrogen oxides emissions can be reduced by the use of low NO_X burners. Sulfur dioxide emissions can be reduced by the removal of sulfur from the fuel before combustion, by the removal of sulfur dioxide during the combustion process, or by the removal of sulfur dioxide from the flue gases after combustion.[1] The pre-combustion controls comprise selection of low sulfur fuels and fuel desulfurization. The combustion controls are mainly for conventional coal-fired plants and involve in-furnace injection sorbents. The post-combustion controls are the flue gas desulfurization (FGD) processes.[2]

Although recent concerns on emissions control suggest the diversification from current fossil fuels technologies to more sustainable ones—that is, the use of renewable energy sources[3] and the use of hydrogen[4]—such energy system transformation is expected to be completed after 2030–2040. Thus it is necessary in parallel to the integration of the above sustainable technologies to improve the efficiency and to reduce the associated cost of emission control technologies, such as FGD processes.

Based on the extensive literature review carried out for the purpose of this work, except in a few cases reported in Córdoba,[2] Álvarez-Ayuso et al.,[5] and Braden et al.,[6] no concise review of the various FGD technologies is available. In particular, regarding FGD technologies, financial considerations as well as sources of raw materials and disposal of the end products are not available in the literature. In this work the different technologies concerning FGD are reviewed. Also, the associated capital and operating costs are analyzed, and a review of possible sources of raw materials required in each process and the viable means of disposal of the end products are discussed.

In section 2, the sulfur dioxide emissions are reviewed. In section 3, an overview of the different available FGD technologies is presented, including process description as well as economic considerations. In section 4, the raw materials required in each process and end products are discussed. The conclusions are summarized in section 5.

*Address correspondence to: Andreas Poullikkas, Department of Electrical Engineering, Cyprus University of Technology, P.O. Box 50329, 3603 Limassol, Cyprus. Email: apoullikkas@cut.ac.cy

2. Sulfur Dioxide Emissions

Sulfur dioxide enters the air mainly from industrial processes and from the combustion of hydrocarbon fuels with a substantial sulfur (S) content and represents a major source of air pollution. Sulfur dioxide is a colorless, corrosive gas that has a bitter taste, but no smell at low levels. The share of SO_2 emissions comes from the use of coal and oil in fossil-fired power plants, in industrial combustion units, in small combustion units in households, and in vehicles. Although the concentration of SO_2 in stack gases emitted by steam generation plants is usually in the range of 800 to 6000 mg/Nm^3, the volume of gases produced by the utility industry worldwide results in the liberation of a large tonnage of sulfur dioxide into the atmosphere.[7]

Sulfur dioxide is a chemical that can be dangerous in many ways. It is known to be lethal to humans at dose levels higher than 1000 mg/Nm^3 for a period of over 10 minutes. At lower levels, sulfur dioxide has been found to be a corrosive irritant to eyes and skin. Sulfur dioxide has been associated with a variety of respiratory diseases and increased mortality rates. Inhalation of sulfur dioxide can cause increased airway resistance by constricting lung passages. Sulfur dioxide is also one of the main ingredients in acid rain. Acid rain occurs when sulfur dioxide or other gaseous chemicals, such as nitrogen oxides, are released into the air. These gases ascend through the atmosphere, and when they reach upper cloud levels, they react with water, oxygen, and sunlight.[8] Various concentrations of sulfuric acid and nitric acid are then produced. These acids mix with the condensed water vapor in the clouds and fall to the ground with the water in various forms of precipitation. This precipitation with greater acidity is what is known as acid rain. Acid rain is dangerous to many natural ecosystems; it can harm plants, sand, make water undrinkable, and unlivable for many animals. The biggest producers of sulfur dioxide emissions are electrical utilities that produce electricity through the burning of fossil fuels.

Sulfur dioxide emissions, from combustion installations using coal or HFO, can be reduced by fuel desulfurization or by FGD, as well as by a combination of these two measures. Low sulfur content fuels with less than 0.5% S are available on the market, the cost of which is high. Flue gas desulfurization is an effective method in which the sulfur dioxide can be removed by the treatment of flue gases by a means of SO_2 absorber.[5] Although water is partially effective as a medium for SO_2 absorption, with removal efficiencies up to 20%, solutions containing appropriate chemical absorbers or seawater have been shown to be practically effective, and sulfur dioxide removal efficiencies ranging from 50% to 99.8% have been achieved by treating waste gases with a variety of chemical reactants.[9] When using a high sulfur content fuel (e.g., 3.5% S), the required SO_2 removal efficiency to reduce the emission values from 5950 mg/Nm^3 to 400 mg/Nm^3 must be above 93%.

3. Flue Gas Desulfurization Technologies

Flue gas desulfurization is an efficient method for the reduction of the sulfur dioxide emissions.[2] Many processes are available in the market, such as (a) wet scrubbers, (b) spray dry scrubbers, (c) sorbent injection, (d) regenerable processes, and (e) combined SO_2/NO_X removal processes. The different flue gas desulfurization available technologies are compared in Table 1. The distribution of existing FGD plants on coal-fired generating units is shown in Figure 1. In the FGD market throughout the world, limestone wet scrubbers take the lead, the byproduct of which is a marketable gypsum. Most of the second (other wet scrubbing) covers the similar process but produces a disposal product. "Other wet scrubbing" sector shown in Figure 1 includes the seawater scrubbing technology as well, whereas other important sectors include spray dry scrubbers, regenerable processes, and sorbent injection systems. Combined SO_2/NO_x processes have a small share around the world with no expectation for

Table 1. FGD available technologies.

Technology	Raw materials	Byproducts	Max fuel S content	Max SO$_2$ removal efficiency
Wet scrubbers				
− Limestone scrubbing	Limestone and water	Marketable gypsum, sludge, waste water	3.5%	95–99%
− Lime scrubbing	Lime and water	Marketable gypsum, sludge, waste water	3.5%	95–99%
− Seawater scrubbing	Seawater and limestone or lime	Waste treated seawater	3.5%	90–95%
Spray dry scrubbers	Lime or calcium oxide and water	Mixture of calcium-sulfate, sulfite, and flying ash	3.5%	90–95%
Sorbent injection processes	Limestone or hydrated lime	Mixture of calcium-sulfate, sulfite, and flying ash	N/A	50%
Dry scrubbers	Limestone or lime or dolomite	Mixture of calcium-sulfate, sulfite, and flying ash	1%	50%
Regenerable processes	Sodium sulphite or ammonia	S or SO$_2$ or H$_2$SO$_4$ or ammonium sulfate	3.5%	90–99%
Combined SO$_2$/NO$_X$ removal processes				

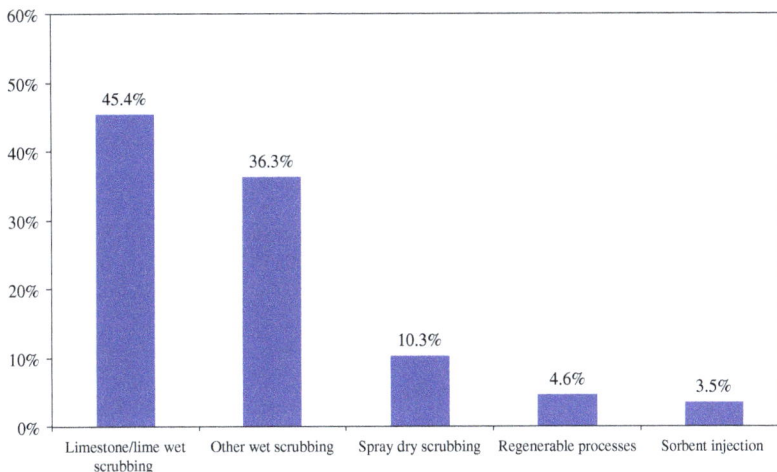

Fig. 1. Installed FGD plants on coal-fired generating units.

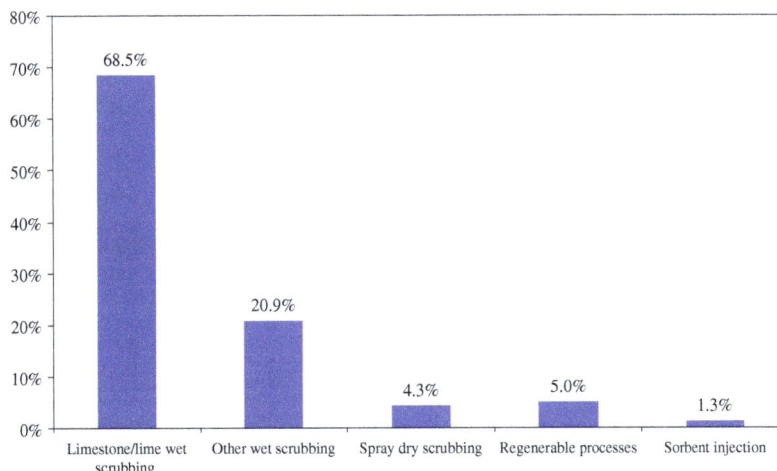

Fig. 2. Planned FGD plants on coal-fired generating units.

this share to change in the future.[8] The planned FGD plants are shown in Figure 2, where the limestone/lime wet scrubbers retains their dominant role. Taken with the next largest sector, which is mainly the wet scrubbing disposal process, the family of wet scrubbers accounts for nearly 90% of new FGD plants.[6] In this section, a review of these methods in terms of the process description, technical characteristics, and capital and operating costs is presented. Operating costs include the fuel cost, staff costs, insurance charges, rates, fixed maintenance, spare parts, chemicals, consumables, and water.

3.1 Wet Scrubbers

Wet scrubbers are the most widely used FGD technology for SO_2 control throughout the world. Calcium-sodium based sorbents have been used in a slurry mixture, which is injected into a specially designed vessel to react with the SO_2 in the flue gas. Commercial wet scrubbing systems are available in many variations and proprietary designs. Systems currently in operation include (a) lime/limestone/sludge wet scrubbers, (b) lime/limestone/gypsum wet scrubbers, and (c) seawater wet scrubbers.

Wet scrubbers can achieve removal efficiencies as high as 99%.[10] The preferred sorbent in operating wet scrubbers is limestone ($CaCO_3$), followed by lime (CaO). In (a), lime or limestone is used as a raw material, and the sludge byproduct produced is disposed of. The second system (b) is the most favored because of its availability, relative low cost, and ability to produce a marketable gypsum. The process is designed to produce a high-quality product (gypsum) suitable for use as raw material in various industries. It is estimated that with the increased cost of land filling in Europe and the introduction of increasingly stricter regulations regarding byproduct disposal, wet scrubbers producing marketable gypsum will overtake all other FGD technologies. The last option (c) is a convenient means for removing sulfur dioxide emissions from coastal power stations. The sorbent used is seawater which, after absorption, is treated with limestone or lime or seawater before discharge back to the sea.

3.1.1 Limestone Scrubbing

In the simplest configuration in wet limestone/gypsum scrubbers, systems (a) and (b), all chemical reactions take place in a single integrated absorber, resulting in reduced capital cost and energy consumption. The integrated single tower system requires less space, thus making it easier to retrofit in existing plants. The absorber usually requires a rubber, stainless steel, or nickel alloy lining as a construction material to control corrosion and abrasion. Fiberglass scrubbers are also in operation.[11] The overall chemical reaction, which occurs with a limestone sorbent, can be expressed in a simple form as

$$SO_2 + CaCO_3 \longrightarrow CaSO_3 + CO_2.$$

In practice, air in the flue gases causes some oxidation, and the final reaction product is a wet mixture of calcium sulfate ($CaSO_4$) and calcium sulfite ($CaSO_3$) as sludge. A forced oxidation step, in the scrubber or in a separate reaction chamber, involving the injection of air produces the saleable byproduct, gypsum ($CaSO_4 \bullet 2H_2O$), by the following reaction:

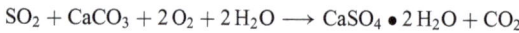

$$SO_2 + CaCO_3 + 2O_2 + 2H_2O \longrightarrow CaSO_4 \bullet 2H_2O + CO_2$$

The typical composition of FGD gypsum, compared with natural gypsum, is shown in Table 2, and the various possibilities for the use of gypsum are tabulated in Table 3. The various scrubber designs available are presented in Table 4.

The principal flow diagram of the limestone scrubbing method is shown in Figure 3. The main stages of the method are the preparation of raw materials, the absorption of the pollutants, the adjustment of the pH of the scrubber suspension, the oxidation of

Table 2. Typical composition (by mass) of natural and FGD gypsum.

Constituent	Gypsum	
	Natural	FGD
Moisture	1%	7–10%
Gypsum	78–95%	94–99%
Cl	<0.001%	<0.01%
Na$_2$O	0.02%	<0.01%
MgO (soluble in water)	–	0.1%
Fe	–	<0.05%
F (soluble in water)	–	<0.05%
SO$_2$	–	<0.05%
CO$_2$	–	1%
K$_2$O (soluble in water)	–	0.5%
pH	6–7	5–8

Table 3. Options for the use of FGD gypsum.

Sector	Application
Gypsum industry	} Gypsum plaster
	} Gypsum wallboard
	} Mortar for mining
Cement industry	Addition agent
Building material industry	Addition agent
Road construction	Aggregate
Landscaping	Aggregate
Mining	Pneumatic packing
Fertilizer	Land improver
Disposal	

Table 4. Available wet scrubber designs.

Scrubber design	Details
Spray tower	Pump pressure and spray nozzles atomize the scrubbing liquid into the reaction chamber providing large particle surface area for efficient mass transfer
Plate tower	Gas is dispersed into bubbles providing large sorbent surface area
Impingement	Vertical chamber incorporates perforated plates with openings partially covered by target plates which are flooded with sorbent. Flue gas and sorbent make contact around the target plate, creating a turbulent frothing zone
Packed tower	Flue gas flows upward through a packing material counter-current to the sorbent that is introduced at the top of the packing through a distributor
Fluidized packed tower	Similar to the packed tower, except that the packing is fluidized

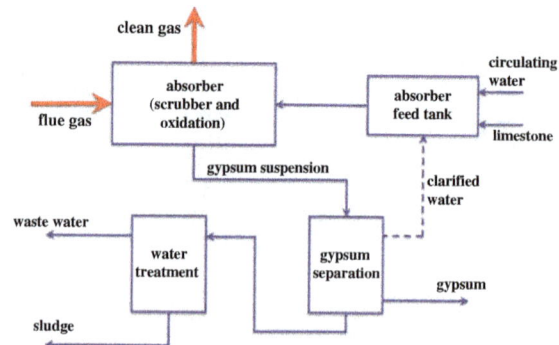

Fig. 3. Flow diagram of the limestone scrubbing method.

sulfite to sulfate, the gypsum separation system, and the waste-water treatment. Further parts of the installation are the storage spaces of limestone and the gypsum.

For the preparation of the absorber suspension, limestone is used. This is mixed with water (demineralized or town water is commonly used) into the absorber feed tank to form a homoge-nous suspension. The suspension, which has a pH of 7 and a $CaCO_3$ content of 20% (by mass), is fed into the absorber. The flue gases pass through the absorber, and the sulfur diox-ide is absorbed by the sorbent suspension. The formed sulfite is oxidized to sulfate by the use of air that is blown from the bottom part of the absorber to obtain enforced oxidation. The absorber efficiency depends on the length of time of direct con-tact between the sorbent suspension and the flue gases, which is directly related to the scrubber design (see Table 4). The clean flue gases are then reheated above 72°C and discharged to the station flue system.

With the absorption of SO_2, the pH value of the suspension drops.[12] The absorption rate depends greatly on the pH value, since for low pH values the SO_2 solubility diminishes; therefore, the suspension is neutralized with fresh feed sorbent suspension in order to keep pH value between 5 and 6.5. By this, there is a sufficient chemical absorption, and the quality of the gypsum produced is maintained.

The gypsum suspension is drawn off from the bottom of the absorber and has a solid matter content of 7–18% (by mass). The gypsum suspension is fed into the gypsum separator, where most of the water content is drained off. Final removal of water is com-pleted by centrifuge, and the gypsum is recovered as fine crystals. Upgrading of gypsum depends on the intended use of the gypsum produced. The amount of gypsum is proportional to the SO_2 mass flow separated from the flue gases. The quality of the gypsum is determined by the share of impurities contained, such as silicates, iron, aluminum, and magnesia compounds (see Table 2). These compounds enter gypsum via fly ash and limestone. The quality of gypsum depends also on the installations for the separation of gypsum. Compared to natural gypsum, the FGD gypsum, in most of the cases, has more constant chemical composition.[13]

The clarified water is returned to the absorber feed tank, where fresh limestone is continuously added to replace that converted to gypsum.[9] The high content of suspended solids and other trace elements in the waste water requires special treatment in a separate unit. The sludge produced is concentrated and then dewatered in a chamber filter press. The resulting press cake is disposed of. The purified water is drawn off in the outfall ditch since it contains heavy metals from the fuel combustion.[14]

Basically, the limestone scrubbing method can be installed downstream of any boiler with a maximum of 2×10^6 Nm^3/h flue gases flow rate. Where there are higher flow rates of flue gases, the FGD is carried out using two or more lines. To achieve high SO_2 efficiencies, removal of the dust from of flue gases is necessary prior to entry into the absorber. The method can be used for fuels with sulfur content of up to 3.5%.

There are many limestone scrubbing FGD suppliers in the market. The sulfur dioxide removal efficiency achieved, in most of the cases, is greater than 95%.[15] The capital cost of the limestone scrubbing method is influenced mainly by the volume flow of the flue gas. Retrofitting a boiler with a FGD installa-tion increases capital costs by 16%, due to differences resulting

from factors of space and boiler recondition. The operating costs are influenced by the annual full load operating hours as well as by the sulfur dioxide mass flow to be separated and by the flue gases flow rate.[16] The capital and operating costs of limestone scrubbing technology are ranging from 191–316 US$/KW for the former and 0.78–1.56 USc/kWh for the latter.

3.1.2 Seawater Scrubbing

A convenient and economical FGD process for removing sul-fur dioxide emissions from coastal power stations is the seawater scrubbing, in which the sorbent used is seawater. Seawater is suit-able for SO_2 absorption for two main reasons. First, it is naturally alkaline, since it contains bicarbonates, which buffer the addition of acid, and, second, the absorbed SO_2 is transformed into sulfate ions, a natural constituent of seawater. Seawater scrubbing offers potential advantages as it is of simple design, requires no bulk chemical in most of the cases, and has low capital and operat-ing costs.[17] The principal flow diagram of the seawater scrubbing method is shown in Figure 4. The main stages of the method are the dust collector, the absorption of the sulfur dioxide, and the adjustment of the pH of the seawater. In this process, the sulfur dioxide is removed from the flue gases by passing them through seawater. The sulfur dioxide reacts with the available water in two stages,

$$SO_2 + H_2O \longrightarrow 2H^+ + SO_3^-,$$

and

$$SO_3 + H_2O \longrightarrow 2H^+ + SO_4^-,$$

to form sulfite, sulfate, and hydrogen ions. The resulting acid-ity is neutralized by the natural alkalinity of the seawater, by the addition of limestone or lime, or by mixing the used seawater in the process with fresh seawater before discharge. The resulting water has a pH value of approximately 7. The process is capable of achieving over 90% of sulfur dioxide removal and can be used for fuels with sulfur content of up to 3.5%.[18]

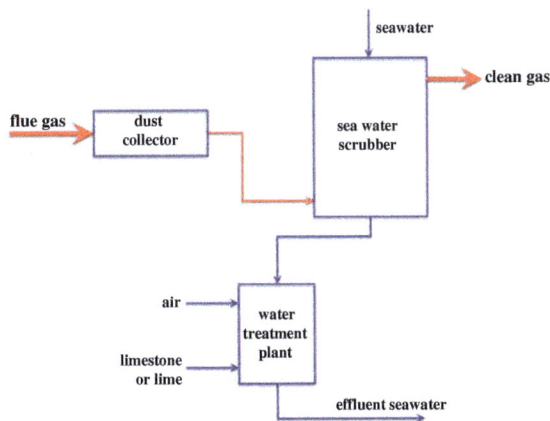

Fig 4. Flow diagram of the sea water scrubbing method.

Seawater scrubbers require the use of an efficient particulate control device such as an electrostatic precipitator (ESP) or a pre-scrubber in which the dust is collected. The flue gases are then forced through a heat exchanger to cool before entering the seawater scrubber, in which they flow upward against seawater flow.[19] The neutral alkalinity of the seawater and any added limestone or lime or seawater neutralizes the sulfur dioxide. From the scrubber the gases return to the heat exchanger in order to heat up before being discharged to the station flue system. The seawater, which has a pH value of 3, passes on to the aeration lagoon, where it is mixed with extra seawater to increase pH and is aerated in order to convert any sulfite to sulfate, to remove the CO_2 produced during acidification, and to raise the oxygen level. After aeration, the water is returned to the sea. The seawater scrubbing method can be installed downstream of any boiler with a maximum of 2.2×10^6 Nm^3/h flue gases flow rate. Where there are higher flow rates of flue gases, the FGD is carried out using two or more lines.

The environmental impact of seawater scrubbing is the effect of the discharged water on the local marine environment. The discharged water, the flow rate of which is 1×10^8 m^3/1000 GWh, has a biochemical oxygen demand because most of the plankton in the intake water is killed, and their breakdown by microorganisms will require oxygen. Biochemical oxygen demand (BOD) is the oxygen required by any organism to cover its oxygen needs.[20] As a result, since the oxygen in the local environment will be less, many organisms will probably die, and the local marine environment will become poorer.

3.2 Spray Dry Scrubbers

Spray dry scrubbers are the second most widely used FGD technology, and their application is limited to flue gases with 0.8 $\times 10^6$ Nm^3/h volume flow rate. Larger plants require the use of several modules to deal with the total flue gas flow. Spray dry scrubbers in commercial use have achieved sulfur dioxide removal efficiency not greater than 90%. The method can be used for fuel with sulfur content up to 3.5%.

Spray dry scrubbing (also referred as semi-dry process) purifies flue gases from SO_2 through the reaction with the lime suspension to the respective salts in a spray absorber. However, spray dry scrubbers require the use of an efficient particulate control device such as an ESP or fabric filter. The sorbent usually used is lime. The absorber construction material is usually carbon steel, making the process less expensive in capital costs compared with wet scrubbers. However, the necessary use of lime in the process increases considerably its operating costs.[21]

The principal flow diagram of the spray dry scrubbing method is shown in Figure 5. The main stages of the method are the pre-dust precipitator (depending on the byproduct utilization), the absorber feed tank, the spray absorber, and the main dust precipitator. Further parts of the installation are the storage spaces of lime and the byproduct.

For the preparation of the absorber suspension, lime is used. This is mixed with water (demineralized or town water is commonly used) into the absorber feed tank to form a homogenous suspension. The spray suspension is injected into the spray absorber through distribution systems. The resulting fine droplets are sprayed into the absorber and are mixed with the hot flue

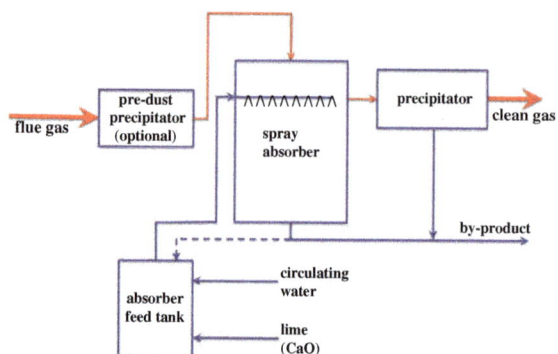

Fig. 5. Flow diagram of the spray dry scrubbing method.

gases. Special flue gas distributors achieve a good mixing of droplets with the flue gas, which is an important requirement for obtaining sufficient SO_2 removal efficiency. On entry, the temperature of the flue gas in the absorber usually varies between 120° and 160°C. The liquid portion of the scrubber suspension evaporates during the residence time in the spray absorber, and the flue gases cool down to temperatures between 65° to 80°C. Therefore, it is not necessary, in most cases, to reheat the flue gases before discharge to the station flue system. The residence time in the absorber normally varies between 10 to 50 seconds, and it is sufficient to allow for the SO_2 to react simultaneously with the hydrated lime to form a dry mixture of calcium sulfate and sulfite. A share of the dry and small-grained byproduct is drawn off at the bottom of the spray absorber. However, a major portion is entrained by the flue gases and precipitated in the main dust precipitator. No waste water results from the purification process because the water is completely evaporated in the spray dry absorber. Factors affecting the absorption chemistry include flue gas temperature, SO_2 concentration in the flue gas, and the size of the atomized or sprayed slurry droplets.[22]

ESP as well as fabric filters may be used as main dust precipitators. Fabric filters offer the advantage of a post reaction of still unreacted sorbent with the remaining SO_2 in the filter cake. This post reaction contributes up to 20% of the total removal efficiency. In the case of an ESP, the efficiency is substantially lower. However, the pressure drop of the ESP is significantly lower than that of the fabric filter. Operation experience has shown that ESP should be corrosion resistant since the outlet flue gas temperature of the spray absorber is 10° to 30°C above the adiabatic saturation temperature. Therefore, a sufficient thermal insulation of the dust precipitator is necessary in order to avoid a temperature drop below due point.

The byproduct also contains unreacted lime. A share of approximately 8% to 15% of the byproduct is thus recirculated to the absorber feed tank and added to the lime suspension. The resulting increased (between 30% and 50%) solid concentration of the sorbent suspension improves the drying process, so that the residual moisture of the byproduct decreases.[23] The mass flow rate of the product depends on the SO_2 concentration in the flue gases as well as on the mass flow of lime suspension used. The byproduct mostly consists of calcium sulfate and calcium

Table 5. Typical composition (by mass) of a spray dry scrubbing method byproduct (without fly ash content).

Constituent	Percentage
$CaSO_3$	40–70%
$CaSO_4$	5–20%
$Ca(OH)_2$	10–20%
$CaCO_3$	1–10%
$CaCl_2$	1–5%
CaF_2	trace amounts

sulfite also including carbonates and unreacted lime. The content of fly ash varies between 1% and 80% (by mass) depending on the existence and type of the pre-dust precipitator. Table 5 shows the composition of a typical byproduct without fly ash. The composition also depends on the quality of the sorbent and of the fuel quality. Normally, the byproduct is disposed of; however, industrial utilization may be possible in the future.

Capital costs of dry spray scrubbers depend mainly on the volume of flue gases but also on the type and the layout of the spray absorber and the injection systems. Different process variants, i.e., the type of the main dust precipitator and flow pattern in the absorber, can bring about a wide range of investment. The operating costs depend mainly on the annual full load operating hours, on the sulfur dioxide concentration, and on the process variant. The capital and operating costs of dry spray scrubbers technology range from 125–216 US$/KW for the former and 0.59–0.70 USc/kWh for the latter.

3.3 Sorbent Injection

Sorbent injection systems are divided into (a) furnace sorbent injection, (b) economizer sorbent injection, (c) duct sorbent injection, and (d) hybrid sorbent injection. The simplest technology is furnace sorbent injection, where a dry sorbent is injected into the upper part of the furnace to react with the sulfur dioxide in the flue gas. The finely grained sorbent is distributed quickly and evenly over the entire cross-section in the upper part of the furnace in a location where the temperature is in the range of 750–1250°C. Commercially available limestone or hydrated lime $(Ca(OH)_2)$ is used as a sorbent.[24] While the flue gas flows through the convective pass, where the temperature remains above 750°C, the sorbent reacts with SO_2 and O_2 to form $CaSO_4$. This is later captured in a fabric filter or ESP together with unused sorbent and fly ash.

In an economizer sorbent injection process, hydrated lime is injected into the flue gas stream near the economizer zone where the temperature is in the range of 300–650°C. In contrast to the furnace sorbent injection process, where the reaction temperature is around 1100°C, $Ca(OH)_2$ reacts directly with SO_2 since the temperature is too low to dehydrate $Ca(OH)_2$ completely. In this temperature range, the main product is $CaSO_3$ instead of $CaSO_4$, and the reaction rate is comparable to or higher than that at 1100°C. The byproduct is captured in a fabric filter or ESP together with unused sorbent and fly ash.

In duct sorbent injection, the aim is to distribute the sorbent evenly in the flue gas duct after the preheater where the

temperature is about 150°C. At the same time, the flue gas is humidified with water if necessary. Reaction with the SO_2 in the flue gas occurs in the ductwork, and the byproduct is captured in a downstream filter.[25]

The hybrid sorbent injection process is usually a combination of the furnace and duct sorbent injection systems aiming to achieve higher sorbent utilization and greater SO_2 removal. Various types of post furnace treatments are practiced in hybrid systems, such as injection of second sorbents, namely sodium compounds, into the duct and humidification in a specially designed vessel.[26]

The sorbent injection systems offer relatively low capital and operating costs, easy to retrofit, easy operation and maintenance with no slurry handling, reduced installation area requirements due to compact equipment, and no waste-water treatment. However, the sulfur dioxide removal efficiency is about 50%.

3.4 Dry Scrubbers

In the dry additive method, which can be viewed as a furnace sorbent injection method, the flue gases are purified by the reaction of the sulfur dioxide with an added pulverized dry sorbent.[27] The resulting byproduct is removed as dry salt in a dust precipitator downstream of the boiler. The sorbent additives commonly used are hydrated lime, limestone, and dolomite $(CaCO_3 \bullet MgCO_3)$. The addition of the additive can be carried out in three different ways, such as (a) addition to fuel, (b) addition to air, and (c) addition to flame area. Such method achieves sulfur dioxide removal efficiency of 50% and can be used for fuels with sulfur content up to 1%.

The principal flow diagram of the dry additive method is shown in Figure 6. The main stages of the method are the additive dosing and injection system and the dust precipitator. Further parts of the installation are the storage spaces of the additive and the byproduct.

The dosing and injection system distributes the additive and creates continuous and homogeneous flow of additive and flue gases. The number of nozzles as well as the injection levels depend on the specific conditions of the boiler.[28] The nozzles are adjusted to give an appropriate additive velocity in order to avoid the additive jets to hit the furnace walls. The additive is injected with a steady quantity of air into the furnace. There, the finely dispersed particles come into contact with the flue gases.[29] The reaction with SO_2 is a dry solid gas reaction that depends

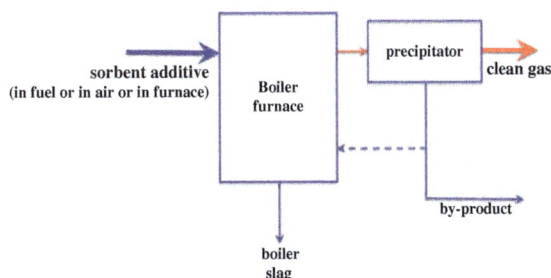

Fig. 6. Flow diagram of the dry additive method.

Table 6. Typical composition (by mass) of a dry additive method byproduct.

Constituent	Additive	
	CaCO$_3$	Ca(OH)$_2$
	Percentage	
Fly ash	20–40%	25–50%
CaCO$_3$	10–20%	5–15%
CaSO$_4$	20–25%	25–30%
CaO	25–35%	25–35%

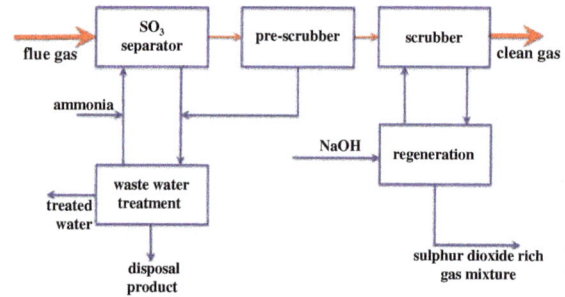

Fig. 7. Flow diagram of the Wellman-Lord method.

on the flue gas temperature, the additive residence time, the additive type and particle size, and the quantity of the additive. The byproduct is removed by a dust precipitator downstream of the boiler. All installations operating with the dry additive process use ESP as a de-dusting system. A share, up to 50%, of the byproduct, may be recirculated to the boiler to increase the efficiency of the process by diminishing the content of unreacted additive. Flue gases need no reheating before being fed to the stack. The composition of the byproduct depends directly on the ash content of the fuel and on the type of the additive used. Table 6 shows the composition of a typical byproduct. The industrial utilization of the product is limited to bedding material for road construction, land fill, and neutralization agent in industrial processes.[30] The disposal of such a byproduct is not permitted.

Variations in the boiler load and, thus, variations on the temperature profile within the furnace have a negative influence on the desulfurization efficiency. To counteract these effects, the sorbent additive is injected in different dosing according to the boiler load. The dry additive method lowers the boiler efficiency from about 0.7% to 1%.

The capital and operating costs are fairly low, ranging from 29–7.74 US$/kW for the former and 0.39–0.70USc/kWh for the latter.

3.5 Regenerable Processes

In regenerable processes, the sorbent is regenerated chemically or thermally and reused. Elemental sulfur or sulfur dioxide or sulfuric acid can be recovered. The revenue from these byproducts can compensate partially for the higher capital costs required in such FGD systems. Systems currently in operation include (a) the Wellman-Lord process and (b) the Walther process.

Regenerable processes generally require no waste disposal, produce little waste water, and have low sorbent makeup requirements.[31] In most such systems, a pre-scrubber is essential to control chlorides. Although these processes can achieve high SO$_2$ removal efficiencies (>95%), they have in general high capital costs and power consumption. The main influencing parameters regarding the capital cost are the flue gas flow rate as well as the quality and composition of the used fuel. In comparison to wet scrubbing systems, the capital cost is higher due to the extra regeneration units, which is in the range of 383–650 US$/kW. The operating costs are determined mainly by the annual full load operating hours and by the quality and composition of the fuel.

3.5.1 Wellman-Lord Process

The Wellman-Lord process, system (a), uses a sodium sulfite (Na$_2$SO$_3$) solution as a sorbent to remove the sulfur dioxide emissions. In the regeneration unit, the sulfur dioxide is desorbed from the sorbent liquid. The resulting SO$_2$ rich gas can be processed to elemental sulfur, liquid SO$_2$, or sulfuric acid. The principal flow diagram of the Wellman-Lord process is shown in Figure 7. The main stages of the method are the sulfite separation, the pre-scrubbing of the flue gases, the absorption of the sulfur dioxide, the regeneration, the sulfur dioxide rich gas processing, and the waste treatment.

Ammonia (NH$_3$) water is injected into the flue gases stream at a temperature between 165° and 170°C. Ammonium sulfate ((NH$_4$)$_2$SO$_4$) is then formed, and it is precipitated with the dust. The residue is mixed with the waste water of the pre-scrubber and purified in the waste-water treatment unit. In the pre-scrubber, which is situated at the bottom of the scrubber, the chlorides and the fluorides are eliminated. For high removal efficiency, the pre-scrubber sorbent liquid is acidified with hydrochloric acid reaching a pH value of 2. Also in the pre-scrubber the flue gases temperature is reduced to 65°C, which is necessary for absorption. In the scrubber, the sulfur dioxide is absorbed by a circulating concentrated sodium sulfite sorbent, where a sodium bisulfite (Na$_2$S$_2$O$_3$) is produced. In a secondary reaction, sodium sulfite is partially oxidized by the oxygen contained in the flue gases to form sodium sulfate (Na$_2$SO$_4$). The four-staged absorber is of the counter flow type.[31] The absorber usually consists of three packed beds with a height of approximately 1.6 m. The mean pressure drop in the scrubber is between 40 and 70 mbar, which is comparable to other wet scrubber systems. In order to increase the contact time between the scrubber sorbent and the flue gases, the sorbent is recirculated at each stage. The recirculated quantity is about five times the amount of fresh liquid required. The overflow of the bottom absorber stage is pumped to the buffer tanks and then fed to the regeneration unit.

The scrubbing liquid is regenerated in a regeneration unit where fresh sorbent liquid is regained by the removal of sulfur dioxide. The separated SO$_2$ rich gas consists of 92% vapor and 8% SO$_2$. The vapor content of the SO$_2$ rich gas is then reduced to 3%, so that, after compression, it can be processed to end products. The losses of the scrubber are compensated by adding NaOH.

The catalytic reduction of sulfur dioxide, using natural gas as a reducing agent, takes place in a two-stage process. In the first stage, the sulfur dioxide reacts with methane (CH_4), using an aluminum and calcium oxide catalyst, and forms hydrogen sulfite and sulfur. In the second stage, the hydrogen sulfite and the rest of sulfur dioxide are converted to elemental sulfur. The gaseous sulfur is condensed and stored at 150°C as liquid in sulfur tanks. In the waste-water treatment unit, the waste water containing ammonia is adjusted with a lime suspension to pH values of 10. Ammonia is recovered in a stripper and refed to the sulfite separator.[32]

The Wellman-Lord process can be used to purify flue gases, with a maximum flow rate of $0.6 \times 10^6 Nm^3/h$, from the combustion of fuels with sulfur content up to 3.5%, and achieves sulfur dioxide removal efficiency of 95%. Where there are higher flow rates of flue gases, the FGD is carried out using two or more lines.

3.5.2 Walther Process

In the Walther process, system (b), flue gas desulfurization is achieved by the reaction of the main sulfur dioxide component with ammonia water, leading to ammonium sulfite, which is then oxidized to ammonium sulfate. The principal flow diagram of the Walther process is shown in Figure 8. The main stages of the method are the scrubbing system, with two scrubbers in series, and the byproduct treatment, which consists of an oxidization unit, a buffer tank, and an evaporator.

The scrubbing system consists of two scrubbers in series. The de-dusted flue gases (dust content approximately 50 mg/Nm^3) are fed to the top of the first scrubber, with a temperature of 70°C, and react with the recirculated sorbent solution, at a pH value of 6, and ammonium sulfite is formed. The sorbent solution, which is composed of 25% ammonia, is produced by mixing water with ammonia. The content of 200 mg/Nm^3 of SO_2 in the clean gases is considered a technically upper limit since for less SO_2 content the emissions of ammonia within the clean gases increase drastically. The ammonium salts produced in the first scrubber remain at the scrubber bottom. The cleaned gases are then fed to the second scrubber and, upon exit, are heated up from 50°C to 90°C before discharge to the station flue system.[33] The scrubbing solution withdrawn from the first scrubber is thoroughly mixed with air in an oxidation vessel, leading to oxidation of ammonium sulfite to ammonia sulfate. The pulverized ammonia sulfate

is granulated. The end product is marketable fertilizer with a nitrogen content of 20% to 25%.

The Walther process can be used to purify flue gases, with a maximum flow rate of $0.23 \times 10^6 Nm^3/h$, from the combustion of fuels with sulfur content up to 1.2%, and achieves sulfur dioxide removal efficiency of 90%. Where there are higher flow rates of flue gases, the FGD is carried out using two or more lines.

3.6 Combined SO₂/NOₓ Removal Processes

Combined SO_2/NO_x removal processes remain fairly complex and costly. However, emerging technologies have the potential to reduce SO_2 and NO_x emissions for less than the combined cost of conventional FGD for SO_2 control and selective catalytic reduction for NO_x control. Most processes are in the development stage, although some of them are commercially used on low to medium sulfur coal-fired power plants.

4. Raw Materials Selection

The selection of FGD raw materials, such as lime, limestone, and ammonia, to supply new or retrofitted FGD plants is complex. Determination of the through-life project cost requires an assessment of both the capital and operating costs. The cost of the feed stock must include both the ex-works price of the raw material and all the transport, handling, and storage costs. The quality of the raw material can have a significant influence on process performance, operating costs, and byproduct quality. The least cost and security of supply of raw material may conflict with technical quality requirements, such as chemical purity, silica content, and physical characteristics.[7] The delivered price is influenced by the differing logistics and costs for delivery from different sources.

The FGD processes require a range of raw materials. Some raw materials are naturally occurring minerals, the properties of which tend to be more variable than for manufactured chemicals.[1] The actual quantity of raw material required is determined by the concentration of sulfur and flue gas mass flow, the removal efficiency required, the process design, and feed stock quality. In many projects, the quality of the byproducts is also vital (e.g., for wallboard use, it has to meet a stringent specification set by the purchaser, such as moisture and particle size).

4.1 Limestone

Limestone is a naturally occurring mineral found throughout the world. Chemical composition in terms of calcium carbonate ($CaCO_3$) content varies by source and can also vary significantly within a source. The majority of limestone is extracted from open cast quarries, using conventional blasting, crushing, and screening techniques. Limestone is used in the steel and chemical industries; however, the majority is used in the construction industry as an aggregate. Limestone is readily transported as a bulk product by road, with transportation by rail and ship, where the facilities exist, being technically and economically feasible.[34]

In determining an appropriate source of feed stock limestone, it is necessary to consider limestone characteristics and how these affect the design and performance of the FGD plant. Calcium carbonate is the prime source of alkalinity for effecting sulfur

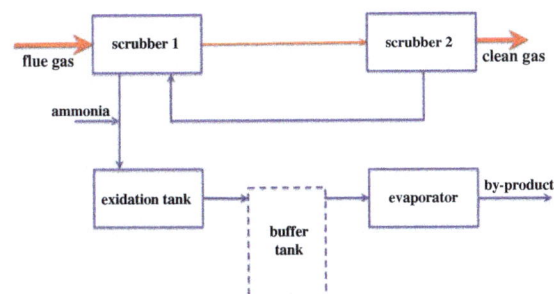

Fig. 8. Flow diagram of the Walther method.

dioxide neutralization. Magnesium carbonate ($MgCO_3$) is also an effective line material for neutralization of SO_2, is more reactive than calcium carbonate, and is found in variable amounts in many sources of limestone.[35] Limits are generally set by gypsum product manufacturers on magnesium compounds as they tend to be soluble. They can play a significant role interacting with chloride and, in the case of wallboard applications, impair the adhesion of the gypsum to the outer layers of paper.

Limitations on permissible levels of silica are set by gypsum users and by the FGD plant design, because silica is abrasive and high levels will cause extra wear in pumps, piping, and other process equipment items.[36] The presence of aluminum together with fluoride may cause a phenomenon known as limestone blinding, in which limestone particles are rendered nonreactive due to the formation of insoluble compounds.

Iron, which is present in limestone, can result in undesirable discoloration problems with gypsum-based products. If the iron level in gypsum is too high for plaster use, the gypsum is sent for use as a cement additive where the color effect is not a disadvantage.[9] Limestone may also be a significant source of trace elements in the process, and the amount is are usually a specific feature of the geology of the quarry.

Since, for wet limestone gypsum systems, the limestone enters the process as a slurry (often specified as 90% with <40 μm in particle size), the size, design, and power of the limestone milling plant is affected by the size distribution and hardness of the incoming stone. To achieve the necessary size grading of limestone, the stone must undergo crushing and milling. The energy required to grind limestone to a size suitable for subsequent feeding to the FGD plant is a function of the initial size of the limestone and its hardness.

4.2 Lime

Lime (CaO) is produced by calcination (1 kg of lime is produced from 1.8 kg of limestone) of limestone in kilns of various designs—for example, rotary or vertical shaft kilns. During calcination, the carbon dioxide is driven off to leave calcium oxide. Impurities in lime generally arise from the limestone feed and contamination from the hot gases produced by the fuel source in the kiln. The reactivity of lime is an important criterion and is influenced by the purity of the limestone feed, the kiln design, and kiln operating temperature, and regime. Physically, lime is produced in various size ranges, with median diameters from 0.5 to about 5 mm.

The systems selected for storage and use of lime require careful attention because of its irritant properties, which can cause serious damage to unprotected eyes or skin. The occupational exposure standards are stringent, so the material is normally kept confined in enclosed systems, silos, pipe work, or vessels, rather than in open storage and bunkers, which are acceptable for the more innocuous limestone material. Where personnel may be exposed to the material, comprehensive protective clothing and breathing masks are likely to be needed.[34] The material is an effective desiccant, attractive moisture from the air, which is another reason for use of closed containment systems, which also limit any dust emissions. Lime is generally transported in bulk using vacuum discharge tankers.

4.3 Hydrated Lime

Hydrated lime ($Ca(OH)_2$), is normally produced by mixing lime with water before full reaction takes place in a reaction vessel at atmospheric pressure. Hydrated lime is generally produced as a dry powder and is white or off-white in color. The compacted bulk density of hydrated lime is around half that of quicklime.[34] Hydrated lime is stored and transported in a similar way to lime; however, it is less hazardous for personnel.

4.4 Ammonia

Commercial grades of ammonia are available for FGD applications either as a solution with approximately 65% water at atmospheric pressure or in pure anhydrous form, as a pressurized refrigerated liquid. Ammonia is widely used in the manufacture of nitrogen fertilizers. In Europe, Japan, and the United States, significant amounts of ammonia are also used in FGD installations that are located at power plants.[37]

Anhydrous ammonia (NH_3) at atmospheric pressure is a colorless, lighter than air gas, with a strong distinctive odor, and is flammable in the presence of air. It is strongly alkaline when dissolved in water and highly corrosive (particularly on copper and zinc). Carbon steel is suitable for storage vessels and pipe work. Bulk delivery can be by road in tankers with a capacity for loads up to 18 t, with rail or ship delivery preferred where the facilities exist.

5. Byproducts Utilization

The byproducts quality depends mainly on the quality of the raw material. Performance constraints are also imposed on the quality of waste-water discharges, particularly in respect to trace element compounds. Trace elements present in FGD raw materials feature significantly in the total trace element inventory of the process.[1] This can be important for elements not easily removed by the waste-water treatment plant.

5.1 Gypsum

FGD gypsum can be produced at purity levels well over 90%, to equal or surpass the quality of many sources of natural gypsum. FGD gypsum is now being successfully used in many countries, particularly by plaster and wallboard manufacturers. The specific properties of an individual FGD gypsum are dependent on the composition of the fuel and limestone used, as well as the process design. Moisture content and crystal size and shape are also important to the gypsum processors.[38]

The value of FGD gypsum to gypsum users depends on the chemical and physical properties and delivery cost compared to supplies of natural gypsum. The ability of the wallboard facility to accept FGD gypsum without the need for significant plant modifications will also influence the price they will be prepared to offer. Limitations on chemical composition and other constraints such as moisture are generally dictated by end use. In Europe, FGD gypsum is used for wallboard manufacture, bag plaster, plaster products, and as a cement additive.[8] In addition to conventional plaster products, other applications have been dentified that offer alternative markets or potential future markets. These include their use in mining mortars, as soil additives to correct sulfur deficiency), fillers, floor screen, road base, and artificial reefs.

Gypsum is suitable for dry compaction using conventional land-filling methods. FGD gypsum has been successfully landfilled to suitably engineered and licensed sites in Europe. In the United States there are some installations where gypsum is pumped to a disposal site, where the water drains naturally for recovery or treatment.[39]

5.2 Ammonium Sulfate

Ammonium sulfate, which is widely used as fertilizer, does not exist as a natural resource, and 60% of world production arises as a byproduct from other processes, particularly from production of various organic chemicals used in the manufacture of synthetic fibers and plastics. There is a significant level of ammonium sulfate production in all of the major continental regions. However, in all regions the overall level of production has stagnated or fallen since the late 1980s.

In this situation, a significant proportion of the ammonium sulfate output from a large European FGD unit might have to be shipped to more distant markets in Africa, Latin America, or the large Asian market.[38] The additional freight and marketing costs associated with supplying more distant markets would tend to offset the likely sales revenues. The material, usually in the form of fine granules, can be handled in bulk by conventional means, although to minimize corrosion of containment vessels, it is advisable to ensure good-quality sealing against moisture ingress.

6. Conclusions

In this work, measures available for the reduction of sulfur dioxide stack emissions were discussed. The various FGD technologies available in the market for the reduction of sulfur dioxide emissions were presented. The process descriptions have been discussed, and the capital and operating costs the various methods were presented. Also, possible sources of raw materials required in each process and the viable means of disposal of the end products were discussed. In the FGD market throughout the world, limestone wet scrubbers take the lead, the byproduct of which is a marketable gypsum. Most of the second (other wet scrubbing) covers the similar process but produces a disposal product. This sector also includes the seawater scrubbing process. Other significant sectors include spray dry scrubbers, regenerable processes, and sorbent injection systems. Combined SO_2/NO_x processes have a small share, and the trend is not expected to change.

References

1. Wirsching, F.; Hüller, R.; Olejnik, R. FGD Gypsum Definitions and Legislation in the European Communities, in the OECD and in Germany. *Stud. Environ. Sci.* **1994**, *60*, 205–216.

2. Córdoba, P. Status of Flue Gas Desulphurisation (FGD) Systems From Coal-Fired Power Plants: Overview of the Physic-Chemical Control Processes of Wet Limestone FGDs. *Fuel* **2015**, *144*, 274–286.

3. Fokaides, P. A.; Kylili, A.; Pyrgou, A.; Koroneos, C. J. Integration Potentials of Insular Energy Systems to Smart Energy Regions. *Energ. Tech. Policy* **2014**, *1*, 70–83.

4. Mousavi Ehteshami, S. M.; Chan, S. H. Techno-Economic Study of Hydrogen Production via Steam Reforming of Methanol, Ethanol, and Diesel. *Energ. Tech. Policy* **2014**, *1*, 15–22.

5. Álvarez-Ayuso, E.; Querol, X.; Tomás, A. Environmental Impact of a Coal Combustion-Desulphurisation Plant: Abatement Capacity of Desulphurisation Process and Environmental Characterisation of Combustion Byproducts. *Chemosphere* **2006**, *65*, 2009–2017.

6. Braden, J.; Kolstad, C.; Woock, R.; Machado, J. Is Coal Desulphurisation Worthwhile? Evidence From the Market. *Energ. Policy* **2001**, *29*, 217–225.

7. Büke, T.; Köne, A. Ç. Estimation of the Health Benefits of Controlling Air Pollution From the Yatağan Coal-Fired Power Plant. *Environ. Sci. Policy* **2011**, *14*, 1113–1120.

8. Feuerborn, H. J. Coal Ash Utilisation Over the World and in Europe. In *Workshop on Environmental and Health Aspects of Coal Ash Utilization*, International Workshop, Tel-Aviv, Israel, Nov 23–24, 2005.

9. Córdoba, P.; Castro, I.; Maroto-Valer, M.; Querol, X. The Potential Leaching and Mobilization of Trace Elements From FGD-Gypsum of a Coal-Fired Power Plant Under Water Recirculation Conditions. *J. Environ. Sci.* **2015**, *32*, 72–80.

10. Akiho, H.; Ito, S.; Matsuda, H. Effect of Oxidizing Agents on Selenate Formation in a Wet FGD. *Fuel* **2010**, *89*, 2490–2495.

11. Borah, D.; Baruah, M. K. Kinetic and Thermodynamic Studies on Oxidative Desulphurisation of Organic Sulphur From Indian Coal at 50–150°C. *Fuel Process. Technol.* **2001**, *72*, 83–101.

12. Córdoba, P.; Font, O.; Izquierdo, M.; Querol, X.; Tobías, A.; López-Antón, M. A.; Ochoa-Gonzalez, R.; Díaz-Somoano, M.; Martínez-Tarazona, M. R.; Ayora, C.; et al. Enrichment of Inorganic Trace Pollutants in Re-Circulated Water Streams From a Wet Limestone Flue Gas Desulphurisation System in Two Coal Power Plants. *Fuel Process. Technol.* **2011**, *92*, 1764–1775.

13. Leiva, C.; García Arenas, C.; Vilches, L. F.; Vale, J.; Gimenez, A.; Ballesteros, J. C.; Fernández-Pereira, C. Use of FGD gypsum in fire resistant panels. *Waste Manag.* **2010**, *30*, 1123–1129.

14. Xiong, Y.; Niu, Y.; Tan, H.; Liu, Y.; Wang, X. Experimental Study of a Zero Water Consumption Wet FGD System. *Appl. Therm. Eng.* **2014**, *63*, 272–277.

15. Dou, B.; Pan, W.; Jin, Q.; Wang, W.; Li, Y. Prediction of SO_2 Removal Efficiency for Wet Flue Gas Desulfurization. *Energy Convers. Manag.* **2009**, *50*, 2547–2553.

16. Stergarek, A.; Horvat, M.; Frkal, P.; Stergarek, J. Removal of Hg0 from flue gases in Wet FGD by Catalytic Oxidation With Air—An Experimental Study. *Fuel* **2010**, *89*, 3167–3177.

17. Bilinski, H.; Matkovič, B.; Kralj, D.; Radulović, D.; Vranković, V. Model Experiments with CaO, MgO and Calcinated Dolomite for Fluoride Removal in a Wet Scrubbing System With Sea Water in Recirculation. *Water Res.* **1985**, *19*, 163–168.

18. Antoniou, C. Vasilikos Power Station, Electricity Authority of Cyprus, Larnaca, Cyprus. Personal communication, May 2014.

19. Xin, M.; Gustin, M. S.; Ladwig, K. Laboratory Study of Air-Water-Coal Combustion Product (Fly Ash and FGD Solid) Mercury Exchange. *Fuel* **2006**, *85*, 2260–2267.

20. Knutzen, J. Effects of Decreased pH on Marine Organisms. *Mar. Pollut. Bull.* **1981**, *12*, 25–29.

21. Tilly, B.; Griffiths, A.; Golland, E. Flue Gas Desulphurisation at Longannet Power Station, Scotland; A Regulators View of the BPEO Assessment. In *The Institute of Energy's Second International Conference on Combustion & Emissions Control*, Proceedings of The Institute of Energy Conference, London, UK, Dec 4–5, 1995; Elsevier: Oxford, 1995; 9–60.

22. Ma, X.; Kaneko, T.; Xu, G.; Kato, K. Influence of Gas Components on Removal of SO2 From Flue Gas in the Semidry FGD Process With a Powder–Particle Spouted Bed. *Fuel* **2001**, *80*, 673–680.
23. Elseviers, W. F.; Verelst, H. Transition Metal Oxides for Hot Gas Desulphurisation. *Fuel* **1999**, *78*, 601–612.
24. Álvarez-Rodríguez, R.; Clemente-Jul, C. Hot Gas Desulphurisation With Dolomite Sorbent in Coal Gasification. *Fuel* **2008**, *87*, 3513–3521.
25. Ochoa-González, R.; Córdoba, P.; Díaz-Somoano, M.; Font, O.; López-Antón, M. A.; Leiva, C.; Martínez-Tarazona, M. R.; Querol, X.; Fernández Pereira, C.; Tomás, A.; et al. Differential Partitioning and Speciation of Hg in Wet FGD Facilities of Two Spanish PCC Power Plants. *Chemosphere* **2011**, *85*, 565–570.
26. Wolff, E. H. P.; Gerritsen, A. W.; Verheijen, P. J. T. Attrition of an Aluminate-Based Synthetic Sorbent for Regenerative Sulphur Capture From Flue Gas in a Fluidised Bed. *Powder Tech.* **1993**, *76*, 47–55.
27. Graf, R. First Operating Experience with a Dry Flue Gas Desulfurization (FGD) Process Using a Circulating Fluid Bed (FGD - CFB). In *Circulating Fluidized Bed Technology: Proceedings of the First International Conference*. Proceedings of the First International Conference on Circulating Fluidized Beds, Halifax, Nova Scotia, Canada, Nov 18–20, 1985; Elsevier: Oxford, 1986; 317–327.
28. Gutiérrez Ortiz, F.; Ollero, P. A Pilot Plant Technical Assessment of an Advanced In-Duct Desulphurisation Process. *J. Hazard. Mater.* **2001**, *83*, 197–218.
29. Nygaard, H. G.; Kiil, S.; Johnsson, J. E.; Jensen, J. N.; Hansen, J.; Fogh, F.; Dam-Johansen, K. Full-Scale Measurements of SO2 Gas Phase Concentrations and Slurry Compositions in a Wet Flue Gas Desulphurisation Spray Absorber. *Fuel* **2004**, *83*, 1151–1164.
30. Zheng, Y. Use of Spray Dry Absorption Product in Wet Flue Gas Desulphurisation Plants: Pilot-Scale Experiments. *Fuel* **2002**, *81*, 1899–1905.
31. Kiel, J. H. A.; Prins, W.; Van Swaaij, W. P. M. Modelling of Non-Catalytic Reactions in a Gas-Solid Trickle Flow Reactor: Dry, Regenerative Flue Gas Desulphurisation Using a Silica-Supported Copper Oxide Sorbent. *Chem. Eng. Sci.* **1992**, *47*, 4271–4286.
32. van der Grift, C. J. G.; Geus, J. W. Reactions of Silica-Supported Copper Oxide as a Regenerable Sorbent for Flue Gas Desulphurisation. *Thermochim. Acta* **1990**, *161*, 131–146.
33. Matthews, C. Flue Gas Desulphurisation - total design. In *Case Studies in Engineering Design*. Elsevier: Oxford, 1998, 217–231.
34. Caselles-Moncho. A.; Ferrandiz-Serrano, L.; Peris-Mora, E. Dynamic Simulation Model of a Coal Thermoelectric Plant With a Flue Gas Desulphurisation System. *Energy Policy* **2006**, *34*, 3812–3826.
35. Wirsching, F.; Huller, R.; Olejnik, R. Gypsum from Flue Gas Desulphurisation Plants. *ZKG Int.* **1994**, *47*, 65–69.
36. Kılıç, O.; Acarkan, B.; Ay S. FGD Investments as Part of Energy Policy: A Case Study for Turkey. *Energy Policy* **2013**, 1461–1469.
37. Pysh'yev, S. V.; Gayvanovych, V. I.; Pattek-Janczyk, A.; Stanek, J. Oxidative Desulphurisation of Sulphur-Rich Coal. *Fuel* **2004**, *83*, 1117–1122.
38. Galos, K. A.; Smakowski, T. S.; Szlugaj, J. Flue-Gas Desulphurisation Products From Polish Coal-Fired Power-Plants. *Appl. Energy* **2003**, *75*, 257–265.
39. Pasini, R.; Walker, H. W. Estimating Constituent Release from FGD Gypsum Under Different Management Scenarios. *Fuel* **2012**, *95*, 190–196.

6

Optimal Sizing of C-Type Passive Filters under Non-Sinusoidal Conditions

ISLAM F. MOHAMED[1], SHADY H. E. ABDEL ALEEM[2*], AHMED M. IBRAHIM[1], and AHMED F. ZOBAA[3]

[1] Faculty of Engineering, Cairo University, Giza, Cairo, Egypt
[2] 15th of May Higher Institute of Engineering, 15th of May City, Helwan, Cairo, Egypt
[3] School of Engineering and Design, Brunel University, Uxbridge, Middlesex, United Kingdom

Abstract: In the literature, much attention has been focused on power system harmonics. One of its important effects is degradation of the load power factor. In this article, a C-type filter is used for reducing harmonic distortion, improving system performance, and compensating reactive power in order to improve the load power factor while taking into account economic considerations. Optimal sizing of the C-type filter parameters based on maximization of the load power factor as an objective function is determined. The total installation cost of the C-type filter and that of the conventional shunt (single-tuned) passive filter are comparatively evaluated. Background voltage and load current harmonics are taken into account. Recommendations defined in IEEE standards 519-1992 and 18-2002 are taken as the main constraints in this study. The presented design is tested using four numerical cases taken from previous publications, and the proposed filter results are compared with those of other published techniques. The results validate that the performance of the C-type passive filter as a low-pass filter is acceptable, especially in the case of lower short-circuit capacity systems. The C-type filter may achieve the same power factor with a lower total installation cost than a single-tuned passive filter.

Keywords: Harmonic distortion analysis, optimization, passive filters, power quality, power systems harmonics

1. Introduction

Harmonic pollution of electrical distribution systems is not new; it can be found in all industrial facilities. Nonlinear loads seem to be the main source of harmonic pollution in a power distribution system. Degradation of the load power factor, increases in transmission line losses, and reduction of the transmission network efficiency are all expected due to the advance and sophistication of nonlinear loads; the level of voltage harmonic distortion in distribution networks will also increase significantly.[1–8]

There is a set of conventional solutions to the harmonic distortion problems that have existed for a long time. Among these solutions, shunt passive filters are the most frequently employed in power-quality markets because of their low cost, which represents the main interest of most users, in the case of power factor

*Address correspondence to: Shady H. E. Abdel Aleem, 15th of May Higher Institute of Engineering, 15th of May City, Helwan, Cairo 11721, Egypt. Email: engyshady@ieee.org

correction and harmonic filtering.[6,9–11] The shunt filters work by short-circuiting harmonic currents as close to the source of distortion as is practical. This keeps the currents out of the supply system.

Practically, the C-type filter has been in operation for years. However, a convenient algorithm for sizing its parameters has rarely been discussed in a simple way; when one asks about designing a C-type filter or a third-order filter, one discovers the problem. Recently, more researchers have tried to present the basic theory of C-type filters and to seek more data about their performance and impact on the power system network within which they are intended to operate.[11] Abdel Aleem et al.[1] introduces an optimal design of the C-type passive filter based on minimization of the total harmonic voltage distortion. The C-type filter has good suppression at the tuned frequencies and efficiently damps the resonance instead of shifting it to a lower harmonic order; also, it offers lower losses when tuned to low frequencies. However, additional features of the C-type filters, especially their installation cost and competitiveness with traditional shunt harmonic passive filters, mainly the single-tuned one, need to be introduced in a simple and convenient manner in order to advance their use in a wide practical domain and to maintain the main advantage of the simplicity of passive filters as a whole.

The optimal design of the passive filters under multi-constraints and multi-objectives is complex since it often involves inconsistent objectives and constraints. It is always treated as a nonlinear programming problem. Consequently, the Genetic Algorithm toolbox (GA) provided by MatLab software is selected to determine the required optimal design and to establish the suitability and effectiveness of the C-type passive filter. GA has several advantages, such as its ability to work with numerical values to build up objective functions without difficulty and its ability to be easily formulated for multi-objective optimization problems for many practical problems.[7,12]

In this article, the optimal design of C-type passive filters based on maximization of the power factor when the filter under non-sinusoidal conditions is presented, while taking into account compliance with the following constraints:

- Maintaining the load power factor PF in an acceptable specified range (90% $\leq$ PF $\leq$ 100%).
- Maintaining the total voltage harmonic distortion $VTHD$ at the point of common coupling between the supplier and the customer in an acceptable specified range (VTHD $\leq$ 5%) with each individual harmonic component being limited to 3%.[13]
- The total current demand distortion (TDD) should be limited to a standard percentage according to the system strength or simply the (I_{SC}/I_L) ratios given in [13].
- Compliance with IEEE Standard 18-2002[14] for shunt power capacitor specifications for its continuous operation.

2. C-Type Filter Configuration

The C-type filter is included in the category of broadband filters. Generally, broadband filters dampen commutation notches more

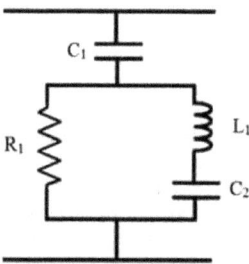

Fig. 1. The configuration of the C-type filter.

effectively than single-tuned filters, as they have a much broader bandwidth. Also, they can eliminate inter-harmonics generated by static frequency converters. Besides, they have good ability to dampen the resonance that may occur.[1] Figure 1 illustrates a circuit model of an installed C-type filter. In what follows, the equations that express the filter components R_1, L_1, C_1 (main capacitor), and C_2 (auxiliary capacitor) will be investigated. The C-type filter has different behaviors with various categories of frequencies and acts as various types of passive filters[1,9] because at fundamental frequency, it acts as a stand-alone capacitor (C_1), where the resistor is bypassed due to the tuned arm (series L_1–C_2); this circuit exhibits much lower losses. As the frequency increases, the filter acts as a single-tuned filter with a damping resistor where the inductor starts to resonate with ($C_1 + C_2$). At higher frequencies, the filter acts as a first-order filter where the inductor magnitude has higher magnitude than the magnitude of ($C_1 + C_2$).

The impedance of the C-type passive filter Z_{Ch} varies with the harmonic order h, where X is the magnitude of the auxiliary capacitive reactance, at fundamental frequency, which is equal to the magnitude of the inductive reactance; it can be illustrated as follows:

$$Z_{Ch} = R_{Fh} + jX_{Fh} = \frac{R * \left[jX\left(h - \frac{1}{h}\right)\right]}{R + \left[jX\left(h - \frac{1}{h}\right)\right]} - j\frac{X_{C1}}{h} \quad (1)$$

where R_{Fh} and X_{Fh} are given as

$$R_{Fh} = \frac{RX_{LCh}^2}{R^2 + X_{LCh}^2} \quad (2)$$

$$X_{Fh} = \frac{R^2 X_{LCh}}{R^2 + X_{LCh}^2} - \frac{X_{C1}}{h} \quad (3)$$

so that

$$X_{LCh} = hX_{L1} - \frac{X_{C2}}{h} = X\left(h - \frac{1}{h}\right) \quad (4)$$

3. System Description

Figure 2 demonstrates a single-phase equivalent circuit of a sample of a distribution system network. It shows the principle of

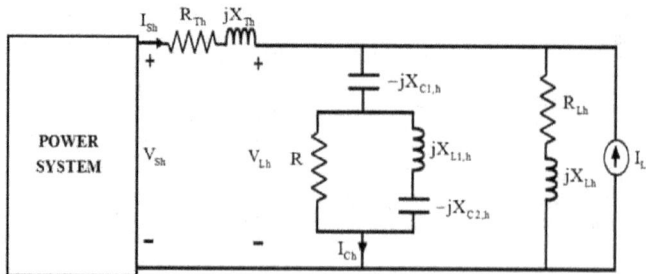

Fig. 2. The system under study.

operation of a *C*-type passive filter dedicated to a harmonic-current-source load (nonlinear load). Also, a background harmonic voltage distortion at harmonic number *h* exists.

Recalling Zobaa et al.,[6] the voltage source representing the utility supply voltage and the harmonic current source representing the nonlinear load are given as functions of time (*t*) as follows:

$$v_s(t) = \sum_h v_{Sh}(t) \tag{5}$$

$$i_L(t) = \sum_h i_{Lh}(t) \tag{6}$$

The *h*th harmonic Thevenin impedance Z_{Th} and load impedance Z_{Lh} are given as

$$Z_{Th} = R_{Th} + jX_{Th} \tag{7}$$

$$Z_{Lh} = R_{Lh} + jX_{Lh} \tag{8}$$

After some derivations using circuit analysis, the *h*th supply current and *h*th compensated load voltage for the system using a *C*-type filter are given, respectively, as

$$I_{Sh} = \frac{NI_{RE} + jNI_{IM}}{D_{RE} + jD_{IM}} \tag{9}$$

$$V_{Lh} = \frac{NV_{RE} + jNV_{IM}}{D_{RE} + jD_{IM}} \tag{10}$$

where

$$NI_{RE} = V_{Sh}(R_{Fh} + R_{Lh}) + I_{Lh}(R_{Fh}R_{Lh} - X_{Lh}X_{Fh}) \tag{11}$$

$$NI_{IM} = V_{Sh}(X_{Fh} + X_{Lh}) + I_{Lh}(R_{Fh}X_{Lh} + R_{Lh}X_{Fh}) \tag{12}$$

$$NV_{RE} = \mu V_{Sh} - I_{Lh}((R_{Th}R_{Lh} - X_{Lh}X_{Th})R_{Fh} - \alpha X_{Fh}) \tag{13}$$

$$NV_{IM} = \delta V_{Sh} - I_{Lh}((R_{Th}R_{Lh} - X_{Lh}X_{Th})X_{Fh} + \alpha R_{Fh}) \tag{14}$$

$$D_{RE} = R_{Fh}(R_{Th} + R_{Lh}) - X_{Fh}(X_{Lh} + X_{Th}) + \beta \tag{15}$$

$$D_{IM} = R_{Fh}(X_{Th} + X_{Lh}) + X_{Fh}(R_{Lh} + R_{Th}) + \alpha \tag{16}$$

so that

$$\alpha = (R_{Lh}X_{Th} + R_{Th}X_{Lh})$$

$$\beta = (R_{Th}R_{Lh} - X_{Lh}X_{Th})$$

$$\mu = (R_{Fh}R_{Lh} - X_{Lh}X_{Fh})$$

$$\delta = (R_{Fh}X_{Lh} + R_{Lh}X_{Fh})$$

Hence, the rms values of the compensated load voltage in volts and the compensated supply current in amperes are given as

$$V_L = \sqrt{\sum_{h=1}^{13} V_{Lh}^2} \tag{17}$$

$$I_S = \sqrt{\sum_{h=1}^{13} I_{Sh}^2} \tag{18}$$

For the system under study shown in Figure 2, the main compensated system indices demonstrating the system performance would be given as the following:

The compensated load power factor *PF* is given as

$$PF = \frac{P_L}{V_L I_S} = \frac{\sum_h G_{Lh}V_{Lh}^2}{\sqrt{\sum_h I_{Sh}^2 \sum_h V_{Lh}^2}} \tag{19}$$

Additionally, in terms of fundamental values of the load active and apparent powers, the compensated load displacement power factor *dPF* is given as

$$dPF = \frac{P_{L1}}{V_{L1}I_{S1}} \tag{20}$$

The transmission loss P_{LOSS} is given as

$$P_{LOSS} = \sum_h I_{Sh}^2 R_{Th} \tag{21}$$

Considering the voltage and current harmonics, which are found from (17) and (18), the total voltage harmonic distortion *VTHD* for the load voltage and total current harmonic distortion *ITHD* for the supply current are given as

$$VTHD = \frac{\sqrt{\sum_{h>1} V_{Lh}^2}}{V_{L1}} \tag{22}$$

$$ITHD = \frac{\sqrt{\sum_{h>1} I_{Sh}^2}}{I_{S1}} \tag{23}$$

Finally, the total current demand distortion *TDD* for the supply current based on the maximum demand current, which is chosen to be equal to the rated load current (I_L), is given as

$$TDD = \frac{\sqrt{\sum_{h>1} I_{Sh}^2}}{I_L} \tag{24}$$

The resistances R_{Lh} and R_{Th} are assumed to be frequency-independent (i.e., $R_{Lh} = R_L$ and $R_{Th} = R_T$).[15]

4. Problem Formulation

4.1 Passive Filter Design

The optimal design of the passive filters under multi-constraints and multi-objectives is complex since it often involves inconsistent objectives and constraints. In this article, the optimal design of C-type passive filters based on maximization of the power factor under non-sinusoidal conditions and calculation of their installation cost are presented, while taking into account compliance with the following constraints:

- Individual harmonics and total harmonic distortions of the voltage and current measured at the PCC are considered constraints for the proposed optimal design approach due to the harmonic limitations placed in IEEE Standard 519-1992. Accordingly, maintaining the total voltage harmonic distortion $VTHD$ and total current demand distortion TDD at the PCC between the supplier and the customer in acceptable ranges are the main constraints regarding harmonic suppression with the C-type filter.
- Maintaining the load power factor at an acceptable specified range ($\geq 90\%$) is an important constraint controls amount of the compensated reactive power with the C-type filter.
- Compliance with IEEE Standard 18-2002[14] for shunt power capacitor specifications for its continuous operation. To achieve this target, capacitors will be capable of continuous operation provided that none of the following limitations are exceeded[14]:

1. 135% of nominal rms filter current ($I_{nominal}$) based on rated kVA ($S_{nominal}$) and rated voltage ($V_{nominal}$), so that

$$100 * \frac{\sqrt{\sum_h I_{Ch}^2}}{I_{nominal}} \leq 135\% \tag{25}$$

where I_{Ch} is the capacitor current at harmonic number h and is given as

$$I_{Ch} = \frac{[V_{Sh}R_{Lh} - \beta I_{Ln}] + j[V_{Sh}X_{Lh} - \alpha I_{Lh}]}{D_{RE} + jD_{IM}} \tag{26}$$

2. 110% of the rated rms voltage, so that

$$100 * \frac{\sqrt{\sum_h V_{Ch}^2}}{V_{nominal}} \leq 110\% \tag{27}$$

where V_{Ch} is the main capacitor voltage at harmonic number h and is given as

$$V_{Ch} = \frac{X_{C1}}{h}\left[\frac{[V_{Sh}R_{Lh} - \beta I_{Lh}] + j[V_{Sh}X_{Lh} - \alpha I_{Lh}]}{D_{RE} + jD_{IM}}\right] \tag{28}$$

3. 120% of rated peak voltage ($V_{C, Peak}$), so that

$$V_{C,Peak} = \sum_h (I_{Ch})\left(\frac{X_{C1}}{h}\right) \tag{29}$$

4. 135% of nameplate kVA

$$100 * \frac{V_C I_C}{S_{nominal}} \leq 135\% \tag{30}$$

As a result, according to the proposed approach, optimal design problem of the C-type passive filter, provided that none of the IEEE Standard 18-2002 limitations are exceeded, can be formulated as follows:

Maximize PF (X_{C1}, X, and R)
subjected to:
$90\% \leq PF$ (X_{C1}, X, and R) $\leq 100\%$,
$VTHD$ (X_{C1}, X, and R) $\leq 5\%$,
TDD (X_{C1}, X, and R) $\leq$ Maximum TDD (defined in IEEE 519).

Genetic Algorithm toolbox (GA) provided by MatLab software is selected to determine the required optimal design and to establish the suitability and effectiveness of the C-type passive filter. Algorithm is initiated with a set of random solutions named population. An individual solution is represented in an encoded form called chromosome. Each chromosome consists of individual structures called genes. Solutions from one population are used to create a new one.[12] In order to create a new population, GA uses selection procedure. Selection is the procedure of maintaining and ignoring bad solutions from both the current and the next population. In the selection process, the solutions are selected according to their values of objective function (fitness). The more fitness, the more chance of being selected. The algorithms will repeat until a termination condition is satisfied. The best solution is returned to represent the optimum (global) solution.

Real-valued representation of parameters is used because real representations give faster, more consistent, and more accurate results. The goal of genetic algorithms is to search for the C-type filter main capacitive reactance (X_{C1}) in order to compensate the system's reactive power, determine the parallel resistance value (R) to ensure the required performance and effectiveness of the proposed filter, and determine reactance value (X) to ensure the required quality of the proposed filter. The optimization fitness (objective function) is based on maximizing load power factor for the system under study with the C-type filter, while complying with the IEEE standards 519-1992 and 18-2002 recommendations.

The lower and upper boundaries of the main capacitive reactance (X_{C1}) are selected as the reactance values of a free capacitor that improve the system's actual dPF from 71.65 to 100%. The range of variability of the inductive reactance (X) is chosen based on the lower and upper boundaries of the tuning harmonic order which is considered between 2 and 12. This search interval of the tuning harmonic order covers all harmonic orders that are observed in the simulated system. The range of variability of the parallel resistance (R) is chosen between 1 to 100 ohms based on a wide search interval of quality factor of the the C-type filter. Population size is chosen to be 100 individuals. Crossover probability equals 0.8. Mutation probability equals 0.01. GA termination condition is 30 generations. Additionally, roulette-wheel selection and shuffling crossover are used in the search algorithm.

4.2 Installation Cost Calculation of the C-Type Filters

In most cases, the main objective of a filter is that the values of filter parameters must be optimized to ensure that they are the most well-fitted solution for the specified techno-economic target. The following relations were determined to calculate the total installation cost of the C-type passive filters. Richards et al. and Zobaa et al.[16–17] summarize the methods used to calculate the cost of parameters of a conventional shunt (single-tuned) passive harmonic filter. The formulation of the harmonic passive filter cost (COST) is arranged in terms of the filter size that represents the reactive power supplied by the capacitors as follows:

$$COST = C_{C1} + C_{C2} + C_L + C_E \qquad (31)$$

where C_{C1}, C_{C2}, C_L and C_E are the main capacitor, auxiliary capacitor, reactor, and energy loss costs, respectively. In order to maintain simplicity, the capacitor and inductor costs are assumed to be equal and proportional to their ratings. So the capacitor unit cost in Egyptian pounds per kilovar $W_C = W_L = 50$ L.E./kvar so that

$$C_{C1} = W_C * V_{C,Peak} * I_C \qquad (32)$$

$$C_{C2} = W_C * \sqrt{I_{C1}^2 + \sum_{h>1} I_{ah}^2} * \sum_h \left(I_{ah} \frac{X_{C2}}{h} \right) \qquad (33)$$

$$C_L = W_L * \sqrt{I_{C1}^2 + \sum_{h>1} I_{ah}^2} * \sum_h h I_{ah} X_L \qquad (34)$$

$$C_E = 8760 * F_V * U_V * \frac{(1+T)^K - 1}{T(1+T)^K} * \sum_h \left(I_{Ch}^2 R_{Fh} \right) \qquad (35)$$

It must be mentioned that the constants included in the previous equations are chosen based on the Egyptian tariff. They can be summarized as follows: interest rate $T = 0.05$, filter lifetime $K = 15$ years, filter utilization factor $U_V = 1.0$, and the cost of power loss per kilowatt hour $F_V = 0.20$ L.E./kwhr [17], where I_{ah} is the hth auxiliary capacitor current in amperes.

5. Results and Discussion

Four cases of an industrial plant (Table 1) were simulated using the GA optimization method. The numerical data were taken from an example in IEEE Standard 519-1992.[13,18] Two short-circuit capacity expressed in mega volt-amperes system capacities are used in the study cases, where the 80 MVA$_{SC}$ is used in Cases 1 and 2, representing a weak system, while the 150 MVA$_{SC}$ is used in Cases 3 and 4, representing a stiff one. The three-phase active-power is 5100 kW and the three-phase inductive reactive-power is 4965 kvar. The 60-cycle supply line voltage is 4160 V. It is assumed that the load harmonics are not sufficiently serious to employ a harmonic filter, but when combined with source harmonics the usage of a pure capacitor would degrade the power factor and overload equipment. Consequently, a C-type passive filter is selected.[19]

Table 2 shows the uncompensated system results to be defined and compared with the proposed filter results. Table 3 shows a summary of the results for the four case studies. Table 4 shows

Table 1. Four cases of an industrial plant under study and their simulation results.

Parameters	Case 1	Case 2	Case 3	Case 4
Short-circuit capacity (MVA)	80		150	
V_{S1} (kV)	2.40	2.40	2.40	2.40
R_{T1} (Ω)	0.02163	0.02163	0.01154	0.01154
X_{T1} (Ω)	0.2163	0.2163	0.1154	0.1154
R_{L1} (Ω)	1.7421	1.7421	1.7421	1.7421
X_{L1} (Ω)	1.696	1.696	1.696	1.696
V_{S5} (V)	72	96	72	96
V_{S7} (V)	48	72	48	72
V_{S11} (V)	24	48	24	48
V_{S13} (V)	12	24	12	24
I_{L5} (A)	33	33	33	33
I_{L7} (A)	25	25	25	25
I_{L11} (A)	9	9	9	9
I_{L13} (A)	8	8	8	8

Table 2. Simulated results of the uncompensated system.

Parameters and cases	Case 1	Case 2	Case 3	Case 4
PF (%)	71.53	71.48	71.53	71.49
dPF (%)	71.65	71.65	71.65	71.65
I_S (A)	923.42	923.47	952.87	952.93
V_L(V)	2245.07	2246.59	2316.19	2317.83
P_{Loss} (kW)	18.45	18.45	10.48	10.48
VTHD (%)	3.71	4.48	3.19	4.09
ITHD (%)	4.35	4.45	4.45	4.55
TDD (%)	4.02	4.11	4.24	4.33

Table 3. Simulated results in the four cases for the optimization process for the C-type filter.

Parameters and cases	Case 1	Case 2	Case 3	Case 4
X_{C1} (Ω)	3.485	3.483	3.484	3.481
X (Ω)	0.668	0.788	0.801	0.498
R (Ω)	93.839	89.093	88.090	79.071
PF (%)	99.800	99.720	99.750	99.520
dPF (%)	99.990	99.990	99.990	99.990
I_S (A)	702.520	702.890	705.970	707.370
V_L(V)	2381.100	2382.150	2391.760	2392.560
P_{Loss} (kW)	10.680	10.690	5.750	5.770
VTHD (%)	2.560	3.240	2.640	3.010
ITHD (%)	5.410	6.070	6.280	8.340
TDD (%)	3.800	4.260	4.420	5.870

the expected installation cost of the C-type filter parameters. Comparison of the simulated results given in Tables 2 and 3 shows that the general performance of the method is satisfactory, providing improvement of the system's overall performance. Additionally, Table 3 shows that the proposed technique results in reduction of the supply current, lower transmission loss, higher transmission efficiency, and higher load power factor compared to the uncompensated system results shown in Table 2.

Table 4. Cost of the C-type filter parameters.

Parameters and cases	Case 1	Case 2	Case 3	Case 4
C_{C1} (thousands of L.E.)	82.104	82.253	83.751	82.7350
C_{C2} (thousands of L.E.)	15.738	18.609	11.981	18.9970
C_L (thousands of L.E.)	21.819	26.683	20.080	24.9580
C_E (thousands of L.E.)	1.450	2.545	3.545	1.6366
COST (thousands of L.E.)	121.049	130.159	119.367	128.3480

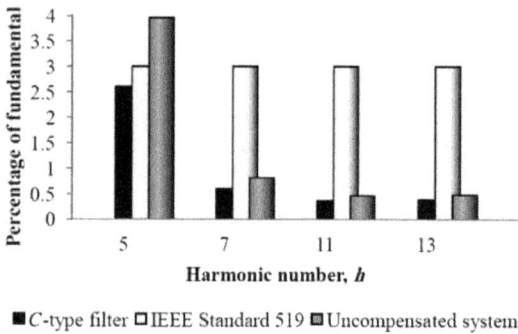

■ C-type filter ☐ IEEE Standard 519 ▨ Uncompensated system

Fig. 3. Harmonic content of the load voltage, Case 2.

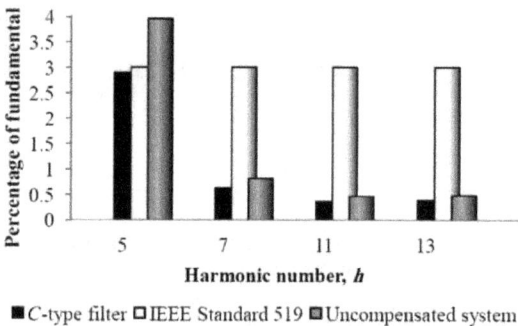

■ C-type filter ☐ IEEE Standard 519 ▨ Uncompensated system

Fig. 4. Harmonic content of the load voltage, Case 4.

As shown in Table 3, the $VTHD$ is dramatically reduced, satisfying the main concern of the IEEE standard 519. Figures 3 and 4 show the values of the load harmonic voltage after compensation where the proposed filter is introduced compared with values for the uncompensated system and the IEEE 519 limits.

Figure 5 shows the harmonic content of the supply current, before and after compensation, for the lower short-circuit capacity system (Case 1), while Figure 6 shows the harmonic content of the supply current, before and after compensation, for the higher short-circuit capacity system (Case 3), both with the same input data for harmonic contents. It has been shown that under distorted supply voltages, any trial to increase the power factor results in an increase in the $ITHD$.[6,18] However, the individual harmonic supply currents and the TDD percentage meet the standard limits defined in IEEE 519.

Table 5 shows the calculated capacitor limits for all cases. Comparison of the calculated and standard limits shows that all

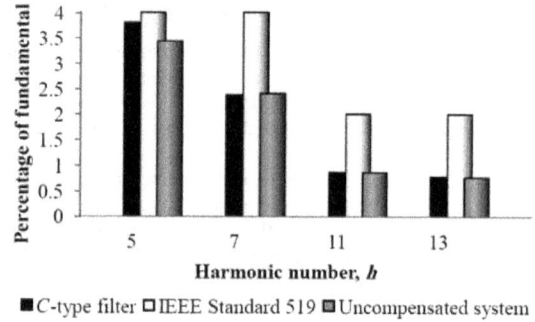

■ C-type filter ☐ IEEE Standard 519 ▨ Uncompensated system

Fig. 5. Harmonic content of the supply current, Case 1.

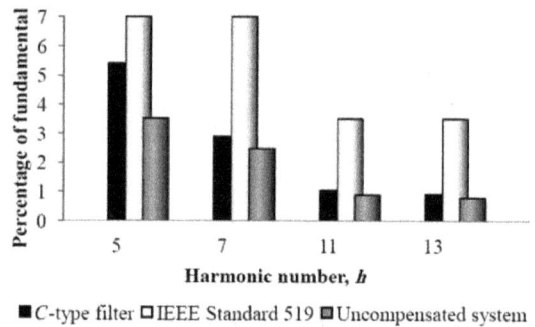

■ C-type filter ☐ IEEE Standard 519 ▨ Uncompensated system

Fig. 6. Harmonic content of the supply current, Case 3.

Table 5. Capacitor loading duties.

Parameters	Case 1	Case 2	Case 3	Case 4
Rms voltage (%)	99.17	99.17	99.61	99.62
Peak voltage (%)	100.13	100.24	100.44	101.40
Rms current (%)	99.22	99.24	99.65	99.83
Apparent power (%)	99.35	99.48	100.09	101.23

Table 6. Comparison of calculated and simulated $VTHD$ and $ITHD$ percentages.

Parameters and cases	Frequency-domain analysis		Time-domain analysis	
	$VTHD$ (%)	$ITHD$ (%)	$VTHD$ (%)	$ITHD$ (%)
Case 1	2.56	5.41	2.56	5.26
Case 2	3.24	6.07	3.26	6.33
Case 3	2.64	6.28	2.62	6.34
Case 4	3.01	8.34	3.15	8.10

values lie within the standard limits in the optimal filter design for all studied cases.

The numerical results for the system under study with time-domain simulation are presented based on Simulink of MatLab software. Table 6 shows the total harmonic distortion percentages for the load voltage and the supply current measured at the point of common coupling based on both frequency and time-domain simulations, respectively. It is obvious the reasonable agreement between them.

Figures 7 and 8 show the linear impedance-frequency scan for the system under study: Cases 2 and 4, respectively, both seen from the harmonic-source side.[17] It is obvious from both figures that C-type filters are more durable to parallel resonance hazards, especially in the case of lower short-circuit capacity systems.[1]

For equal short-circuit capacity systems, the additional supply voltage harmonic contents for the same harmonic load will result in an increase in the line current passing through the compensated load and thus a decrease in the load power factor.[20] Also, an increase in the voltage and current total harmonic distortions will be obvious. The transmission efficiency decreases because of the higher transmission losses and thus higher transmission voltage drop. Furthermore, the compensator components rating increases, which will be reflected in their cost.[21]

Considering the total installation costs of the C-type filter and the conventional single-tuned passive filter presented in Abdel-Aziz et al.,[15] Figures 9 and 10 show installation cost

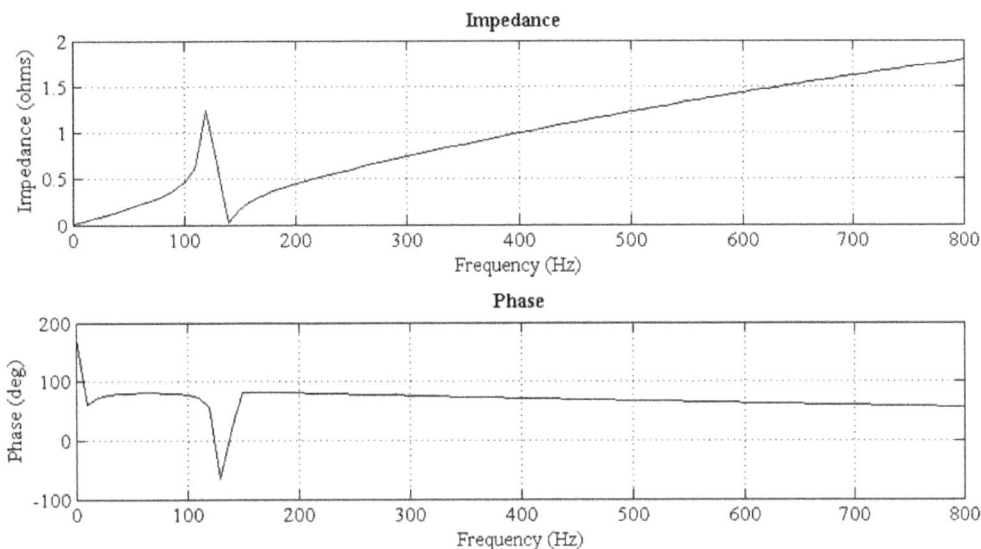

Fig. 7. Linear impedance-frequency scan, Case 2.

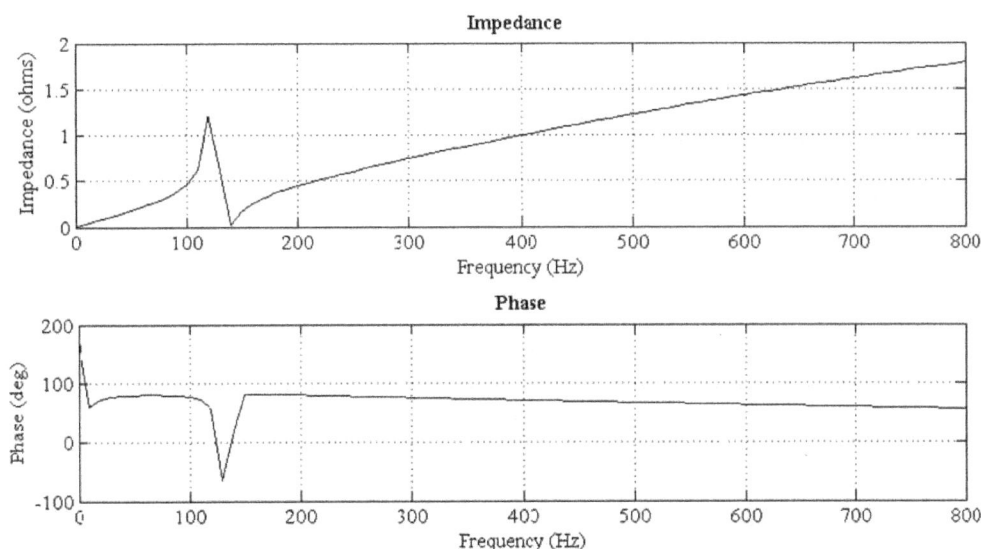

Fig. 8. Linear impedance-frequency scan, Case 4.

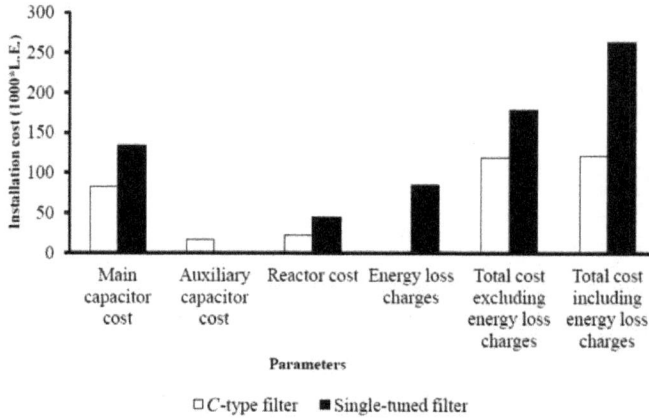

Fig. 9. Cost comparison of C-type and single-tuned passive filters, Case 1.

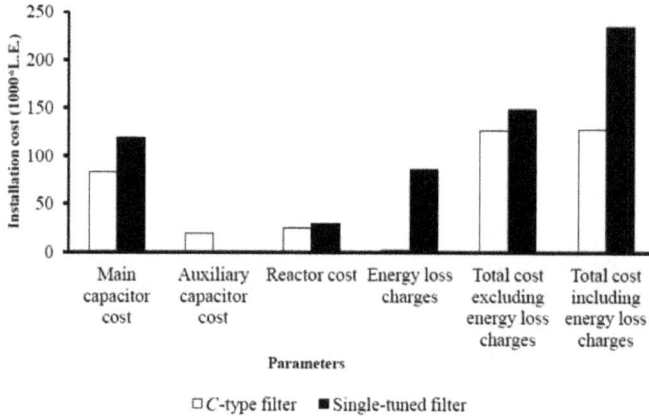

Fig. 10. Cost comparison of C-type and single-tuned passive filters, Case 3.

comparison between the two filters for the system under study in Cases 1 and 3, respectively. The conventional single-tuned passive filter data are taken from Zobaa et al.[6] and Abdel-Aziz et al.[15] However, it must be mentioned that the equivalent resistance of the reactor at fundamental frequency was ignored in Zobaa et al.[6] and Abdel-Aziz et al.[15]; thus, in the comparison it is assumed to have the smallest acceptable value of 1% of the fundamental value of the compensator reactor. This allows it to be taken into account in the results rather than being neglected. Additionally, Table 7 presents a comparison of the present method and the method presented in Abdel-Aziz et al.,[15] for all cases under study. It is obvious that results of the proposed filter are similar to those presented in Abdel-Aziz et al.[15]

Tables 3 and 7 show that the proposed filter provides higher *PF* percentage when compared to the conventional filter. It can also be observed that I_S, V_L, and *dPF* values obtained by the proposed filter and the conventional filter presented in Abdel-Aziz et al.[15] are very close to each other. Finally, it can be mentioned that the proposed filter has lower *VTHD, ITHD,* and P_{LOSS} values than those presented in Abdel-Aziz et al.[15]

Table 7. Simulated results in the four cases for the optimization process for the single-tuned passive filter presented in Abdel-Aziz.[15]

Parameters and cases	Case 1	Case 2	Case 3	Case 4
X_{C1} (Ω)	4.73	4.73	4.15	4.15
X (Ω)	1.25	1.25	0.66	0.66
R (Ω)	0.012	0.012	0.006	0.006
PF (%)	99.67	99.59	99.50	99.37
dPF (%)	100	100	100	100
I_S (A)	702.93	703.11	707.32	707.8
V_L(V)	2383.44	2384.84	2393.59	2395.06
P_{Loss} (kW)	10.64	10.64	5.74	5.74
VTHD (%)	6.59	7.82	6.24	7.55
ITHD (%)	6.54	6.93	8.93	9.66
COST (thousands of L.E.)	178	181	149	151.5

Additionally, it is obvious that the installation cost of the C-type passive filter is reasonable compared to the single-tuned passive filter, which gives the proposed filter a great advantage as a reasonably competitive cost solution. It is also evident that the C-type passive filter is cheaper than the conventional shunt filter when the energy loss cost is included because of its lower power loss.

In conventional harmonic filtering methods, single-tuned and high-pass filters are employed to suppress harmonic currents injected in the power system. Single-tuned filters provide strong reduction of harmonic currents at a specific tuning harmonic number but suffer from series and parallel resonance. High-pass filters attenuate high-order harmonic currents; however, they cannot be tuned to a lower-order harmonic due to their expected large power losses and installation cost. In this article, the C-type filter is employed as a different approach in low-pass filtering to avoid the previous disadvantages of conventional shunt passive filters.[1] It is obvious that the C-type passive filter can work effectively as a low-pass filter and mitigates low-order harmonics with a good harmonic attenuation at its tuned frequency, especially in the case of lower short-circuit capacity systems compared to systems with higher short-circuit capacity. Besides, the C-type filter promises low ohmic losses compared to the conventional low-pass and high-pass passive filter configurations. Finally, the C-type filter may achieve the same power factor with a lower total installation cost than a single-tuned passive filter.

6. Conclusions

For nonlinear loads, it is necessary to use harmonic filters. Such filters have dual purposes: the first is to improve the associated poor power factor, and the second is to prevent the harmonic load currents from being injected into the network. In this article, the general system performance when using C-type filters as low-pass filters is implemented and discussed. Four cases with two different system short-circuit capacities have been tested, and the general performance of the method used is satisfactory, providing improvement of distortion levels and power factor correction compared with other published results. The required investment cost is estimated for the C-type passive filter, and it was evident that the C-type filter may achieve the same load power factor with a lower installation cost than a conventional single-tuned passive filter.

Appendix: Nomenclature

C_1:   Main capacitor of the C-type filter in microfarads
C_2:   Auxiliary capacitor of the C-type filter in microfarads
C_{C1}:   Main capacitor cost of the C-type filter in Egyptian pounds
C_{C2}:   Auxiliary capacitor cost of the C-type filter in Egyptian pounds
C_E: The present value cost of energy losses in Egyptian pounds
C_L: Reactor cost of the C-type filter in Egyptian pounds
COST: Total C-type filter installation cost in Egyptian pounds
dPF: Displacement power factor as a percentage
F_V: Filter utilization factor

G_{Lh}:   Load conductance at harmonic number h in mhos
H   Harmonic number
I_{ah}:   Auxiliary capacitor current at harmonic number h in amperes
I_{Ch}:   Main capacitor current at harmonic number h in amperes
I_C: Rms value of the main capacitor current in amperes
I_L: Rms value of the rated load current in amperes
I_{Lh}:   Load current in amperes at harmonic number h
I_S: Root-mean-square (rms) value of supply current in amperes
I_{SC}:   Rms value of the short-circuit current in amperes
I_{Sh}:   Supply current in amperes at harmonic number h
ITHD: Current total harmonic distortion as a percentage
K: Filter lifetime in years
L_1:   Inductance of the C-type filter in henrys
PF: Load power factor as a percentage
P_L: Load active power per phase in watts
P_{Loss}:   Transmission power loss in watts
P_S: Supply active power per phase in watts
Q_C: Main capacitor rating in reactive volt-amperes
R_1:   Damping resistor of the C-type filter in ohms
R_F, X_F: Fundamental values of the equivalent resistance and reactance of the C-type filter in ohms
R_{Fh}, X_{Fh}: C-type filter resistance and reactance in ohms at harmonic number h
R_{Lh}, X_{Lh}: Load resistance and reactance in ohms at harmonic number h
R_{Th}, X_{Th}: Thevenin source resistance and reactance in ohms at harmonic number h
TDD: Total demand distortion as a percentage
U_V: Cost of power loss in Egyptian pounds per kilowatt hours
V_C: Rms value of the main capacitor voltage in volts
V_{Ch}:   The main capacitor voltage at harmonic number h in volts
V_L: Rms value of load voltage (line to neutral) in volts
V_{Lh}:   Load voltage in volts at harmonic number h
V_S: Rms value of supply voltage (line to neutral) in volts
V_{Sh}:   Supply voltage at harmonic number h in volts
VTHD: Voltage total harmonic distortion as a percentage
W_C, W_L: Capacitor and reactor unit costs in Egyptian pounds per kilovar
X: Magnitude of the auxiliary capacitive reactance or the inductive reactance of the C-type filter at fundamental frequency in ohms
X_{C1}:   Magnitude of the main capacitive reactance of the C-type filter at fundamental frequency in ohms
Z_{Ch}:   The hth harmonics impedance of the C-type filter in ohms
Z_{Lh}:   The hth harmonics load impedance in ohms
Z_{Th}:   The hth harmonics transmission impedance in ohms

References

1. Abdel Aleem, S. H. E.; Zobaa, A. F.; Abdel Aziz, M. M. Optimal C-Type Passive Filter Based on Minimization of the Voltage Harmonic Distortion for Nonlinear Loads. *IEEE Trans. Ind. Electron.* **2012**, *59*, 281–289.

2. Ma, J.; Mi, C.; Zheng, S.; Wang, T.; Lan, X.; Wang, Z.; Thorp, J. S.; Phadke, A. G. Application of Voltage Harmonic Distortion Positive Feedback for Islanding Detection. *Electr. Pow. Compo. Sys.* **2013**, *41*, 641–652.

3. Xiao, Y.; Zhao, J.; Mao, S. In *Theory for the Design of C-type Filter*, 11th Int. Conf. Harmonics and Quality of Power, ICHQP 2004, Lake Placid, New York, Sept 12–15, 2004; IEEE: New York, pp 11–15.

4. Das, J. C. Passive Filters—Potentialities and Limitations. *IEEE Trans. Ind. App.* **2004**, *40*, 232–241.

5. F. D. Garza. *Mitigation of Harmonic Currents and Conservation of Power in Non-Linear Load Systems.* U.S. Patent 20110121775, May 26, 2011.

6. Zobaa, A. F.; Abdel-Aziz, M. M.; Abdel Aleem, S. H. E. Comparison of Shunt-Passive and Series-Passive Filters for DC Drives Loads. *Electr. Power Compon. Syst.* **2010**, *38*, 275–291.

7. Zacharia, P.; Menti, A.; Zacharias, T. Genetic Algorithm Based Optimal Design of Shunt Compensators in the Presence of Harmonics. *Electr. Power Syst. Res.* **2008**, *78*, 728–735.

8. Nassif, A. B.; Xu, W.; Freitas, W. An Investigation on the Selection of Filter Topologies for Passive Filter Applications. *IEEE Trans. Power Deliv.* **2009**, *24*, 1710–1718.

9. Dugan, R. C.; Granaghan, M. F. M.; Santoso, S.; Beaty, H. W. *Electric Power Systems Quality*; McGraw-Hill: New York, 2002.

10. Abdel Aleem, S. H. E.; Elmathana, M. T.; Zobaa, A. F. Different Design Approaches of Shunt Passive Harmonic Filters Based on IEEE Std. 519-1992 and IEEE Std. 18-2002. *Recent Pat. Electr. Eng.* **2013**, *6*, 68–75.

11. Balci, M. E.; Karaoglan, A. D. Optimal Design of C-type Passive Filters Based on Response Surface Methodology for Typical Industrial Power Systems. *Elec. Pow. Comp. Syst.* **2013**, *41*, 653–668.

12. Raymond, C.; Chapra, S., *Numerical Methods for Engineers: with Software and Programming Applications;* McGraw-Hill: New York, 2002.

13. IEEE Standard 519–1992, *IEEE Recommended Practices and Requirements for Harmonic Control in Electrical Power Systems*, 1992.

14. IEEE Standard 18-2002, *IEEE Standard for Shunt Power Capacitors*, 2002.

15. Abdel-Aziz, M. M.; El-Zahab, E. E. A.; Zobaa, A. F.; Khorshied, D. M. Passive Harmonic Filters Design Using FORTRAN Feasible Sequential Quadratic Programming. *Electr. Pow. Syst. Res.* **2007**, *77*, 540–547.

16. Richards, G. G.; Tan, O. T.; Klinkhachorn, P.; Santoso, N. I. Cost-Constrained Power Factor Optimization with Source Harmonics Using LC Compensators. *IEEE Trans. Ind. Electron.* **1987**, *IE-34*, 266–270.

17. Zobaa, A. F.; Abdel Aleem, S. H. E. A New Approach for Harmonic Distortion Minimization in Power Systems Supplying Nonlinear Loads. *IEEE Trans. Ind. Inf*. **2014**, *10*, 1401–1412.

18. Zeineldin, H. H.; Zobaa A. F. Particle Swarm Optimization of Passive Filters for Industrial Plants in Distribution Networks. *Electr. Pow. Compo. Sys.* **2011**, *39*, 1795–1808.

19. Abdel Aleem, S. H. E.; Balci, M. E.; Zobaa, A. F.; Sakr, S. Optimal Passive Filter Design for Effective Utilization of Cables and Transformers under Non-sinusoidal Conditions. *Proceedings of 16th Int. Conf. Harmonics and Quality of Power*, ICHQP '14, Bucharest, Romania, May 25–28, 2014; IEEE: Bucharest, Romania, pp 626–630.

20. Lia, X.; Bhatb, A. K. S. A Fixed-frequency Series-parallel Resonant Converter with Capacitive Output Filter: Analysis, Design, Simulation, and Experimental Results. *Electr. Pow. Compo. Sys.* **2014**, *42*, 746–754.

21. Elmathana, M. T.; Zobaa, A. F.; Abdel Aleem, S. H. E. Economical Design of Multiple-Arm Passive Harmonic Filters for an Industrial Firm—Case Study. In *15th Int. Conf. Harmonics and Quality of Power*, ICHQP '12, Hong Kong, China, June 17–20, 2012; IEEE: Hong Kong, pp 438–444.

7

Investigation of Aluminum Primary Batteries Based on Taguchi Method

AMIR ERFANI, MILAD MUHAMMADI, SOHEIL ASGARI NESHAT, MOHAMMAD MASOUD SHALCHI, and FARSHAD VARAMINIAN*

School of Chemical, Gas and Petroleum Engineering, Semnan University, Semnan, Iran

Abstract: In this study, calcium hypochlorite, sodium hypochlorite, potassium ferricyanide, and hydrogen peroxide are experimentally investigated as electrolytes for aluminum batteries. Different factors such as NaOH concentration, electrolyte concentration, current density, and temperature are experimentally investigated using the Taguchi statistical method. Also ANOVA is utilized to determine the importance of each parameter on performance of the battery. For each of these 4 chemistries, a L9 orthogonal array with the 4 factors of control at three levels each was employed. Optimization of aluminum batteries' performance in low concentration electrolytes using commercial, inexpensive aluminum with the purity according to ASTM 1050 standard (purity 99.5%) is investigated. At the optimum operating conditions, specific energies up to 2386 WH/kg and power densities up to 77.6 mW/cm^2 are obtained.

Keywords: Primary battery, aluminum anode, Taguchi, optimization

1. Introduction

Experimental investigations, optimization, and modeling of candidate electrochemical systems have been important subjects of electric vehicles studies.[1] High theoretical energy capacity, voltage, and specific energy have attracted many researchers to develop aluminum-based batteries.[2-6] Applications of these batteries for submarine vehicles have been recommended by many researchers and inventors.[7,8] Aluminum batteries have been explored using different kinds of electrolytes and anode purity grades.[9-12]

Aluminum primary batteries are expressed by aluminum oxidation and electrolyte reduction at cathode surface. Principally because of higher activity of an oxidizing agent, they have higher current density than aluminum/air batteries. This makes them a favorable choice when an abundant oxidizing agent such as oxygen in air is not available (e.g., for submarine applications). Important parameters in controlling performance are temperature, electrolyte type and concentration, caustic concentration,

*Address correspondence to: Farshad Varaminian, School of Chemical, Gas and Petroleum Engineering, Semnan University, 35131-19111 Semnan, Iran. Email: fvaraminian@semnan.ac.ir

anode purity, foaming characteristics, use of corrosion inhibitors, and current density.[4,9,12] Self-corrosion of aluminum is the most important and most studied drawback of these electrochemical systems. Use of anionic exchange membranes or ionic liquids as solvents have been newly discussed as novel ways to lower self-corrosion of aluminum.[13] Reaching extremely high current densities in high concentrated electrolytes with reasonable columbic efficiencies demands highly pure aluminum as anode, which is not economically feasible.[14] Earlier research has focused on obtaining high columbic efficiency through using highly pure aluminum anodes. A new approach is to use commercial-grade aluminum in an optimized condition and enhancing overall performance by utilizing a fuel cell to make use of produced hydrogen.[15]

Statistical based designs of experiments are applied to diminish the number of experiments in a multi-parametric system. These methods are usually utilized for evaluation of chemical formulations, structures, materials, etc.[16] The most widely used methods include factorial, response surface, and Taguchi.[17-19] Among these designs, the Taguchi method has gained more attention because it suggests fewer experiments for a defined system because of its fast converging characteristics. The Taguchi method searches the optimum conditions through quality analysis in orthogonal matrices.[20,21] It has been successfully used in both engineering and electrochemical applications.[22-27] The Taguchi method uses signal to noise ratio (SNR) as criteria to evaluate the optimum condition for any defined optimization problem. SNR can maximize, minimize, or nominalize responses of a system to applied changes. Inherently, this method includes a series of assumptions (e.g., no interaction between parameters). Applying the Taguchi method in a system that does not satisfy these assumptions is in vain. One must consider the validity of the Taguchi method assumptions by a comparison between experimental data and the ones predicted by the Taguchi method.

In this research, by utilizing the Taguchi method, optimization of aluminum batteries' performance in low concentration electrolytes using commercial and inexpensive aluminum anodes are investigated. This research aims to examine whether or not a design of an experiment with assumption of no interactions between parameters can effectively and correctly optimize thermo physical variables of aluminum primary batteries. There have been very few studies on calcium hypochlorite and potassium ferricyanide as electrolytes for aluminum batteries. Another goal of this research is to examine different chemistries proposed elsewhere in a comparable condition and discuss their performance and characteristics.

2. Materials, Procedures, and Methods

2.1 Testing Setup

To conduct experiments at different constant temperatures, an electrochemical setup consisting of the following main parts was devised and built in our laboratory: (1) reactor and stirring system, (2) pump and piping system, (3) hot water batch, (4) temperature control unit, and (5) voltage and amperage meters. In this setup, the thermocouple immersed in electrolyte measures temperature and transmits the value to the temperature controller. The controller turns the pump on and off accordingly and helps maintain a constant temperature in the reactor by circulating hot water from the water batch into the reactor's jacket. Anode and cathode electrodes are placed in the middle of the reactor as two parallel plates with a 0.5-cm gap in between. The reference electrode is located right in this gap. Figure 1 illustrates the experimental setup.

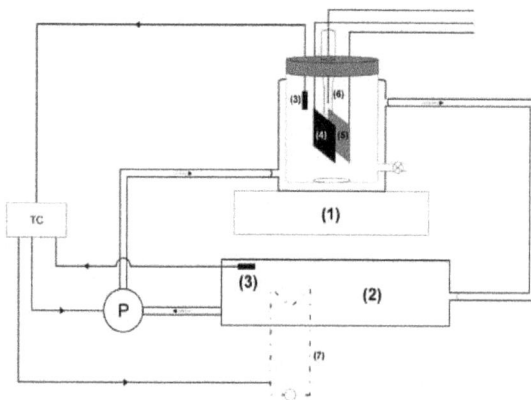

Fig. 1. Diagram of the testing setup, P: Pump, TC: Temperature Controller, (1) Stirrer, (2) Hot water batch, (3) Thermocouple, (4) Anode, (5) Cathode, (6) Reference electrode, (7) Heating element.

2.2 Experiment Conditions, Materials, and Measurements

A solution of 3.5% NaCl (seawater salinity) was used to dilute solutions to the specified concentration of each experiment. Electrolytes of each experiment were individually made and poured in the 500-cc reactor and then heated to experiment temperature. 99.5% purity aluminum (ASTM 1050) was selected as anode. Anodes were cut out of 0.3-mm thick plates with all of its surface except the area facing cathodes covered by epoxy. The active surface of anodes and cathodes were each 4.5 × 4.5 cm; cathode were out of planar nickel, which, according to reactions in aluminum primary batteries, only works as an electric current collector.[11-13] After removal of a thin layer of aluminum surface, anodes were weighted. Experiments were carried out in 20-minute intervals. To prevent accumulation of hydrogen bubbles on the anode surface, a magnetic stirrer with a constant rate of 300 RPM was used. During the experiments, current and voltage were measured, and after 20 minutes, the anode was taken out of reactor, washed, dried, and again weighted. According to difference in mass, anodic columbic efficiency was calculated. All experiments were repeated twice, and if there were more than 5% error in responses, the experiment was repeated for a third time. To gather data, which could be applied in designing of aluminum primary batteries, all materials used were of commercial grade with their inherent impurities.

For columbic efficiency calculations, initial and final mass of anodes were measured using AND GR 200 with 0.1 mg accuracy. Amperage between electrodes was measured by FOTEK DA-24T, AC current meter with 0.01A accuracy. A FOTEK DV-24T voltage meter was used to measure the potential difference between each electrode and an Ag/AgCl reference electrode. Principally due to polarization losses, open and closed circuit voltage of cells are not equal. The very first value shown by the voltage meter, after cutting the circuit, was assumed as the nearest value to the actual closed circuit voltage. An example of these measurements is presented in Table 1.

2.3 Design of the Experiments

Prior to Taguchi design, tests were carried out to find which parameters affect battery performance. Preliminary tests have

Table 1. Measurements example, Ferricyanide run8 NaOH: 1 M Ferricyanide concentration: 0.2 M, Load minimum, temperature: 50°C.

Time (Min)	Current (A)	Al vs. Ag/AgCl potential (mV)	Ferricyanide vs. Ag/AgCl potential (mV)
0	0.62	−1633	−1082
2	0.62	−1632	−1082
4	0.62	−1630	−1083
6	0.54	−1630	−1083
8	0.54	−1629	−1085
10	0.54	−1629	−1085
12	0.54	−1628	−1085
14	0.54	−1628	−1083
16	0.54	−1628	−1081
18	0.53	−1627	−1085
20	0.52	−1626	−1090

hown that the performance of cells is dependent mainly on the following items: (1) chemistry of the electrolyte, (2) electrolyte concentration, (3) electrolyte temperature, (4) current density, and (5) alkalinity of electrolyte. The performance optimization of the cells using the conventional method of changing one variable at a time while keeping all others constant (full factorial) requires a considerable number of tests. Therefore, we applied the Taguchi method. In our study, NaOH concentration (alkalinity), electrolyte concentration, current density, and temperature were considered, and each of these four factors was tested at three different levels. To obtain three different levels of current densities, different loads were applied. These loads were r_0, r_0+5 and r_0+10 Ohm—by (r_0) we mean that it was tried that electrochemical cell generates electrical energy with highest amperage per surface. This was achieved by lowering the load as low as possible, but it is noteworthy that because of the ampere meter and other electrical resistance, there was always a minimum load on the system. The design of experiments as well as data analysis was carried out using MINITAB16. Degrees of freedom are calculated using Eq. 1; this optimization problem has 8 degrees of freedom. So a matrix of Taguchi design consisting of 9 experiments is chosen.

$$\text{Degrees of freedom} = (\text{number of levels} - 1) \\ \times \text{number of factors} \quad (1)$$

As there are four factors each with three levels, classic design of experiments suggests 3^4 experiments. Utilizing the Taguchi method has lowered the required number of the tests by a factor of 9. The Taguchi method not only optimizes parameters based on an objective function but also could predict the value of this function at specified conditions. While the Taguchi method helps us with conducting a very limited number of experiments, this method lacks the ability to give us a model to predict goal function at points between studied levels.

There are three scenarios to optimize S/N ratios available depending on the type of objective function; lower is better (LB), nominal is best (NB), or higher is better (HB). Eq. (2), Eq. (3), and Eq. (4) give SNR for each scenario accordingly.[18]

$$SNR = -10Log\left(Y^2\right) \quad (2)$$

$$SNR = -10Log\left(\frac{Y^2}{S^2}\right) \quad (3)$$

$$SNR = -10Log\left(\frac{1}{Y^2}\right) \quad (4)$$

where Y is the resulting value calculated under experiment No. i and S indicates the standard deviation of the resulting values for all experiments. An aluminum primary battery works very similarly to a fuel cell. Aluminum provided from an aluminum source enters the electrochemical reactor and produces electricity due to aluminum consumption. The first challenge is to maximize generated energy per consumption of a kilogram of aluminum (MJ/kg Al), which is called specific energy. For this objective function, higher is better is the choice for calculating SNR. Using Eq. (4), signal to noise ratio for specific energy objective

function is rewritten (Eq. 5). Furthermore, in a real aluminum battery, there are limitations on battery volume and electrode surface; therefore, the second challenge is to maximize energy per anode surface, which is called power density (mW/cm^2). Eq. (6) gives the signal to noise ratio for second objective function. First and second objective functions (Eqs. 5 and 6) should be optimized simultaneously as the most significant objective function. In order to achieve this purpose, Eq. (7) is optimized.

$$SNR1 = -10Log\left(\frac{1}{Specific\ Energy^2}\right) \quad (5)$$

$$SNR2 = -10Log\left(\frac{1}{Energy\ density^2}\right) \quad (6)$$

$$SNR3 = -10Log\left(\frac{\dfrac{1}{Specific\ Energy^2} + \dfrac{1}{Energy\ density^2}}{2}\right) \quad (7)$$

For understanding the effect of the studied parameters on the cell performance, analysis of variance (ANOVA) was carried out. As a result of ANOVA analysis, the influential degree on the performance characteristics of parameters are determined by calculating the percent contribution of design parameters. To illustrate the contribution of each factor, analysis of sums of squares was applied, and it is shown in Eq. (8) to Eq. (11).

$$\text{Percentage contribution} = \frac{SS\ factor}{SS\ total} \times 100 \quad (8)$$

$$SS_{mean} = \frac{\left(\sum SNR\right)^2}{n} \quad (9)$$

$$SS_{factor} = \frac{\sum SNR^2_{factor-i}}{N} - SS_{mean} \quad (10)$$

$$SS_{total} = \sum SNR^2 - SS_{mean} \quad (11)$$

where SS_{total} is the total sums of squares, SS_{mean} is the sums of square due to the mean, SS_{factor} is the sums of squares due to factor, $SNR_{factor-i}$ the sum of i level of factor, and N is the repeating number of each levels of factors. Among all the oxidizing agents, based on their stability, solubility in water, availability, and price, we chose calcium hypochlorite, sodium hypochlorite, potassium ferricyanide, and hydrogen peroxide. From different commercially available aluminum grades, 99.5% pure aluminum was chosen due to relatively high purity and reasonable price. In selecting upper and lower levels of temperature, chemical activity, safe distances from electrolyte boiling point, and absence of foaming were taken into account.

3. Results and Discussion

3.1 Optimization

In all different chemistries for generation of electrical energy in aluminum batteries, there are three major reactions—two of them are anodic and cathodic reactions related to electrolyte, and the third reaction is self-corrosion of aluminum in caustic solution as shown in Eq. 12.

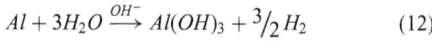

$$Al + 3H_2O \xrightarrow{OH^-} Al(OH)_3 + {}^3/_2 H_2 \tag{12}$$

Increasing temperature results in a higher reaction rate but also a lower potential difference. To obtain a higher energy efficiency, one must increase both chemical potential and reaction rate. Therefore, finding the temperature associated with highest columbic efficiency could be considered as an optimization problem.

The anodic reaction consists of three major steps that control reaction rate. In the first step, hydroxide ions move to the anode interface (mass transfer); in the second step, electrochemical reaction occurs; and in the third step, the products move away from the anode (mass transfer). While stirring the solution can lower or possibly diminish the mass transfer resistance, the electrochemical reaction is dependent on reactivity of the electrolyte and nature of the anode. Electrochemical reaction can also produce intermediate products (Eq. 13). The $Al(OH)_4^-$ gives a gray shadow color to the electrolyte solution.

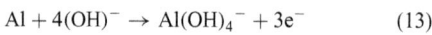

$$Al + 4(OH)^- \rightarrow Al(OH)_4^- + 3e^- \tag{13}$$

3.1.1 Ca(ClO)$_2$

For Al/Ca(ClO)$_2$ cell anodic and cathodic reactions are as follows:

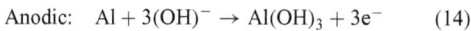

$$\text{Anodic:} \quad Al + 3(OH)^- \rightarrow Al(OH)_3 + 3e^- \tag{14}$$

Cathodic:

$$3/4\,Ca(ClO)_2 + 3/2H_2O + 3e^- \rightarrow 3/4CaCl_2 + 3(OH)^- \tag{15}$$

Overall: $Al + 3/4Ca(ClO)_2 + 3/2H_2O \rightarrow 3/4CaCl_2 + Al(OH)_3$

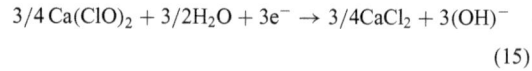

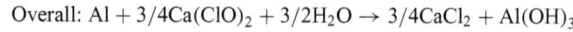

$$E^\circ_{cell} = 2.9V \tag{16}$$

where E°_{cell} is the standard cell potential. For Al/Ca(ClO)$_2$ cell matrix of the experiments and the corresponding results are given in Table 2. In preliminary experiments, using Ca(ClO)$_2$ as electrolyte, it was noticed that excessive foaming occurs in high temperatures and concentrations. To avoid this excessive foaming, the upper level of temperature chosen was 50°C. For this case main effects plot for signal to noise ratios are illustrated in Figure 2. Optimum conditions for each of three responses and contribution of each factor are illustrated in Table 3.

Although Figure 2 might suggest that a load of more than 10 ohm might be more efficient, it should be noticed that current density would be too small to be practical. The same discussion is also true for lowering Ca(ClO)$_2$ concentration below 25 g/L. All selected ranges for study parameters have been chosen very carefully to be practical for an aluminum battery that uses commercial (not highly pure) aluminum anode.

3.1.2 Al/NaClO

Al/NaClO cell, anodic, cathodic, and overall reactions are as follows:

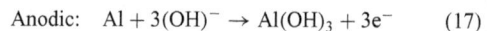

$$\text{Anodic:} \quad Al + 3(OH)^- \rightarrow Al(OH)_3 + 3e^- \tag{17}$$

Table 2. Matrix of the experiments and corresponding experimental results for Al-Ca(ClO)$_2$.

Test #	NaOH (M)	Ca(ClO)$_2$ (gr/lit)	Load (Ohm)	Temperature (°C)	Current density (mA/cm^2)	Delta V (mV)	Efficiency (%)	Specific energy (WH/Kg Al)	Energy density (mW/cm^2)
1	0.1	25	r_0	30	2.96	900	56	1500.0	2.67
2	0.1	50	r_0+5	40	4.79	1166	73	2647.3	5.60
3	0.1	75	r_0+10	50	2.96	1087	40	1278.9	3.22
4	0.3	25	r_0+5	50	4.84	1650	17	816.0	7.99
5	0.3	50	r_0+10	30	3.56	1315	81	3094.1	4.68
6	0.3	75	r_0	40	20.25	1432	62	2286.0	30.00
7	0.5	25	r_0+10	40	9.38	1742	12	625.3	16.20
8	0.5	50	r_0	50	2.96	1078	23	798.5	3.20
9	0.5	75	r_0+5	30	2.76	899	47	1259.0	2.47

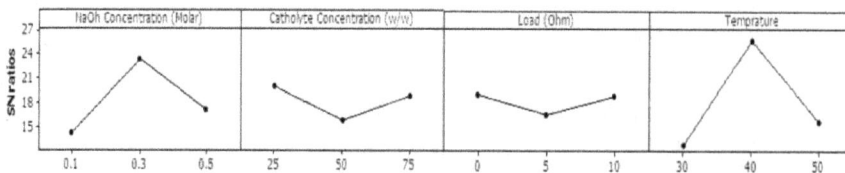

Fig. 2. Al/Ca(ClO)$_2$ main effects plot for signal to noise ratios, specific energy, and energy density as response.

Table 3. Al/Ca(ClO)$_2$ Optimum conditions for each of three responses and percentage contribution of each factor.

	Specific energy		Energy density		Specific energy and energy density
	Optimized value	Percentage contribution	Optimized value	Percentage contribution	Optimized value
NaOH (M)	0.3	40	0.3	29	0.3
Ca(ClO)$_2$ (g/L)	50	33	25	6	25
Current density (mA/cm^2)	4.32	1	18.4	2	18.4
Temperature (°C)	30	26	40	63	40

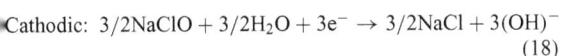

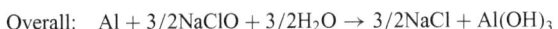

Cathodic: $3/2NaClO + 3/2H_2O + 3e^- \rightarrow 3/2NaCl + 3(OH)^-$ (18)

Overall: $Al + 3/2NaClO + 3/2H_2O \rightarrow 3/2NaCl + Al(OH)_3$

$E^\circ_{cell} = 2.8V$ (19)

For NaClO matrix of the experiments and the corresponding results are given in Table 4. Optimum conditions for each of the three responses and contribution of each factor are illustrated in Table 5. For this case main effects plot for signal to noise ratios are illustrated in Figure 3.

According to Figure 3 and Table 5, increasing resistance, which is analogous to decreasing current density, lowers signal to noise ratio and, consequently, efficiency of the cell. Explanation for this effect is that although higher resistance has no impact on self-corrosion, it limits anodic and cathodic reactions rate, and, consequently, columbic efficiency deteriorates. Results suggest that the higher the current density, the higher the efficiency. The physical meaning is that in a practical aluminum battery, surface in contact with electrolyte should be a minimum. This minimum should be calculated from current densities at optimum conditions, e.g., for a Al/NaClO battery, if a system needs 77.6 mW electrical energy, only a surface of 1 cm^2 of anode should be in contact with electrolyte, which is the optimum current density of this chemistry, and more surface should be introduced if more power is desired.

When current density and potential difference are of importance, sodium hypochlorite surpasses calcium hypochlorite, but this cell has lower columbic efficiency and specific energy. One problem associated with sodium hypochlorite is that it is not available in a concentrated solution or solid form, which leads to some issues in handling.

Table 4. Matrix of the experiments and corresponding experimental results for Al/NaClO.

Test #	NaOH (M)	NaClO (W/W)	Load (Ohm)	Temperature (°C)	Current density (mA/cm^2)	Delta V (mV)	Efficiency (%)	Specific energy (WH/Kg Al)	Energy density (mW/cm^2)
1	0.1	0.02	r_0	30	14.42	1612	20	877.6	21.63
2	0.1	0.03	r_0+5	40	8.89	1790	10	488.8	15.58
3	0.1	0.04	r_0+10	50	4.94	1712	4	174.4	8.51
4	0.3	0.02	r_0+5	50	10.37	1887	7	357.0	19.60
5	0.3	0.03	r_0+10	30	4.94	1660	6	268.6	8.20
6	0.3	0.04	r_0	40	32.00	1800	24	1269.0	57.60
7	0.5	0.02	r_0+10	40	5.53	1830	5	270.0	10.10
8	0.5	0.03	r_0	50	42.96	1805	21	1087.4	77.60
9	0.5	0.04	r_0+5	30	7.21	1738	10	482.2	12.53

Table 5. Al/NaClO Optimum conditions for each of three responses and percentage contribution of each factor.

	Specific energy		Energy density		Specific energy and energy density
	Optimized value	Percentage contribution	Optimized value	Percentage contribution	Optimized value
NaOH (M)	0.5	2	0.5	6	0.5
NaClO (w w-1)	0.03	1	0.03	2	0.03
Current density (mA/cm^2)	38.4	93	77.6	80	77.6
Temperature (°C)	40	4	40	12	40

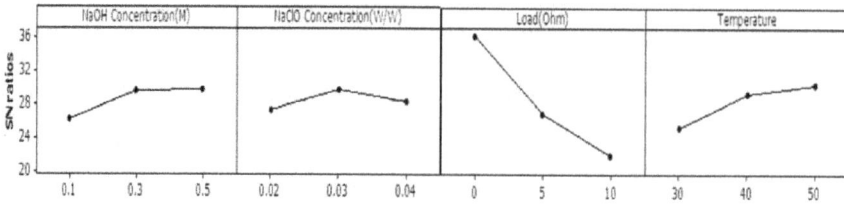

Fig. 3. Al/NaClO main effects plot for signal to noise ratios, specific energy, and energy density as response.

3.1.3 Al/K₃Fe(CN)₆

For Al/K$_3$Fe(CN)$_6$ cell anodic and cathodic reactions are as follows:

$$\text{Anodic:} \quad Al + 3(OH)^- \rightarrow Al(OH)_3 + 3e^- \tag{20}$$

$$\text{Cathodic:} \quad 3FeCN_6{}^{3-} + 3e^- \rightarrow 3FeCN_6{}^{4-} \tag{21}$$

Overall:

$$Al + 3\left(OH^-\right) + 3(FeCN_6)^{3-} \rightarrow Al\left(OH_3\right) + 3(FeCN_6)^{4-}$$

$$E°_{cell} = 2.8V \tag{22}$$

Using potassium ferricyanide as electrolyte, it was observed that concentration of NaOH has to reach a minimum threshold before anodic and cathodic reactions start. Although a potassium ferricyanide solution is red, adding sufficient amount of NaOH

changes its color to yellow. This color change also suggests concentrations in which anodic and cathodic reactions occur. For K$_3$Fe(CN)$_6$ matrix of the experiments and the corresponding results are given in Table 6. Optimum conditions for each of three responses and importance of each factor are illustrated in Table 7. For this case main effects plot for signal to noise ratios are illustrated in Figure 4.

3.1.4 Al/H₂O₂

Cell reactions for Al/H$_2$O$_2$ are as follows:

$$\text{Anodic:} \quad Al + 3(OH)^- \rightarrow Al(OH)_3 + 3e^- \tag{23}$$

$$\text{Cathodic:} \quad 3/2H_2O_2 + 3e^- \rightarrow 3(OH)^- \tag{24}$$

$$\text{Overall:} \quad Al + 3/2H_2O_2 \rightarrow Al(OH)_3 \quad E_{cell} = 2.3 \tag{25}$$

Table 6. Matrix of the experiments and corresponding experimental results for Al/K$_3$Fe(CN)$_6$.

Test #	NaOH (M)	Catholyte (M)	Load (Ohm)	Temperature (°C)	Current density (mA/cm²)	Delta V (mV)	Efficiency (%)	Specific energy (WH/Kg Al)	Energy density (mW/cm²)
1	0.50	0.1	r_0	30	7.40	585	96	1339.3	3.70
2	0.50	0.2	r_0+5	40	0.99	378	10	103.9	0.38
3	0.50	0.3	r_0+10	50	0.00	0	0	0.0	0
4	0.75	0.1	r_0+5	50	2.96	570	5	81.1	1.68
5	0.75	0.2	r_0+10	30	0.99	506	54	816.1	0.50
6	0.75	0.3	r_0	40	3.95	439	22	284.1	1.73
7	1.00	0.1	r_0+10	40	1.48	555	5	71.4	0.82
8	1.00	0.2	r_0	50	28.64	543	64	1042.0	12.70
9	1.00	0.3	r_0+5	30	0.99	490	90	1289.5	0.49

Table 7. Al/K$_3$Fe(CN)$_6$ Optimum conditions for each of three responses and percentage contribution of each factor.

	Specific energy		Energy density		Specific energy and energy density
	Optimized value	Percentage contribution	Optimized value	Percentage contribution	Optimized value
NaOH (M)	1	24	0.5	25	1
K₃Fe(CN)₆ (M)	0.2	20	0.25	28	0.1
Current density (mA/cm²)	12.1	28	13.3	36	13.3
Temperature (°C)	30	28	50	11	30

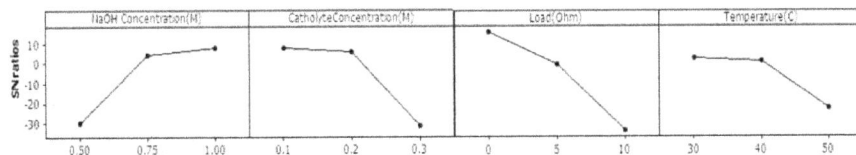

Fig. 4. Potassium ferricyanide main effects plot for signal to noise ratios, specific energy and energy density as response.

Table 8. Matrix of the experiments and corresponding experimental results for Al/H_2O_2.

Test #	NaOH (M)	Catholyte (M)	Load (Ohm)	Temperature (°C)	Current density (mA/cm²)	Delta V (mV)	Efficiency (%)	Specific energy (WH/Kg A1)	Energy density (mW/cm²)
1	0.1	0.25	r_0	30	4.94	734	72	986.5	3.61
2	0.1	0.5	r_0+5	40	2.47	876	33	862.2	2.16
3	0.1	0.75	r_0+10	50	1.48	875	60	1562.5	1.30
4	0.03	0.25	r_0+5	50	3.95	1035	11	341.0	4.10
5	0.03	0.5	r_0+10	30	1.23	710	19	399.7	0.90
6	0.03	0.75	r_0	40	4.94	795	64	1517.2	3.90
7	0.05	0.25	r_0+10	40	27.16	1200	15	523.3	3.56
8	0.05	0.5	r_0	50	21.73	1212	20	588.7	21.00
9	0.05	0.75	r_0+5	30	1.98	680	40	906.6	1.44

Table 9. Al/H_2O_2 Optimum conditions for each of three responses and percentage contribution of each factor.

	Specific energy		Energy density		Specific energy and energy density
	Optimized value	Percentage contribution	Optimized value	Percentage contribution	Optimized value
NaOH (M)	0.1	28	0.5	16	0.5
H_2O_2 (M)	0.75	5	0.25	11	0.25
Current density (mA/cm²)	4.5	61	24.2	48	24.2
Temperature (°C)	40	6	50	25	50

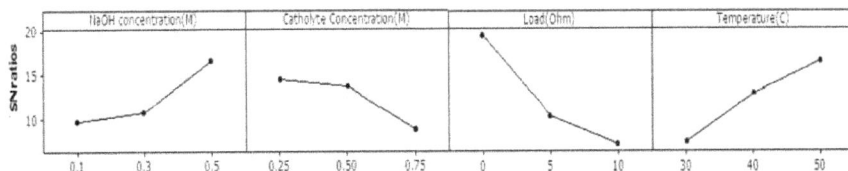

Fig. 5. H_2O_2 main effects plot for signal to noise ratios, specific energy and energy density as response.

For H_2O_2, matrix of the experiments and the corresponding results are given in Table 8. Optimum conditions for each of three responses and importance of each factor are illustrated in Table 9. For this case main effects plot for signal to noise ratios are illustrated in Figure 5.

3.2 Analysis of the Experiment Results

For every studied chemistry, at optimum condition a test was carried out. Table 10 summarizes characteristics of examined aluminum electrolyte at optimum conditions.

Results show that $Al/Ca(ClO)_2$ cell has a high potential difference and specific energy. It also has reasonable current density. In optimized temperature, no foaming occurs. Columbic efficiency of this cell is 57%, which shows that 57% of aluminum is consumed in anodic reaction, and 43% is consumed in hydrogen generation reaction, which generates 48 g hydrogen per kilogram of aluminum consumed. It was observed that although NaClO has the lowest columbic efficiency in comparison to the other three electrolytes studied, it has the highest open/closed circuit potential difference and power density.

Table 10. Characteristics of examined aluminum catholyte at optimum conditions.

Pair	Cell voltage (V)	Specific energy of metal (WH/Kg Al)	Energy density (mW/cm^2)	Current density (mA/cm^2)	Columbic efficiency (%)	Hydrogen generation (gr H$_2$/kg Al)
Al/Ca(ClO)$_2$	1.4	2386	18.4	12.83	57	48
Al/NaClO	1.8	1087	77.6	42.96	21	88
Al/K$_3$Fe(CN)$_6$	0.6	1067	13.3	23.7	63	41
Al/H$_2$O$_2$	1.3	1250	24.2	18.08	33	74

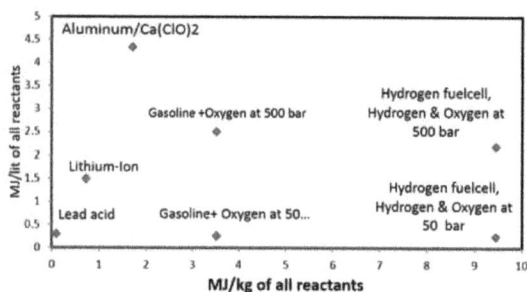

Fig. 6. Specific energy of Al/Ca(ClO)$_2$ and other fuels for submarine propulsion applications; in calculation of reactants weight, it was assumed that oxygen must be carried.

A desirable fuel needs to have high energy both per mass (energy density) and volume. Hydrogen fuel cells have the highest energy per mass, but even at high pressures, hydrogen gas has a very low density. On the other hand, aluminum batteries have a high energy per volume of the fuel. Figure 6 illustrates specific energy of Al/Ca(ClO)$_2$ and other fuels. Figure 6 is prepared from the submarine application point of view where weight and volume of oxygen as co-reactant in internal combustion or hydrogen fuel cell has been taken into account.

Using Eq. 3, based on experiment results at optimum conditions, a corresponding SNR can be found. Verification of the model was achieved by comparing these experimental SNRs with SNRs predicted by the model. These experimental and predicted results are given in Table 11.

For Al-Ca(ClO)$_2$ Al-NaClO and Al-H$_2$O$_2$, experimental and predicted signal to noise ratios show a better agreement than for Al-K$_3$Fe(CN)$_6$. It suggests that for these chemistries it is reasonable to assume there is no significant interaction between the parameters, and the Taguchi method is valid. For

Table 11. Predicted and experimental signal to noise ratios for specific energy and power density as goal functions.

Cell chemistry	Predicted SNR3	Experimental SNR3	Absolute relative error percentage
Al-Ca(ClO)$_2$	33.6	27.8	20%
Al-NaClO	40.8	40.8	0%
Al-K$_3$Fe(CN)$_6$	52.4	25.5	100%
Al-H$_2$O$_2$	30.2	29.0	4%

Al-K$_3$Fe(CN)$_6$, predicted and experimental results do not match. This is due to interactions between sodium hydroxide and potassium ferricyanide in electrolyte. High dependency of K$_3$Fe(CN)$_6$ activity on caustic concentration has led to deviation of predicted and experimental signal to noise ratio for this system. Understanding the mechanism of this dependency would require further research.

4. Summary and Conclusions

In this study, four commercial oxidizing agents with suitable properties were selected as electrolytes to be used in aluminum batteries. The Taguchi method and ANOVA were applied for design of experiments and analysis of results accordingly. While optimization of signal to noise ratio led to highest specific energy (energy per weight of aluminum consumed) and energy density (energy per anode surface), the ANOVA helps to find the contribution of each factor on the performance of the battery. Based on the Taguchi matrix of experiments, tests were carried out and signal to noise ratios for three appropriate objective functions were calculated. Aluminum batteries investigated in this research were of commercial-grade aluminum and low concentration of sodium hydroxide. Most important, this study aimed to examined the ability of the Taguchi optimization method to predict aluminum battery efficiency in response to different combinations of factors/levels through conducting a very limited number of experiments with an acceptable level of accuracy. The results showed that the Taguchi method is able to model three studied chemistries well and can therefore be applied in future studies in other electrochemical systems. The best chemistry and operation conditions could generate 2386 Wh/kg Al energy. In this condition, the battery is capable of producing 48 gr H$_2$/kg Al. Using a hydrogen fuel cell alongside the battery results in increasing the performance of system by a factor of 2. Although NaClO has the lowest columbic efficiency in comparison to the other three electrolytes studied, it has the highest open/closed circuit potential difference and power density. Although potassium ferricyanide can act as a high-performance electrolyte for highly pure aluminum and concentrated sodium hydroxide solutions, our research shows that in low concentrated solutions and commercial-grade aluminum, potential differences are very low and not practical. Al-CaClO$_2$ at its optimum conditions has the highest specific energy, but current density and potential difference are of importance; sodium hypochlorite surpasses other electrolytes. As current density increases, the importance factor of resistance with respect to other electrolytes increases ıs well. For instance, in the highest current density of sodium ıypochlorite and calculations based on sum of squares shows that ʼesistance has an importance factor of 96%. Although studied ɔhemistries have shown reasonable specific energies and energy ɟensities, the study of cell voltage components, cost, longevity, ınd safety consideration will require additional research.

Funding

This research is funded by Semnan University.

References

1. He, H.; Xiong, R.; Guo, H.; Li, S. Comparison Study on the Battery Models Used for the Energy Management of Batteries in Electric Vehicles. *Energ. Conv. Manag.* **2012**, *64*, 113–121.
2. Linden, D.; Reddy, T.B. *Handbook of Batteries;* McGraw-Hill: New York, 2002.
3. Vlaskin, M.; Shkolnikov, E.; Bersh, A.; Zhuk, A.; Lisicyn, A.; Sorokovikov, A.; Pankina, Y.V. An Experimental Aluminum-Fueled Power Plant. *J. Power Sources* **2011**, *196*, 8828–8835.
4. Li, Q.; Bjerrum, N.J. Aluminum as Anode for Energy Storage and Conversion: A Review. *J. Power Sources* **2002**, *110*, 1–10.
5. Yang, S.; Knickle, H. Design and Analysis of Aluminum/Air Battery System for Electric Vehicles. *J. Power Sources* **2002**, *112*, 162–173.
6. Zaromb, S. Electrochemical Power Generation, U.S. Patents, 1983.
7. Hunter, J.A.; Hamlen, R.P. Aluminium Battery With Electrolyte Circulation Means, U.S. Patents, 1994.
8. Dow, E.G.; Yan, S.G.; Medeiros, M.G.; Bessette, R.R. Separated Flow Liquid Catholyte Aluminum Hydrogen Peroxide Seawater Semi Fuel Cell, U.S. Patents, 2005.
9. Licht, S. Novel Aluminum Batteries: A Step Towards Derivation of Superbatteries. *Colloids Surf. A Physicochem. Eng. Aspects* **1998**, *134*, 241–248.
10. Licht, S.; Myung, N. A High Energy and Power Novel Aluminum/ Nickel Battery. *J. Electrochem. Soc.* **1995**, *142*, L179–L182.
11. Rusek, J. Direct Hydrogen Peroxide Fuel Cell, U.S. Patents, 2008.
12. Doche, M.; Novel-Cattin, F.; Durand, R.; Rameau, J. Characterization of Different Grades of Aluminum Anodes for Aluminum/Air Batteries. *J. Power Sources* **1997**, *65*, 197–205.
13. Egan, D.R.; Ponce de León, C.; Wood, R.J.K.; Jones, R.L.; Stokes, K.R.; Walsh, F.C. Developments in Electrode Materials and Electrolytes for Aluminium–Air Batteries. *J. Power Sources* **2013**, *236*, 293–310.
14. Tang, Y.; Lu, L.; Roesky, H.W.; Wang, L.; Huang, B. The Effect of Zinc on the Aluminum Anode of the Aluminum–Air Battery. *J. Power Sources* **2004**, *138*, 313–318.
15. Wang, L.; Wang, W.; Yang, G.; Liu, D.; Xuan, J.; Wang, H.; Leung, M.K.H.; Liu, F. A Hybrid Aluminum/Hydrogen/Air Cell System. *Int. J. Hydrog. Energy.* **2013**, *38*, 14801–14809.
16. Turgut, E.; Çakmak, G.; Yıldız, C. Optimization of the Concentric Heat Exchanger with Injector Turbulators by Taguchi Method. *Energ. Conv. Manag.* **2012**, *53*, 268–275.
17. Demirtas, M.; Karaoglan, A.D. Optimization of PI Parameters for DSP-based Permanent Magnet Brushless Motor Drive Using Response Surface Methodology. *Energ. Conv. Manag.* **2012**, *56*, 104–111.
18. Montgomery, D.C. *Design and Analysis of Experiments*; Wiley: New York, 2006.
19. Bilanovic, D.; Andargatchew, A.; Kroeger, T.; Shelef, G. Freshwater and Marine Microalgae Sequestering of CO2 at Different C and N Concentrations—Response Surface Methodology Analysis. *Energy Conv. Manag.* **2009**, *50*, 262–267.
20. Roy, R. K. *Design of Experiments Using the TAGUCHI Approach*; Wiley Interscience: New York, 2001.
21. Yang, H.-T.; Peng, P.-C. Improved Taguchi Method Based Contract Capacity Optimization for Industrial Consumer With Self-Owned Generating Units. *Energ. Conv. Manag.* **2012**, *53*, 282–290.
22. Sadeghi, S.H.; Moosavi, V.; Karami, A.; Behnia, N. Soil Erosion Assessment and Prioritization of Affecting Factors at Plot Scale Using the Taguchi Method. *J. Hydrol.* **2012**, *448*, 174–180.
23. Yusoff, N.; Ramasamy, M.; Yusup, S. Taguchi's Parametric Design Approach for the Selection of Optimization Variables in a Refrigerated Gas Plant. *Chem. Eng. Res. Design* **2011**, *89*, 665–675.
24. Zhang, J.Z.; Chen, J.C.; Kirby, E.D. Surface Roughness Optimization in an End-Milling Operation Using the Taguchi Design Method. *J. Mater. Proc. Tech.* **2007**, *184*, 233–239.
25. Medeiros, M.G.; Bessette, R.R.; Deschenes, C.M.; Atwater, D.W. Optimization of the Magnesium-Solution Phase Catholyte Semi-Fuel Cell for Long Duration Testing. *J. Power Sources* **2001**, *96*, 236–239.
26. Bao, Z.; Yang, F.; Wu, Z.; Nyallang Nyamsi, S.; Zhang, Z. Optimal Design of Metal Hydride Reactors Based on CFD–Taguchi Combined Method. *Energ. Conv. Manag.* **2013**, *65*, 322–330.
27. Yao, A.W.L.; Chi, S.C. Analysis and Design of a Taguchi–Grey Based Electricity Demand Predictor for Energy Management Systems. *Energ. Conv. Manag.* **2004**, *45*, 1205–1217.

8

Interoperable Framework for IEC61850-Compliant IEDs and Noncompliant Energy Meters with SCADA

MINI S. THOMAS, IKBAL ALI, and NITIN GUPTA*

Department of Electrical Engineering, Faculty of Engineering and Technology, Jamia Millia Islamia, New Delhi, India

Abstract: Modern substation automation involves integrating IEC 61850 compliant IEDs and noncompliant energy meters with an already-existing multi-vendor SCADA at the station level. Due to the lack of knowledge toward the IEC 61850 standard, it is difficult to write applications for reading and processing data from IEDs. Therefore, utilities either have to replace the older applications with IEC 61850 compliant ones or use protocol converters to acquire data from IEDs and to exchange the data with other applications. This article presents an approach for an interoperable framework of SCADA with IEC61850 compliant IEDs and non-IEC61850 energy meters for accessing data. The data are stored in an open file format that can be accessed by any other application at the station level with minimal engineering efforts. To integrate a non-IEC61850 energy meter with the SCADA, an AMI head-end application is developed that also stores the data in an open file format. The implemented methodology allows retaining the SCADA, a non-IEC61850 compliant application, while opting for IEC61850-based substation automation systems.

Keywords: AMI, IEC61850, IED, interoperability, MMS, SCADA, Web service, XML

Abbreviations

ACSI	Abstract Communication Service Interface
AMI	Advanced Metering Infrastructure
DO	Data Object
HMI	Human–machine Interface
IEC	International Electrotechnical Commission
IED	Intelligent Electronic Device
LN	Logical node
MMS	Manufacturing Message Specification
MVL	MMS-Virtual-Lite
OPC	Object Linking and Embedding (OLE) for Process Control
SCADA	Supervisory Control and Data Acquisition
VMD	Virtual Manufacturing Device
XML	Extensible Markup Language
XSLT	Extensible Style Sheet Language Transformations

*Address correspondence to: Nitin Gupta, Research Scholar, Jamia Millia Islamia University, New Delhi 110025, India. Email: nitin_ias@yahoo.co.in

1. Introduction

Today's substation automation systems are equipped with IEDs and energy meters to provide operational and nonoperational data to the utilities. The data collected from these IEDs or energy meters are utilized for identifying and locating faults in a power system, determining the load profile, and detecting outages.[1,2] The integration of multi-speak SCADA and metering data plays a vital role in data collection and is possible only if all the meters/IEDs are multi-speak compliant or vendor neutral.[3] Considering the smart grid objectives for data exchange, utilities are more concerned in choosing the most appropriate device for their applications in order to provide better services, and, as a result, the whole infrastructure comprises smart devices from multiple vendors.[4] Each vendor has their own applications to read data from the smart devices.[5] This poses a challenge to integrate and interoperate the applications and devices at the station level. Further, there are high chances that the advanced devices purchased from one vendor do not operate with another vendor application because of differences in communication protocols.[6] Currently, the whole industrial and power segment is suffering from the problem of interoperability.[7] This is because of the dynamic need of customers, changes in communication protocols, and inclination toward modernized energy infrastructure.

The standard IEC61850 is one of the advanced communication standards that resolve the interoperability issues among the IEDs/devices at the bay level of the substation.[8,9] Integration of IEC61850 compliant IEDs/devices is possible only if they re compatible with each other. It depends on the knowledge of the IED manufacturer toward the IEC61850 standard. Each vendor has their own way of using IEC61850 in their devices.[10] As a result, most of the utilities face challenges n accessing data from IEC61850-based IEDs manufactured by different vendors. Experimentation with IEC61850 IEDs s in full swing to test the functional features, including nteroperability at the bay level. However, still less attention has been paid at the station level, where a plurality of multi-vendor applications is running. Some of these applications, which are not compliant to IEC61850, need to access data from IEC61850 IEDs and noncompliant energy meters for further processing. In such a scenario, the application uses protocol converters or OPC server to fetch data from these IEC61850 devices and noncompliant energy meters.[12,13,14,15,16] Customizing OPC client-server interface for multi-vendor IEDs requires much engineering time and effort, and internal knowledge of the IEC61850 standard for providing interaction with non-IEC61850 compliant applications. There is no universal client-server OPC interface available that can integrate multivendor IEC61850 compliant IEDs and non-IEC 61850 compliant meters. As a result, multiple OPC servers are required for different vendor IEDs or energy meters. Further, some of these applications store data in their proprietary database format, which is complex and hidden from the users. As an alternative, protocol converters[12,13] can be used for acquiring data for non-IEC61850 compliant application/SCADA from IEC61850 compliant IEDs/meters wherein the protocol converter converts data from a IEC61850 format to a SCADA/application readable format. This solution entails extra cost to achieve interoperability at the station level. Therefore, utilizing the non-IEC61850 compliant application for accessing data from IEC61850 compliant IEDs, without using any protocol converter or OPC server, demonstrated in this article is a novel solution and is different from the solutions described in literature.[12-16] Considering the cost, time, and effort and to strengthen the limited knowledge toward accessing data from IEC61850-based devices using a non-IEC61850 compliant SCADA application, this article describes an IEC61850 client-server profile that can help engineers, researchers, and students to use the source code of IEC61850 and integrate noncompliant SCADA applications at the station level.

This article presents an approach for providing an interoperable framework of SCADA, i.e., a noncompliant IEC61850 application, with IEC 61850 compliant IEDs and noncompliant energy meters, where SCADA fetches data from IEDs and energy meters. The approach uses a MMS protocol of IEC61850 standard for establishing the client-server communication to access data from IEC61850-based IED and stores the data in an open file format (i.e., extensible markup language (XML)). Client-server communication is established by customizing an MMS client program, i.e., IEC61850-based application source code. For non-IEC61850 energy meter, SCADA is further integrated with another data source containing values of electrical parameters acquired by the AMI head-end application. Moreover, an IEC61850 client-server profile is also created to demonstrate the fetching of IED data and storing it in open standard format (XML) to support interoperability at the station level. The methodology of this article allows retaining the

SCADA application while opting for IEC61850-based substation automation.

The article is organized as follows. Section 2 describes the role of MMS in IEC61850 for establishing a client/server communication. Section 3 depicts a hierarchical model of IEC61850-based IED and describes how logical devices, logical nodes, and their data attributes are used in the MMS client application. Section 4 describes the interoperable framework for integrating a non-IEC 61850 SCADA application with a IED data source created by a MMS client program and energy meter data source created by AMI head-end application to validate interoperability at the station level. The section further describes a communication profile established between the IEC61850-based client program and the IEC61850-based IED for storing data in an open format, and Web service is also implemented to access the data remotely. The conclusion is given in section 5.

2. Manufacturing Message Specification in IEC61850

MMS is an international application-layer protocol for exchanging or transferring information among real devices such as IEDs and the computer applications.[17] MMS plays an important role in mapping the services of abstract communication service interface (ACSI) models of IEC61850 to the lower level and vice versa so that all the IEDs respond in the same fashion from the network point of view. This is because services provided by ACSI in IEC 68150-7-2 cannot be used directly to make communication with another device over a network. Therefore, the objective of MMS is to specify how physical devices shall operate when the messages are received to make the communication between the devices vendor independent. For example, services provided by MMS enable the MMS-compliant device to read the variables from different MMS-compliant devices because the MMS services (e.g., read, write) and messages exchanged are identical.[18]

Using a range of services, MMS provides an ability to define and explore the functionalities and capabilities of IEDs for exchanging data between the intelligent devices and other utility applications. For better clarity, a visual representation depicting an analogy between MMS and telephone communication is created in Figure 1. A caller, as shown in Figure 1, speaking in the language "French" is assumed as an "MMS client" who wants to communicate with an international receiver in the United States. The communication between the French caller and the U.S. receiver is possible only if both understand a common language, French or English. If not, then one of them has to use a local language translator, which is equivalent to a virtual manufacturing device (VMD) to ensure the correct delivery of messages between them.

Therefore, VMD plays the role of language translator that provides an abstraction layer to hide the internal functionalities of real physical devices from the external environment. The main objective of VMD is to define the following three things[19]:

1. *Objects:* MMS objects are the variables defined in the server device (IED). These objects provide capabilities for accessing operational, control, and other parameters defined in a

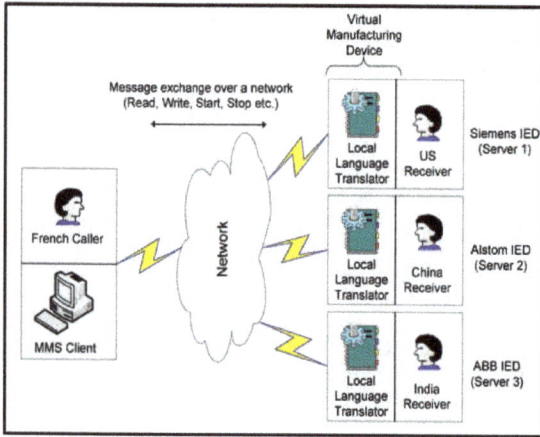

Fig. 1. MMS client/server analogy with telephone communication.

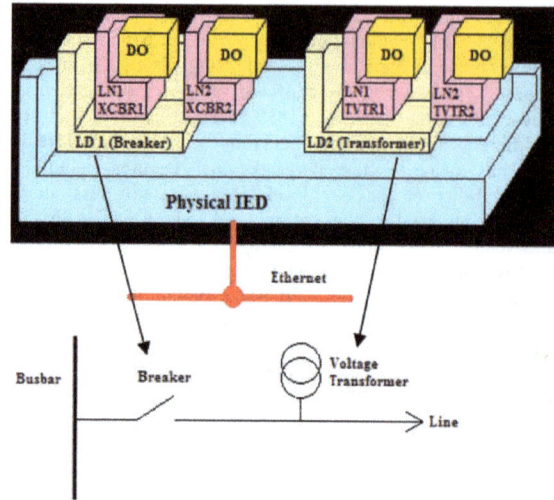

Fig. 2. Pictorial of LDs, LNs, and DOs inside a physical IED.

physical IED. In short, MMS objects enable the client application to see what is happening inside the IED.

2. *Services:* Client device or application uses various MMS services such as Read, Write, Start, Stop, and so on to manipulate the objects or to obtain the status information of the objects defined in the physical IED.

3. *Behavior:* When the client device sends a service request for an object to the server, the way the server responds upon receiving the service request is called behavior.

For exchanging the information between the IEC61850-compliant devices and applications, a client/server scheme is used in MMS for non-time-critical applications, which is a connection-oriented protocol. With the client/server scheme, a client application can perform read/write operations with an IEC61850 IED only after establishing a connection with the IED described in section 4. Some of the advantages gained by connection-oriented protocol are as follows:

• It provides an acknowledgment message upon exchanging the information successfully;
• Because of the acknowledgment message, read/write/start/stop and other services are more reliable; and
• It supports encoding and decoding of information from security point of view.

3. Hierarchical Data Model of IEC61850-Based IED

To establish communication between a client program and an IEC61850-based IED, it is important to understand the data model of the IED. As the standard IEC61850 follows an object-oriented approach, it is not possible to read data of a particular object until their root nodes and subsequent nodes nomenclature is not known. The IEC61850 standard defines a number of logical nodes (LNs) to provide various virtual functionalities for representing the operations of physical devices. A virtual representation of logical devices and logical nodes of an IED correlating the physical breaker and voltage transformer is depicted

Table 1. List of logical node groups.

Logical group	Description
A	Automatic Control
C	Control
G	Generic
I	Interfacing and Archiving
L	System Logical Node
M	Metering and Measurement
P,R	Protection Related
S	Sensor Monitoring
T	Instrument Transformer Related
X	Switch-gear Related
Y	Power Transformer
Z	Further Power System Equipment

in Figure 2. Instances of a circuit breaker (XCBRs) are represented as LNs, which further comprises data objects (DOs) representing variables related to status, measured and control value, and some other information of a physical IED. These data objects further provide the actual value of status, measurement settings, and other information.

The standard IEC61850 has categorized all the LNs into 13 logical groups as shown in Table 1.[20] Thus, by reading the initial first character of a logical node, it is easy to identify the functionalities of that logical node. This article is focused on metering aspects and that is why the logical node MMXU is selected. For example, in an instance MMXU1, as shown in Figure 3, "M" indicates that the logical node is meant for measurement related functions. This logical node is used for measuring voltage, current, power, frequency, impedance, and so on in a three-phase system. Similarly, based on the application functional requirements, a particular logical node can be selected from Table 1.

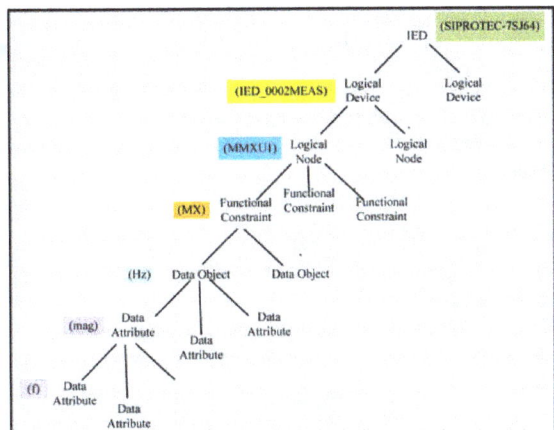

Fig. 3. Hierarchical data model of SIPROTEC-7SJ64.

Considering Siemens IED (SIPROTECT-7SJ64) used in our test setup, a hierarchical representation of logical device, logical node, data object, and data attribute is developed and shown in Figure 3, where for better clarity, each node is tagged with an equivalent name. This hierarchical model will help in establishing the client-server communication described in section 4.

4. Interoperable Framework of SCADA

Figure 4 shows an interoperable framework of SCADA for accessing data from IEC61850-based application/IED and non-IEC61850 based application/energy meter. The configured framework operates on Ethernet layer and employs the following applications/devices:

1. MMS client source code, i.e., MMS-EASE Lite (a Sisco product[21]), is an IEC61850-compliant application. Hereinafter, the IEC61850-based MMS client source code will be called a "client program."

2. IEDs[1-n], as shown in Figure 4, are IEC61850 relay IEDs.

3. A server running an AMI head-end application developed in Ref.[22] for fetching energy parameters from a network of ModSim32 virtual energy meters, which is a simulated version of non-IEC61850 meters supporting MODBUS TCP/IP protocol.[23] The network of energy meters are also connected to the station level.

4. IEDScout application[24] is used to determine various domain names, logical nodes, and their attributes defined in IEC61850 IED so that they can be called correctly inside the client program while establishing the communication with IEC61850 IED.

5. Hardware CMC 256plus[25] plays the role of a field device and is used to provide test signals to IEC61850 relay IEDs.

6. Vijeo Citect SCADA[25] is an industrial process SCADA, i.e., a non-IEC61850 application used for accessing the data from IEC61850 IED and non-IEC61850 energy meter.

Figure 4 shows the schematic of the laboratory setup that is used in implementing the proposed methodology and is available at the Substation Automation Laboratory of Jamia Millia Islamia University, New Delhi, India. The process level in the test setup is represented by the CMC 256plus test set, the bay level is represented by the relay IED, and the station level, where different applications and databases are running, is represented by the computer. The notable functionalities of the proposed methodology to achieve interoperability will be discussed next.

4.1 XML Storage

Data extracted by the client program from IEC61850 IED and the data extracted by an AMI head-end application[22] from non-IEC61850 energy meters are wrapped into XML tags, which is software or hardware independent. The benefit of wrapping the

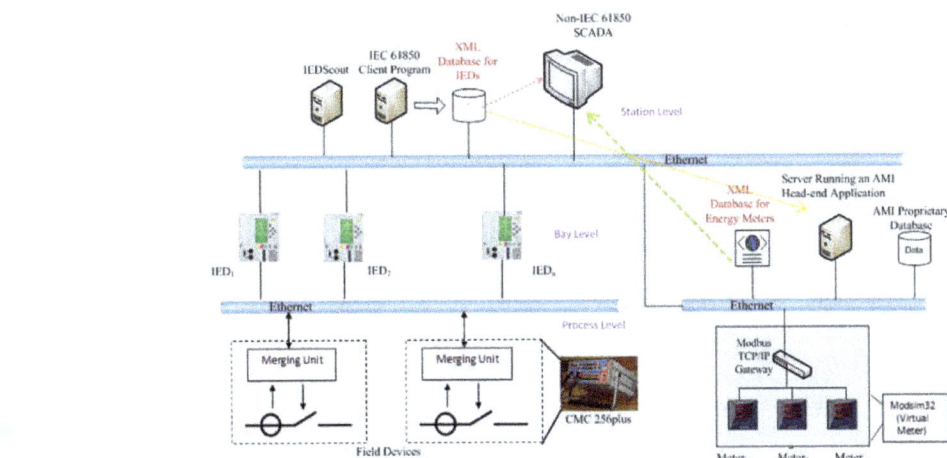

Fig. 4. SCADA laboratory test setup.

```
▼<IED_Database>
  ▼<IED_Reading>
    ▼<DateTime>
        <IED_ID>Siprotect-7SJ64-IED1</IED_ID>
        <Date>6/12/2013</Date>
        <Time>11:15:48AM</Time>
      </DateTime>
    ▼<Voltage>
        <IED_ID>Siprotect-7SJ64-IED1</IED_ID>
        <IED_Voltage_AB>12.47</IED_Voltage_AB>
        <IED_Voltage_AC>12.47</IED_Voltage_AC>
        <IED_Voltage_A>6.03</IED_Voltage_A>
        <IED_Voltage_B>6.03</IED_Voltage_B>
        <IED_Voltage_C>6.03</IED_Voltage_C>
      </Voltage>
    ▶<Current>...</Current>
    ▶<Power>...</Power>
    ▶<IED_OtherParam>...</IED_OtherParam>
    </IED_Reading>
  ▼<IED_Metadata>
      <IED_ID>Siprotect-7SJ64-IED1</IED_ID>
      <IED_Location>Lab1</IED_Location>
    </IED_Metadata>
  </IED_Database>
```

Fig. 5. XML relational database stored by client program.

data in XML tags is that it describes the structure of data in a natural language, which a user can easily understand.

A partial representation of a stored XML database created by the client program for IEC61850 IED is shown in the Figure 5. The stored database refers to very simple XML relational database schema consisting of the following four levels:

1. IED_Database: the root element of the database,
2. IED_Reading and IED_Metadata: represents the elements of the database tables,
3. DateTime, Voltage, Current, and other elements: represents the database records,
4. IED_Voltage_AB, IED_Voltage_AC, and IED_Voltage_A: represents the database fields.

Using this relational structure, the client program can add other tables, records, and fields with values. The format structure of the created storage is simple and easy to understand. Thus it provides access to non-IEC61850 applications, to fetch data from IEC61850 IEDs, without needing the cooperation from different vendors for performing the various operations such as data extraction, data transformation, and data presentation. For fetching the IED parameters, a communication profile needs to be established between the *client program* and IEC61850 IED, which is described in the next section. An XML database for non-IEC61850 energy meters is being created by the AMI head-end application.[22] To store data in a XML format, a data storage routine is being called by the developed AMI head-end application for non-IEC61850 energy meters where data is stored with the following information (i) date and time, (ii) meter model, (iii) location of the installed meter, and (iv) parameter's value. This information helps in tracking the historical values of any parameter of interest. The history of the logged parameters enables the utilities to generate bills, calculate load profiles, or do other analysis for taking certain actions. Also, the same AMI head-end application, which is Modbus TCP/IP compliant, can also access the database of IEC61850 IED without any protocol converter.

Once the values are stored in the XML database, the utilities can share the XML database with their customers, allowing them to use their own extensible style sheet language transformations (XSLT) language for processing the XML database and displaying the parameters in the form of a Web page.

Open database (i.e., XML) is used due to the following advantages:

1. Easy to understand and any kind of data processing/analysis.
2. This helps utilities in retaining SCADA for accessing data from IEC61850 IEDs without using protocol converters or OPC servers.
3. Non-SCADA applications such as XSLT language can be used to present or process the XML data to remote users via a Web page.
4. Using XML storage, applications can query/access the particular parameters directly from the XML database instead of sending the requests to the devices; therefore, it reduces traffic congestion at the bay level.

4.2 MMS Client-Server Communication

The MMS client provides a high-level program interface layer referred to as MMS-Virtual-Lite (MVL), which is closely coupled to the lower layer subsystem components and provides an application framework that is suitable for client applications. For client applications, MVL provides easy-to-use API hooks for performing various operations such as MMS connection control and MMS read/write services. MVL supports VMD to provide a straightforward mechanism by which variables defined in the IED can be accessed. Thus users are not required to write their own MMS programs from scratch for fetching data from IEC61850-based IEDs/devices.

As discussed in section 2, VMD provides an abstraction layer for an external world to hide the internal functionalities of a real physical IED/server. Therefore, a plurality of object models of IEC61850 are mapped to specific MMS objects as shown in Figure 6. For instance, the IED server is mapped to MMS VMD, logical device object is mapped to MMS domain, logical name is mapped to MMS named variable, and data are mapped to named variable type description.

Such a mapping is helpful in establishing the communication between the MMS client application and the IEC61850-based server device/IED. A communication flow for reading the IED energy parameters, e.g., frequency (Hz), using the client program is created in Figure 7. While establishing connection with the remote server device, i.e., SIPROTEC 7SJ64 (IEC61850 IED), the client program uses a function domvar_type_id_create() to send a request to the IED server with following parameters:

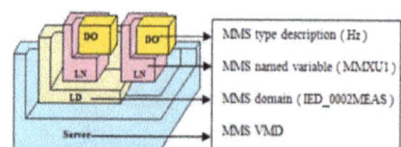

Fig. 6. IEC61850 to MMS object mapping.

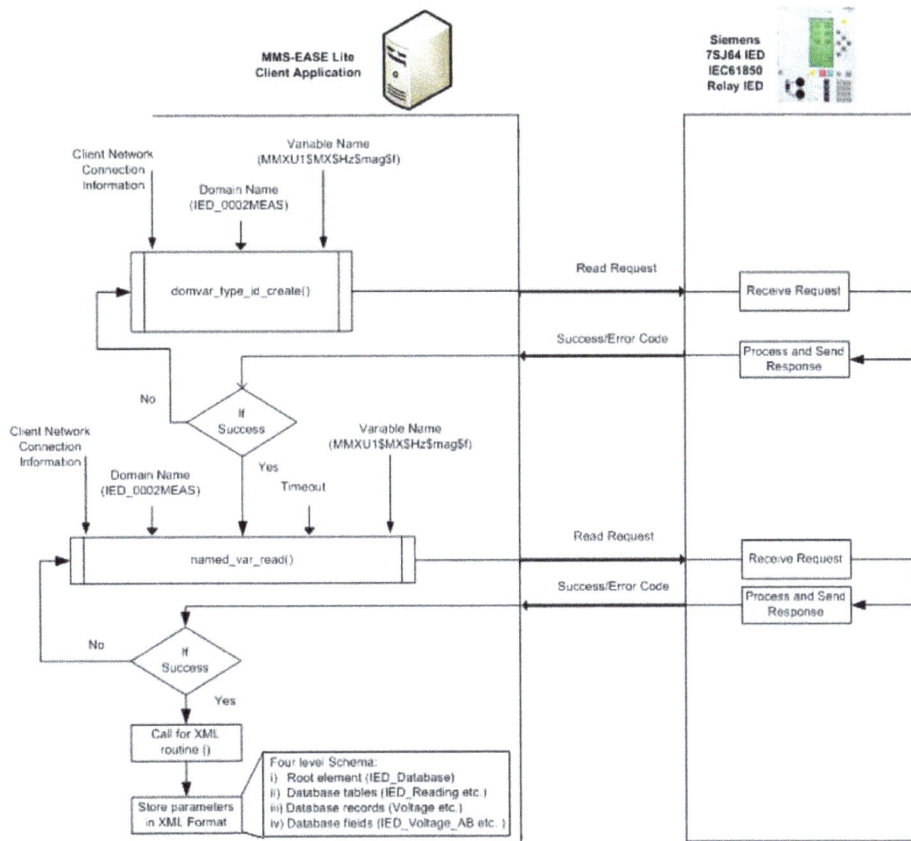

Fig. 7. Communication profile between client (MMS-EASE Lite) and server (Siemens IED).

1. A network connection information, to specify the remote server to which the MMS protocol data unit (PDU) is to be sent, or from which the PDU will be received.
2. A particular domain name, say for specifying the metering operations (e.g., IED_0002MEAS), whose data structure is shown in Figure 8. IEDScout from Omicron can be used for viewing this data structure.
3. A particular variable name, e.g., MMXU1MXHz$mag, whose value need to be accessed. The format of calling the variable, e.g., variable "mag" as shown in Figure 3, in the client program has the following parts:
   - "MMXU1" is an instance of the logical node (LN) class MMXU,
   - "MX" is a functional constraint related with measurement functionality. Functional constraint depicts the services that are possible for a particular DataAttributeComponent (DA).
   - "Hz" is DataObjectComponent (DO), and
   - "mag" is a DA representing the magnitude of the complex value.

The function domvar_type_id_create() called in the client program returns the type for any domain-specific variable and creates a unique type identifier (ID) that is used as an input to

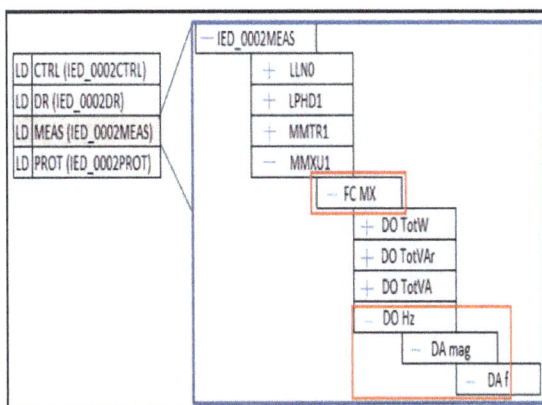

Fig. 8. Data structure of an IEC61580 IED (e.g., Siemens SIPROTEC-7SJ64)

access the particular variables in the function named_var_read(). For every variable, a unique type ID is created. Once the function named_var_read() is called, the client program waits for 1 second

Table 2. MMS-Ease Lite source code output.

Configured parameters	Parameter values
Value of Voltage in Phase AB	12.465000 V
Value of Voltage in Phase A	6.030000 V
Value of current in Phase A	9.750000 A
Value of Frequency	49.100000 Hz

to get the server response. The waiting time for listening the server response can be varied by setting the "Timeout" parameter in the function named_var_read().

The behavior of the IED/server involves receiving the read request from the client program and processing the information to send the response (+ve response) or an error code (-ve response) back to the client program. Once the client program receives a positive response from the server, the client program polls the desired electrical parameters and stores the parameters' value in an interoperable XML data source, as shown in Figure 5, by calling an XML data storage routine similar to as provided in Ref.[2] In the experimental test setup, the client program (coded using MMS-EASE Lite) is customized to read and store the parameter values, i.e., Voltage for phase AB, Voltage for phase A, Current for phase A, and Frequency from IEC61850 IED in an XML data source different from the data source of energy meters. The output of the client program is shown in Table 2.

4.3 Validation of Interoperability Using Non-IEC 61850 SCADA

Vijeo Citect process SCADA,[25] a non-IEC61850 application, is used to validate the interoperability demonstration at the station level. This SCADA application fetches data from IEC61850 IEDs through the XML database and is implemented using Cicode, which is a scripting language similar to C language. For validation of the interoperable framework concept, authors of this article have customized the Cicode in order to monitor the field devices. Cicode supports a multitasking and multithreading environment to interface with the plurality of XML data sources, where the energy data of IEC61850-based IED and non-IEC61850 energy meters are stored, for accessing real-time information. Figure 9 shows an architecture where SCADA application needs to extract parameters from IEC61850 IEDs and also from non-IEC 61850 energy meters. The HMI of process SCADA in a tabular form is demonstrated in Figure 9, where the left portion shows the value of parameters extracted from an XML data source of the IEC61850-based IED, and the right portion shows the value of parameters extracted from another XML data source of the simulated energy meter. Most of the SCADA/HMI vendors have their own scripting languages. Table 3 shows a list of major SCADA vendors, widely used in industrial applications, named with their scripting platform that can be easily customized to parse the XML data sources. Functionality of these scripting platforms is similar to the Cicode scripting language used in this article.

Using the proposed methodology for IEC61850-based IEDs and non-IEC 61850 energy meters, where data from both are stored in XML format for further processing, does not

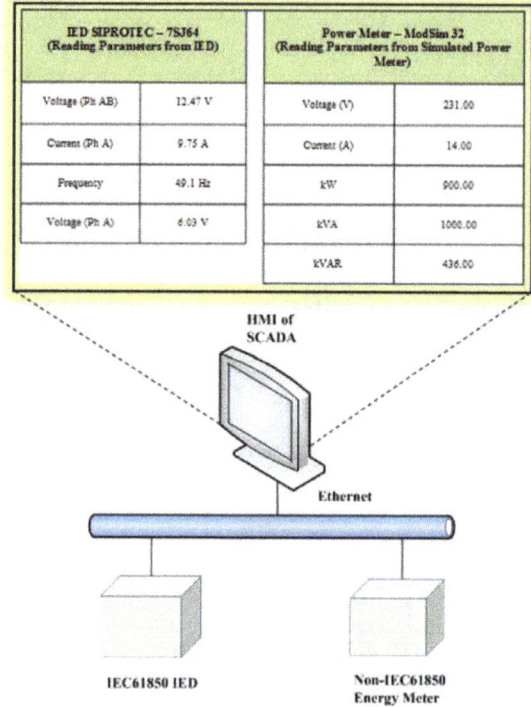

Fig. 9. SCADA accessing XML data source of IEC61850-based IED and noncompliant energy meter.

Table 3. List of SCADA/HMI with scripting platform.

SCADA vendors	SCADA/HMI	Scripting platform
Rockwell Automation	RSView32	Visual Basic for Application (VBA)
Invensys	Wonderware InTouch	QuickScript
GE	iFIX Intellution	VBA
Siemens	SIMATIC WinCC	Visual Basic Scripting (VBS) and VBA
iCONICS	Genesis	ScriptWorX (uses VBA)
Honeywell	Experion	VBS
Schneider Electric	Vijeo Citect	Cicode

require a protocol converter. Therefore, it helps in retaining the existing SCADA, a non-IEC61850 compliant application, while implementing the IEC61850-based substation automation systems.

4.4 Remote Data Access

For accessing the data from the IEDs and energy meters, a Web service is developed using a .Net based server. The Web service, running on an intermediate Web server, accesses the XML data sources of remotely located IEDs and energy meters. The remote

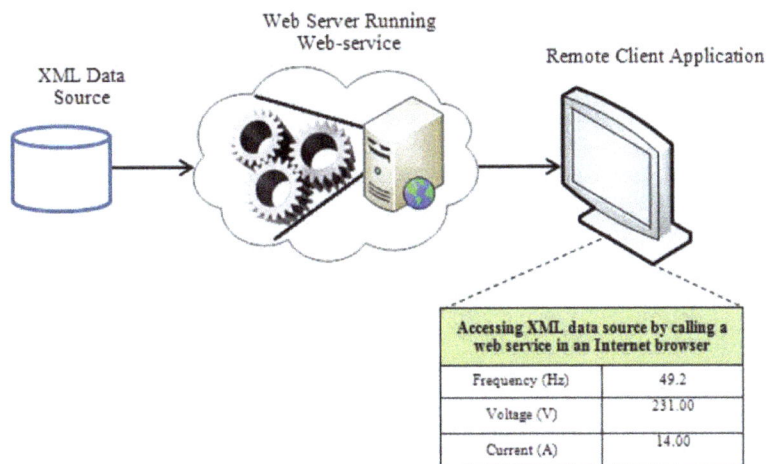

Fig. 10. Remote access to data source using a Web service.

client application, e.g., Internet browser, uses the Web service for displaying the extracted parameters as shown in Figure 10.

The Web service provides an abstraction to the remote client applications. This abstraction helps in making the system more robust in terms of accessing the parameters of IEC61850-based IEDs or non-IEC61850 energy meters with minimal engineering efforts.

Database platforms (e.g. SQL, DB2, MS Access, MySQL, Oracle, Excel) can use schema/data conversion tools to import data from the open file format, made available to the client application using the Web service, for generating bills, reports, etc.

Thus this section demonstrates an interoperable framework for accessing metering data through an open file format and making it interoperable and accessible to multi-vendor utility applications. Following are the advantages of the demonstrated interoperable framework, and they are compared in Table 4 with other published work.

Table 4. A Comparison of demonstrated framework of this article with other published work.

	SYNC[11]	Byong et al.[12]	Crispino et al.[13]	Srinivasan et al.[14]	Jack et al.[15]	Demonstrated framework
Is protocol converter required?	Yes	Yes	NA	NA	NA	No, Protocol converter and OPC interface both are not required
Is OPC client-server interface required?	NA	NA	Yes	Yes	Yes	
Does it support IEC61850 compliant IED and noncompliant meters?	Yes, it supports both the devices	Supports only IEC61850 based IEDs	Supports only IEC61850 based IEDs	Supports only IEC61850 based IEDs	Supports only IEC61850 based IEDs	Yes, the framework is applicable for IEC61850 compliant IEDs and noncompliant meters
Does it support integration (interoperability) with 3rd party applications, e.g., billing application, XSLT?	A detailed understanding of protocol converter is required for integrating the 3rd party applications	No	No	No	No	Yes, 3rd party application can be integrated easily. Users need to understand only XML schema that is user friendly and is not a proprietary schema.
Does it provide remote data access in the form of a Web service?	No, only remote configuration is supported for maintenance purpose	No	No	No	Yes, remote access is provided but not as a Web service	Yes, the demonstrated Web service is generic and any 3rd party application can call this service for accessing data

NA: Not applicable.

- Data from IEC61850-compliant IEDs and energy meters are accessible to traditional SCADA/HMI without needing the protocol converter or OPC interface.
- All types of third-party applications (Industrial SCADA or Enterprise Application) running at the station level can fetch data from XML-based platform independent storage, which was not possible earlier because of proprietary database.
- XML storage can support simultaneous and multiple connections with applications at the station level that help in reducing the traffic of reading the data from IEDs/meters directly.
- For remote metering, Web services can be customized to tailor the data as per the user's need.

5. Conclusion

This article presents an implementation for achieving interoperability among the applications at the station level for extracting data of IEC61850-based IEDs and non-IEC 61850 energy meters. To demonstrate this approach for interoperability, the source code of the IEC61850-compliant application (client program) is used and customized to fetch data from a physical IEC61850 IED and stored in XML format. To integrate a non-IEC 61850 energy meter, an AMI head-end application is developed to fetch data and is used for storing the parameter values in a different XML database.

The approach demonstrated in this article helps non-IEC61850 compliant applications at the station level such as SCADA or any other scripting applications in accessing the open file format to read IED or energy meter data without a protocol converter and thus avoids unnecessary delay and cost. Moreover, a detailed insight of a MMS model and communication profile between client and server is provided to help practicing engineers and researchers to customize the IEC61850 client program as per the requirements. Web service has helped in demonstrating the remote metering, which enable consumers/utilities to design their own reporting templates to display fetched data.

References

1. Duncan, B. K.; Bailey, B. G. Protection, Metering, Monitoring, and Control of Medium-Voltage Power Systems. *IEEE Trans. Indust. Appl.* **2004**, *40*, 33–40.
2. Kezunovic, M. Smart Fault Location for Smart Grids. *IEEE Trans. Smart Grid* **2011**, *2*, 11–22.
3. See, J. D.; Latham, S.; Shirek, G.; Carr, W. C. Report on Real-Time Grid Analysis Pilots. *IEEE Trans. Indust. Appl.* **2012**, *48*, 1170–1176.
4. Li, F.; Qiao, W.; Sun, H.; Wan, H.; Wang, J.; Xia, Y.; Xu, Z.; Zhang, P. Smart Transmission Grid: Vision and Framework. *IEEE Trans. Smart Grid* **2010**, *1*, 168–177.
5. Strasser, T.; Andrén, F.; Lehfuss, F.; Stifter, M.; Palensky, P. Online Reconfigurable Control Software for IEDs. *IEEE Trans. Indust. Inform.* **2013**, *9*, 1455–1465.

6. Ma, R.; Chen, H.-H.; Huang, Y.-R.; Meng, W. Smart Grid Communication: Its Challenges and Opportunities. *IEEE Trans. Smart Grid* **2013**, *4*, 36–46.
7. Von Dollen, D. *Report to NIST on the Smart Grid Interoperability Standards Roadmap*, Technical Report SB1341-09-CN-0031, Electric Power Research Institute (EPRI), **2009**.
8. Manassero, G. Jr.; Lorenzetti Pellini, E.; Senger, E. C.; Nakagomi, R. M. IEC61850-Based Systems—Functional Testing and Interoperability Issues. *IEEE Trans. Indust. Inform.* **2013**, *9*, 1436–1444.
9. Ozansoy, C. R.; Zayegh, A.; Kalam, A. The Application-View Model of the International Standard IEC61850. *IEEE Trans. Power Delivery.* **2009**, *24*, 1132–1139.
10. O'Reilly, R.; Beng, T. C.; Dogger, G. Hidden challenges in the implementation of 61850 in larger substation automation projects. *Proc. Distributech.* **2008**.
11. SYNC 2000, Field IO and Communication Gateway. http://www.ulepl.com/pdf/SYNC%202000.pdf (accessed Nov. 2, 2014).
12. Yoo, B.-K.; Yang, S.-H.; Yang, H.-S.; Kim, W.-Y.; Jeong, Y.-S.; Han, B.-M.; Jang, K.-S. Communication Architecture of the IEC 61850-based Micro Grid System. *J. Elec. Eng. Tech.* **2011**, *6*, 605–612.
13. Crispino, F.; Villacorta, C. A.; Oliveira, P. R. P.; Jardini, J. A.; Magrini, L. C. An Experiment Using an Object-oriented Standard—IEC 61850 to Integrate IEDs Systems in Substations. *IEEE/PES Transmission and Distribution Conference and Exposition: Latin America* **2004**, 22–27.
14. Seshadhri, S.; Kumar, R.; Vain, J. Integration of IEC 61850 and OPC UA for Smart Grid Automation. *IEEE Conference on Innovative Smart Grid Technologies-Asia (ISGT Asia)* **2013**, 1–5.
15. Chang, J.; Vincent, B.; Reynen, M. Protection and control system upgrade based on IEC- 61850 and PRP. *67th IEEE Annual Conference for Protective Relay Engineers* **2014**, 496–517.
16. *Industrial Automation Systems, Manufacturing Message Specification.* Part 1, ISO Standard 9506-1, **2003**.
17. *Communication Networks and Systems in Substations – Part 8-1: Specific Communication Service Mapping (SCSM)—Mappings to MMS (ISO 9506-1 and ISO 9506-2) and to ISO/IEC 8802-3*, IEC 61850-8-1, 2004–05.
18. Sorensen, J. T.; Jaatun, M. G. An Analysis of the Manufacturing Messaging Specification Protocol. *5th International Conference on Ubiquitous Intelligence in Computing*, **2008**, 602–615.
19. *Communication Networks and Systems in Substations – Part 7-4: Basic Communication Structure for Substation and Feeder Equipment—Compatible Logical Node Classes and Data Classes*, IEC 61850-7-4, 2004–05.
20. MMS-EASE Lite: IEC 61850 for Embedded Systems. http://www.sisconet.com/downloads/MktLit_mmslite.pdf (accessed Nov. 2, 2014).
21. Thomas. M. S.; Ali, I.; Gupta, N. Interoperability Framework for Data Exchange between Legacy and Advanced Metering Infrastructure. *Int. J. Energ. Tech. Policy* **2012**, *2*, 49–59.
22. ModSim32-Application Description.http://www.win-tech.com/html/modsim32.htm (accessed Oct. 28, 2014).
23. IEDScout. Versatile Software Tool for Working with IEC 61850 Devices. http://www.omicron.at/en/products/pro/communication-protocols/iedscout/ (accessed Oct. 28, 2014).
24. CMC256plus. High Precision Relay Test Set and Universal Calibrator. http://www.omicron.at/en/products/pro/secondary-testing-calibration/cmc-256plus/ (accessed Oct. 28, 2014).
25. Vijeo Citect. 10 Things You Should Know About SCADA. http://www.schneider-electric.co.za/documents/solutions/energy-efficiency-documents/13.pdf (accessed Nov. 2, 2014).

9

Role of Wind Power in the Energy Policy of Turkey

BİLGİN ŞENEL[1], MİNE ŞENEL[1], and LEVENT BİLİR[2]*

Industrial Engineering Department, Tunceli University, Tunceli, Turkey
[2]*Energy Systems Engineering Department, Atılım University, Ankara, Turkey*

Abstract: Energy used to be produced mostly from fossil fuels until the near past, and the use of these fuels has created many environmental problems. For this reason, renewable energy resources have become important for energy production. As a result of technological developments in the field, electricity production using wind turbines with high capacities is possible in the current era. An increasing tendency to use renewable energy resources in the world has revealed a need to study the current position of wind energy in Turkey also and particularly the policies to increase the usage of wind energy. First, the current position of energy production and energy policy in Turkey is summarized and the details about installed wind power plants are presented. Second, the future projections about the energy power plants are discussed. The details about current and future wind power plants in Turkey are especially given. Incentives for wind power plant constructions are also presented in the study. As a result, it can be concluded that Turkey has great wind energy potential. The use of wind turbines can provide a considerable amount of the country's energy need if bureaucratic and security problems are solved and if effective incentives can be given for renewable energy.

Keywords: Wind energy, renewable energy, energy policy of Turkey, incentives for renewable energy in Turkey

1. Introduction

Wind occurs as a result of the differences in the heat levels on the earth's surface, which create a movement in the air. Two percent of solar energy turns into kinetic energy in the form of wind.[1] It has been discovered that wind could be a powerful resource for energy by humanity ages before. People have used the wind to move sailing boats, to pump water to higher levels, and to grind grains. Although steam engines operating according to thermodynamic processes were used to provide energy especially in the 18th century, still wind energy was needed to pump the water in these steam engines.[2] At the beginning of the 20th century, the use of fossil fuel began to be discussed due to its negative effects on the environment. The use of fossil-based energy resources gives rise to the formation of greenhouse gases (CO_2, SO_2, and NO_x). The amount of CO_2 in the atmosphere is 25% higher today than the period before the Industrial Era. It is expected this amount be doubled by 2050. The emission of acid gases such as SO_2 and NO_x has begun to threat the global climate also. Thus,

it is necessary to take urgent precautions. One of these urgent precautions is to increase the use of renewable energy resources. The United Nations gathered in Kyoto, Japan, to deal with climate change in 1997. A decrease in emission was set as a goal to reduce the greenhouse effect by the Kyoto Protocol, which was accepted at the end of the meeting and signed by 141 countries, including Turkey. As it is shown in Figure 1, the use of renewable energy resources in electricity production increases every year.

Wind energy is commonly used in Denmark, Spain, Germany, Britain, and the US today to produce electricity.[4] The capacity of installed wind energy was 6,100 MW in 1996. It is 33 times higher with 203,500 MW at the beginning of 2010, as shown in Table 1.

According to a scenario study, which has been prepared by the International Energy Agency and which assumes that the current policies continue as they are (WEO2011), electricity production is estimated to increase from 203,500 MW to 285,690 MW in 2020, 354,680 MW in 2030, and 393,680 MW in 2035 with an average of 2.6% increase rate. These numbers indicate a 96.4% increase between 2009 and 2035.[7]

The importance of wind energy is also understood in Turkey, and there are significant numbers of studies investigating wind energy potential in Turkey. Karsli and Geçit[8] investigated wind energy potential in Gaziantep, a city located in the southeast of Turkey, and they found a very encouraging mean power density of 222 W/m^2 at 10 m height. Ozerdem and Turkeli[9] determined the wind characteristics at the Urla region of İzmir, a city that is on the Aegean coast of Turkey. They reported that the region has high wind energy potential. Hepbaşlı and Özgener[10] interpreted the development of wind energy in Turkey: its history,

*Address correspondence to: Levent Bilir, Energy Systems Engineering Department, Atılım University, Kızılcaşar Mah., İncek, Gölbaşı, 06836 Ankara, Turkey. Email: levent.bilir@atilim.edu.tr

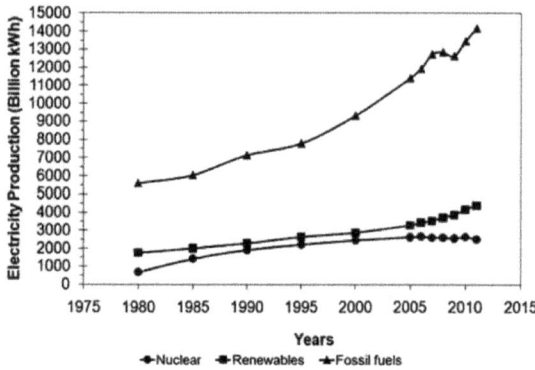

Fig. 1. World electricity production.[3]

Table 1. World installed wind power capacity[5] and the prediction for 2020–2030.[6]

Years	Installed wind power capacity (MW)
1996	6,100
1997	7,600
1998	10,200
1999	13,600
2000	17,400
2001	24,322
2002	31,181
2003	39,295
2004	47,693
2005	59,024
2006	74,122
2007	93,930
2008	120,903
2009	159,213
2010	203,500
2020 (Prediction)	285,590
2030 (Prediction)	354,680

the reconstruction of the Turkish electricity market, and wind energy applications by the end of October 2003. The total theoretical potential of available wind energy in Turkey was examined. It was stated that the annual potential of southeastern Anatolia and the western shore of Turkey is 88,000 MW. Köse[11] studied the possible wind energy potential in Kütahya, Turkey in his research. An observation station was constructed on the campus of Dumlupınar University, and possible electricity power was analyzed. According to the results of the observations, 36.62 W/m^2 of energy density was found in 30 meters height with 4.62 m/s of wind speed in a period of 20 months. Şahin, Bilgili, and Akilli[12] declared that the potential of wind energy on the eastern Mediterranean shores of Turkey is enough to produce electricity, and they found out that the average power density is 500 W/m^2 at 25 meters height in many areas. Akpinar and Akpinar[13] investigated the wind speed and probability distribution of four wind power plants

(Elazığ-Maden, Elazığ-Keban, Elazığ, Elazığ-Ağın) in a period of 8 years (1998–2005).

Gökçek and Genç[14] conducted a pre-study related to the cost of wind energy in Central Anatolia, Turkey. The cost of electricity were attempted to be found out for the seven different chosen regions with 2.5, 5, 10, 20, 30, 50, 100, and 150 kW wind energy conversion systems by using the time series method. According to the calculations, a wind turbine with a capacity of 150 kW and a hub height of 30 meters supplies 120,978 kWh of energy annually in Pınarbaşı. İlkiliç and Turkbay[15] conducted research on the potential for wind energy systems in Turkey. The use of wind energy and wind energy potential in different regions were analyzed via the time series method. As a result, it was found out that the wind energy potential of Turkey is satisfactory. Çelik[16] took the Çanakkale region under consideration. He calculated wind energy potential values for the city center and Bozcaada (an island in the Aegean Sea near Çanakkale city). A mean wind energy density of 354.75 W/m^2 for Bozcaada and 139.18 W/m^2 for Çanakkale at 10 m height was found in the study. Yaniktepe et al.[17] took wind speed data at 10 m height between January 2008 and August 2011 and calculated a wind potential of 24.587 W/m^2 for Osmaniye, Turkey. Akpınar[18] evaluated the wind energy potential at the northeastern coast region (east Black Sea) of Turkey. He used the wind data collected at 10 m height for 9 years and determined the wind power density of 6 locations. He reported that the maximum wind energy density is 59.96 W/m^2 in Sinop, and the density values of Sinop, Hopa, and Trabzon are higher than those of the other 3 locations of the study. The studies that have been done so far reveal that wind energy is a promising alternative in Turkey, and the use of this huge energy resource will increase in the following years, after the reconstruction of the Turkish electricity market.

In this study, current electricity consumption and electricity production in Turkey are analyzed, and also Turkey's future plans for energy production are summarized in detail. Correspondingly, a list of current wind power plants with their production capacities (up to July 2012) is presented in this study in which the present situation of wind energy in Turkey is also discussed. Future wind power plants are also taken into consideration, and the incentives for these renewable energy investments are presented in detail. The study reveals that the wind energy potential of Turkey is significant, and a considerable amount of electricity demand can be met if this potential can be used effectively.

2. Energy Policy of Turkey

The geographic coordinates of Turkey lie at 36°–42° north and 26°–45° east, and it is located in a temperate zone, where four seasons can be experienced. Some of Turkey's land is located in the Balkan Peninsula, which is in the southeastern part of Europe; the rest of its land is located in the southwestern part of Asia. Turkey being close to energy resources of the Middle East and Asia can serve as a bridge for European countries to reach these energy resources.

Unfortunately, Turkey can produce only 2% of its energy need, which has primary importance for its industrialization, by using its own resources. Not being able to use its own energy resources efficiently made Turkey dependent on close areas for energy.

Thus, increasing import outlays have reached 44 billion 294 million dollars in 2011. This is one of the main reasons that increase its foreign trade deficit, one of the most serious problems of Turkey.

When the energy policy of Turkey is reviewed, it can be seen that the public sector has a share of 75% in energy production although important progress has been made by the private sector in electricity distribution. This low ratio of the private sector in energy production is due to bureaucracy. Bureaucratic procedures take too much time—mainly the time required to get a license is a weighty matter for private sector. Additionally, investors have problems in getting environmental effect reports, and they are not encouraged enough to invest. Such reasons make private-sector investors less willing to invest in renewable resources.[19]

This kind of obstacle leads to an increase in foreign trade deficit in countries such as Turkey, which correspondingly results in impairment in the effective usage of renewable energy resources. Turkey, in the awareness of the situation, spurred some new investments, especially on natural gas and petroleum, in order to meet its own needs and to gain a strategic position in the field and to increase its income. First, governments of Turkey and Azerbaijan signed important agreements about the transfer of Azeri natural gas. According to the agreements, Turkey will import annually 6–7 billion cubic meters of gas from the Shah Deniz II project, and it will have the right to re-export natural gas to the third countries. According to the reports, Azerbaijan will pay Turkey a transport fee of $45/1,000 cubic meters—thus Turkey will be able to raise its annual revenue of $300 million from the transport of Azeri natural gas to Europe. With another key agreement, the price of natural gas, which Turkey imports from the Shah Deniz I project, increased to $300/1,000 cubic meters from $120/1,000 cubic meters.[20] With these agreements, Turkey will be a transit country in piping both natural gas and petroleum in the future. In this position, Turkey will be able to import cheap natural gas and petroleum and decrease its current deficit. Natural gas, which is subject to the agreement, has a share of about 38.6%, while hydroelectricity power has a share of 30%, lignite and coal have a share of 13%, and wind energy has a share of 0.04% among the total energy resources.

Hydroelectric power plants in Turkey are able to generate 45 TWh of energy annually by using about 35% of their hydroelectricity potential. It is planned to construct 502 new hydroelectric power plants in order to make use of the full potential until 2020.[21,22] State Hydraulic Works have begun to construct new hydroelectricity power plants that have 550 MW of power capacity.

While the Turkish government plans not to import energy by 2023, the number of new approvals and licenses increases day by day. According to the "No energy import policy," the goals of the Turkish Republic are specified below:[23]

- 125,000 MW of installed power (54,423 MW in 2010)
- Increasing the share of renewable energy to 30%
- 60,717 kilometers of energy transmission line (49,104 kilometers in 2010)
- Capacity of power distribution on the level of 158,460 MVA (98,996 MVA in 2010)
- Decreasing the electricity loss or the use of illegal electricity to 5% and increasing the use of smart grid

- 5 billion m^3 of natural gas storage capacity (2.6 billion m^3 in 2010)
- Establishing a power exchange
- Activating 8 nuclear reactors with a capacity of 10,000 MW
- Constructing 4 nuclear reactors with a capacity of 5,000 MW

If the goals for 2023 are not achieved, Turkey will be drawn into a strategic vortex as its own local energy resources decrease.

When the amount of electricity production in Turkey from 1970 to 2010 (Table 2) is analyzed, it can be seen that a decrease is observed only in 2001 and 2009 if compared to the previous years. The cause of these decreases is severe financial crisis in both years. However, energy production has always increased in the other years. Such factors as the continuous increase in electricity production (and consequent consumption) and the continuous increase in energy prices (due to the ambiguous conjuncture in the world) indicate that Turkey can continue its current growth rate as long as it supports enterprises that use the country's own energy resources.

When Table 3, which shows the shares of different resources used in electricity generation in Turkey, is analyzed, it can be seen that the share of renewable energy resources (other than hydroelectric) increased considerably after 2006. The reason of this significant increase is that a renewable energy law, which supports the use of renewable energy resources, was enacted in 2005. However, obviously the current share of renewable energy resources (in 2012), which is 3.1%, is significantly below the shares of fossil fuel. Another significant problem for Turkey is that coal and mostly imported natural gas still have large share values in electricity generation. This situation points out that Turkey has still not been able to use its own resources.

Some serious problems have begun to be seen about both the environment and human health because of energy production in thermal power plants that use coal as fuel. In a study, it was stated that the amount of radiation in 34 out of 50 villages near the Yatağan Thermal Power Plant was much higher than the level that humans can tolerate, and it was 19 times higher than the acceptable level in the areas where the ashes were left.[26] Respiratory diseases in Yatağan, the town, are two times higher than Muğla, the province.[27] Local coal is burned in this power plant to generate energy. Although there is considerable financial profit by

Table 2. Electricity production in Turkey between 1970 and 2010.[24]

Year	Electricity production (GWh)	Year	Electricity production (GWh)
1970	8,623.0	2001	122,724.7
1975	15,622.8	2002	129,399.5
1980	23,275.4	2003	140,580.5
1985	34,218.9	2004	150,698.3
1990	57,543.9	2005	161,956.2
1995	86,247.4	2006	176,299.8
1996	94,861.7	2007	191,558.1
1997	103,295.8	2008	198,418.0
1998	111,022.4	2009	194,812.9
1999	116,439.9	2010	211,207.7
2000	124,921.6		

Table 3. Shares of resources in electricity production in Turkey.[25]

Year	Coal (%)	Liquid fuels (%)	Natural gas (%)	Hydro (%)	Renewable energy resources (%)
2000	30.6	7.5	37.0	24.7	0.3
2001	31.3	8.4	40.4	19.6	0.3
2002	24.8	8.3	40.6	26.0	0.3
2003	22.9	6.5	45.2	25.1	0.2
2004	22.9	5.1	41.3	30.6	0.2
2005	26.7	3.4	45.3	24.4	0.2
2006	26.5	2.5	45.8	25.1	0.2
2007	27.9	3.4	49.6	18.7	0.4
2008	29.1	3.8	49.7	16.8	0.6
2009	28.6	2.5	49.3	18.5	1.2
2010	26.1	1.0	46.5	24.5	1.9
2011	28.9	0.4	45.4	22.8	2.6
2012	28.4	0.7	43.6	24.2	3.1

Table 4. The inventory of greenhouse gas emission in Turkey.[28]

Years	Greenhouse gas emission (million tons CO_2 equivalent)				
	1990	1995	2000	2005	2010
Sector					
Energy production	132.88	161.50	213.20	242.34	285.07
Industrial processes	15.44	24.21	24.37	28.78	53.94
Agriculture	30.39	29.23	27.85	26.28	27.13
Waste	9.72	23.88	32.79	33.58	35.97
Total	188.43	238.82	298.21	330.98	402.11

this production, coal causes significant damage to the health of people living in that area.

Values related to greenhouse gas (including mainly CO_2) emission, one of the disadvantages of using fossil fuel, can be seen in Table 4. As can be found, energy production has the biggest rate of greenhouse gas emission in Turkey.

3. Wind Energy Policy and Its Development in Turkey

Turkey's geographic location has the advantage of extensive use of many renewable energy resources.[19,28,29] In this regard, renewable energy resources appear to be one of the most efficient and effective solutions for sustainable energy production and environmental pollution prevention in Turkey. Preliminary data show that the Marmara, Aegean, and Southeast Anatolia regions of Turkey are highly suitable for wind energy applications.[11]

In Turkey, electricity production through wind energy was first realized at the Çeşme Altınyunus Facilities in 1986. The hub height of the used turbines is 24.5 m, and the blade diameter is 14 m. Each of these turbines provides electricity power of 55 kW at a wind speed of 12 m/s. 100,000 kWh of electricity energy per year is generated in Çeşme from this wind power plant, and this amount of energy meets 4% of the resorts' annual energy demand.[11]

As a result, the Turkish Wind Energy Association (TWEA) was founded in Turkey as a branch of the European Wind Energy Association in 1992. Electricity production through wind energy connected to a system in Turkey began at the ARES Wind Farm in Alaçatı, Çeşme with twelve 600-kW wind turbines in 1998. The biggest wind power plant in Turkey with a capacity of 10.2 MW was constructed in Bozcaada in 2000.[30] There has been a 100% annual increase in the installed power and energy production since 2005, when law no. 5346 (Renewable Energy Law) was enacted. The installed wind power plant capacity reached 1,329 MW at the end of 2010, 1,805.85 MW at the end of 2011, and 2,041.35 MW at the end of July 2012. 56 out of 157 licenses given to produce electricity through wind energy in Turkey are already installed power plants, and the other 101 wind power plants are planned to be installed in the future. New plants will have an installed power of 6,224.20 MW. The wind power plants in service and their capacities are given in Table 5.

Greenhouse gas emission (especially CO_2 production) caused by energy production is also reduced with wind energy use. The carbon dioxide emission for 1 kWh electricity production is given as 6.89551×10^{-4} metric tons CO_2/kWh.[31] When the installed wind power capacity of Turkey is considered, yearly electricity production from wind power plants can reach 1.788×10^{10} kWh. As a result, a total of 12.33 million tons of CO_2 emission per year can be prevented with the use of wind energy in Turkey. This amount of CO_2 emission reduction represents approximately 3.1% of the total CO_2 emission of Turkey in 2010.

The present power capacity of the installed wind power plants for different cities in Turkey is given in Table 6. There are in total 11 wind power plants being built in 2012 and are to be in service in 2013. The total installed power of the new 11 power plants will be about 452.65 MW when their construction is finished. Thus, 2,495.25 MW of energy will be produced with 56 power plants in service and 11 new power plants that will be in service in 2013. 191 new wind turbines will be constructed in these new 11 power plants. By the end of 2012, the number of turbines with a power between 0.5 MW and 4.8 MW will be 925, and this number will reach 1116 with the 191 new wind turbines by the end of 2013. When the fact that the necessary area for a wind turbine is 100 m^2 is considered, the total area of the installed wind turbines in Turkey will be 111.600 m^2 by 2013.

There is a significant number of wind power plants in Turkey as seen in Table 5, which indicates the progress in the usage of wind energy as a resource. New wind power plant constructions are also in progress, and the increasing capacity can be seen in Figure 2. The highest investment was realized in 2010.

Although capacity increase in 2010 was higher than it was in 2009, there was a descending trend in the years 2011 and 2012. In 2013, an increase in capacity (23% as compared to the previous year) was observed after 3 years. However, it can be still said that the total wind energy potential of Turkey, a country with 77 million citizens and a land area of 783,562 m^2, cannot be used sufficiently and effectively. The issues in renewable energy (consequently, wind energy) investments and the actions taken in order to increase renewable energy use in Turkey will be discussed in the following part of the study.

Sinan Bubik, the Renewable Energy Business Unit Manager at Siemens, says, "It was difficult to invest in Turkey because of various bureaucratic procedures for both domestic and foreign

Table 5. Wind energy power plants in service up to July 2012.[32]

	Name of the Company	Name of the Project	Installed Capacity	Location	Turbine Power (Mw)	Installation Date
1	ABK	Söke-Çatalbük	30.00	Aydın	2.00	2012
2	BAKTEPE	Amasya	40.00	Amasya	2.50	2012
3	ENERJİSA	Dağpazarı	39.00	Mersin	3.00	2012
4	EOLOS	Şenköy	27.00	Hatay	3.00	2012
5	SARAY DÖKMAD	Saray	4.00	İstanbul	2.00	2012
6	SOMA	Soma	2.00	Manisa	0.90	2012
7	GARET	Karadağ	10.00	İzmir	2.50	2012
8	CAN	Metristepe	40.00	Bilecik	2.50	2012
9	AKSU	Aksu	72.00	Kayseri	2.00	2012
10	PEM	Kıllık	40.00	Tokat	2.50	2012
11	AKHİSAR	Akres	45.00	Manisa	2.50	2011
12	ALENTEK	Susurluk	45.00	Balıkesir	2.50	2011
13	AYRES	Ayres	5.00	Çanakkale	1.80	2011
14	DORUK	Seyitali	30.00	İzmir	2.00	2011
15	ENERJİSA	Çanakkale	29.90	Çanakkale	2.30	2011
16	GALATA	ŞahRES	93.00	Balıkesir	3.00	2011
17	KARDEMİR MDNC.	Bozkaya	12.50	İzmir	2.50	2011
18	SOMA	Soma	140.00	Manisa	2.00	2011
19	GARET	ŞaRES	22.50	Çanakkale	2.50	2011
20	ZİYARET RES	Ziyaret	57.50	Hatay	2.50	2011
21	AKDENİZ	Mut	33.00	Mersin	3.00	2010
22	ALİZE	Çataltepe	16.00	Balıkesir	2.00	2010
23	BAKRAS	Şenbük	15.00	Hatay	3.00	2010
24	ALİZE	Kuyucak	25.60	Manisa	2.90	2010
25	AS MAKİNESAN	Bandırma	25.00	Balıkesir	2.50	2010
26	BERGAMA	Aliağa	90.00	İzmir	2.50	2010
27	BİLGİN	Soma	90.00	Manisa	2.50	2010
28	BOREAS	Boreas 1 Enes	15.00	Edirne	2.50	2010
29	KORES KOCADAĞ	Kores Kocadağ-2	17.50	İzmir	2.50	2010
30	SABAŞ	Turguttepe	24.00	Aydın	2.00	2010
31	ROTOR	Gökçeada	135.00	Osmaniye	2.50	2010
32	ÜTOPYA	Düzova	30.00	İzmir	2.50	2010
33	BELEN	Belen	36.00	Hatay	3.00	2010
34	BORASCO	Bandırma	60.00	Balıkesir	3.00	2010
35	ALİZE	Keltepe	20.70	Balıkesir	0.90	2009
36	ALİZE	Çamseki	20.80	Çanakkale	2.80	2009
37	AYEN	Akbük	31.50	Aydın	2.10	2009
38	AKEN	Ayyıldız	15.00	Balıkesir	3.00	2009
39	ALİZE	Sarıkaya	28.80	Tekirdağ	4.80	2009
40	BAKİ	Şamlı	113.40	Balıkesir	3.00	2008
41	DARES DATÇA	Dares Datça	29.60	Muğla	1.70	2008
42	DENİZ	Sebenoba	30.00	Hatay	2.00	2008
43	INNORES	Yuntdağ	57.50	İzmir	2.50	2008
44	DOĞAL	Sayarlar	34.20	Manisa	0.90	2008
45	LODOS	Kemerburgaz	24.00	İstanbul	2.00	2008
46	SANKO	Çatalca	60.00	İstanbul	3.00	2008
47	ANEMON	İntepe	30.40	Çanakkale	0.80	2007
48	DENİZ	Karakurt	10.80	Manisa	1.80	2007
49	DOĞAL	Burgaz	14.90	Çanakkale	1.70	2007
50	MARE MANASTIR	Mare Manastır	39.20	İzmir	1.70	2007
51	SUNJUT	Sunjüt	1.20	İstanbul	0.60	2006
52	TEPERES	TeRES	0.85	İstanbul	0.85	2006
53	YAPISAN	Bandırma	35.00	Balıkesir	1.50	2006
54	BORES BOZCAADA	Bozcaada	10.20	Çanakkale	0.60	2000
55	ARES	Ares	7.20	İzmir	0.60	1998
56	ALİZE	Çeşme	1.50	İzmir	0.50	1998

Table 6. Installed capacity for licensed wind power plants in Turkey.[32]

City	Capacity (MW)	City	Capacity (MW)
Balıkesir	423.10	Mersin	72.0
Manisa	345.70	Kayseri	72.0
İzmir	325.40	Tokat	40.0
Hatay	165.50	Amasya	40.0
Osmaniye	135.00	Bilecik	40.0
Çanakkale	133.70	Muğla	29.6
İstanbul	90.05	Tekirdağ	28.8
Aydın	85.50	Edirne	15.0

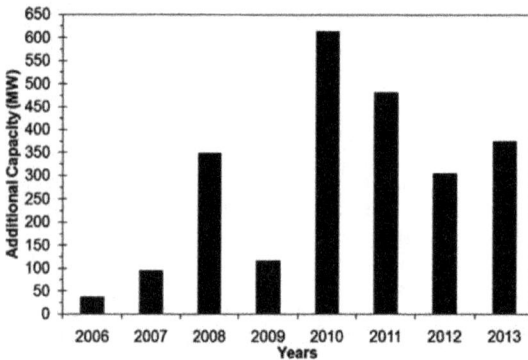

Fig. 2. Wind power plant construction capacity (MW) between 2006 and 2013 in Turkey.[32]

investors. For today, we can say that there is a better investment environment for all sectors, including energy sector."[33] This statement shows that if the incentives are encouraged and a suitable environment is provided, investment in renewable energy resources will increase.

On the other hand, Turkey has the potential to produce 5,000 MW of unlicensed electric energy from wind power plants with a maximum capacity of 500 kW. This rate forms 10% of the available electric power of Turkey, and it shows that there is €7–8 million of investment potential in terms of finance. 238 applications were made to the Unlicensed Electricity Production Association in order to evaluate this potential and produce electricity at home. 130 of these applications were confirmed, and 108 projects are being evaluated. However, the number of applications is still not enough, and they should be increased.

Additional amendments were made to Renewable Energy Law number 5346 (enacted in 2005) in 2007 because energy production through licensed and unlicensed power plants was not enough for the increasing energy demand of Turkey. These amendments aimed to support domestic production, and additional purchase was guaranteed. The incentives for the private sector provided by the electricity market license regulation are[1,34]

- Exercise price for each year is the average wholesale price of the previous year in Turkey determined by the Electricity

Market Regulatory Authority. It cannot be less than in Turkish Liras equivalent to 5 c€/KWh, and it cannot be more than in Turkish Liras equivalent to 5.5 c€/KWh.

- Fixed-price rate is valid for power plants that have been operating for 10 years before December 31, 2011.
- Turkish Electricity Distribution Corporation or licensed distributors will give priority to the system connection of the power plants that use renewable energy sources.
- Public land, excluding the forbidden areas, will be allocated to the wind power investors.
- There will be an 85% discount in permission cost, rent money, easement, and occupancy permission prize for the first 10 years of investment.

Another law amendment (Domestically Produced Equipment Use in Electricity Production Plants from Renewable Energy Sources Law), with some additional incentives, was enacted in 2011 in order to reduce the increasing cost of imported wind turbines. Domestic production was encouraged with this law amendment, which requires at least 55% domestic production of the wind turbine equipment. With these developments, interest in equipment production and wind power plant installation in Turkey have begun to increase day by day.

A new law amendment is being prepared by the Grand National Assembly of Turkey in 2012 in order to increase the licensed energy production limit from 500 kW to 1 MW. This law amendment will certainly affect the increasing number of projects positively.

Despite the amendments on the renewable energy law, it was seen that the obtained power was 2,043.25 MW, and it was below expectations. The reason for this situation is the severe bureaucratic procedures of the government according to the Turkish Electricity Producers Association. Another important reason is the intermittent nature of wind power according to Gökçınar and Uyumaz.[35] If a solution can be found for this problem, the incentives will decrease and the efficiency of wind power investment will increase. The precautions that can be taken in order to overcome these problems are listed below:

- Applying some incentives (such as fixed-price, area allocation, and financial support) to smaller power plants.
- It is necessary to put some applications such as carbon return and incentive payments into effect.
- Use of hybrid systems in wind energy will solve or minimize the intermittent power production problem. Thus, promoting such systems is an adequate way to increase the efficiency of wind energy using power plants.

In Table 7, it can be seen that İzmir has the highest number of licenses to produce electricity by using wind energy when the available and future licenses are analyzed. Malatya and Kastamonu have the minimum number of licenses.

When the cities with a license in Turkey are analyzed, cities that have the highest number of licenses to install a wind power plant are in the Marmara region, but the ones having the lowest number of licenses are in the Eastern Anatolia region. The capacity distribution to the regions of Turkey is tabulated in Table 8.

Table 7. Licensed wind power plant capacity in cities of Turkey.[32]

City	Capacity (MW)	City	Capacity (MW)
İzmir	917.45	Sakarya	70.00
İstanbul	382.50	Uşak	69.00
Balıkesir	372.90	Yalova	66.00
Çanakkale	288.00	Gaziantep	63.00
Kocaeli	281.00	Bolu	60.00
Aydın	275.00	Isparta	60.00
Kırklareli	262.00	Karaman	57.00
Kayseri	237.00	Bursa	50.00
Konya	236.00	Çorum	45.00
Muğla	210.80	Yozgat	45.00
Sivas	187.00	Tokat	42.80
Hatay	162.50	Adıyaman	42.75
Manisa	162.00	Amasya	42.00
Osmaniye	160.00	Aksaray	40.00
Kırşehir	150.00	Eskişehir	39.00
Afyon	148.00	Antalya	37.00
Tekirdağ	134.00	Mersin	35.00
Bilecik	120.00	Sinop	35.00
Zonguldak	120.00	Bartın	27.00
K.maraş	117.50	Düzce	15.00
Edirne	111.00	Erzurum	15.00
Samsun	107.00	Kastamonu	10.00
Denizli	76.00	Malatya	10.00
Ankara	72.00		

Table 8. Licensed wind power plant capacity distribution for the regions of Turkey.[32]

Region (MW)	Capacity
Marmara	2137.40
Aegean	1856.26
Central Anatolia	1063.00
Mediterranean	573.00
Black Sea	503.80
Southeast Anatolia	105.75
East Anatolia	25.00

When the annual average wind speed on 50 meters height in Turkey, given in Figure 3, is analyzed, it is seen that the Eastern Anatolia region, which has the lowest number of licenses, has considerably high wind speed.

Because of terrorist attacks that began in 1984 and became more violent every day in the eastern part of the country, there was not a suitable investment environment in this region. Safety problems of the region and the great amount of money needed for wind turbine investment prevent the private sector from constructing wind power plants in the east. In our neighboring country, Syria, with which Turkey has the longest border, sectarian violence has turned into a civil war. Another neighboring country, Iran, also has some political problems with Turkey. These conditions prevent Turkey from making investments such as wind power plants in many areas of this region. However, if

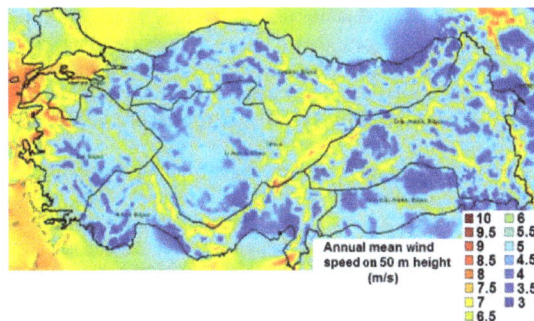

Fig. 3. Annual average wind speed on 50 meters in Turkey.[36]

Turkey can solve these safety problems and can have good relations with its neighbor countries, the unused wind potential in its eastern region can be added to its renewable energy production.

Another important development in the wind energy field in Turkey is the preparation of the Wind Energy Atlas of Turkey. The Electricity Power Resources Survey and Development Administration prepared this Wind Energy Atlas, in cooperation with the Turkish State Meteorological Service and in coordination with the Ministry of Energy and Natural Sources, in order to facilitate wind power resources evaluation. The atlas is very good because it is prepared using measured wind data, ground levels, environmental obstacles, and topography. The investors may get this atlas from the Electricity Power Resources Survey and Development Administration whenever they want.

4. Conclusions

In this study, the present energy production and policy of Turkey is taken under consideration, and the share of wind energy among different energy resources was investigated. The objective of the study is to present the current situation and the future of wind energy in Turkey. Current wind power plants and current wind energy production capacity in Turkey were analyzed. Additionally, new wind power plants under construction and future wind power capacity of Turkey were stated. The obstacles to new wind energy investments in Turkey were described, and some solutions in order to increase these new investments were expressed. First of these solutions are the incentives that the Turkish government gives to renewable energy plants. Second, hybrid system construction is suggested in order to overcome the intermittent nature of the wind. Additionally, the precautions that are taken and that should be taken in order to increase wind energy use in Turkey were presented.

In order to meet the need for energy resources, which increases the foreign trade deficit, the Turkish government brought forward new incentives for both domestic and foreign sectors. The General Directorate of Renewable Energy has been formed within the Ministry of Energy and Natural Sources in order to produce energy from green energy resources. This unit aims to provide both financial and technical aid to install power plants such as wind power plants, solar energy plants, and hydroelectric power plants.

The number of wind power plants in Turkey will be 67 in the very near future. 13 new power plants have started their electricity production recently, and 11 are being constructed. These plants under construction are planned to be in service in 2013. With these power plants, Turkey will get a power of 2,495.25 MW from wind energy at the end of 2013.

As a result, Turkey has encouraged domestic and foreign incentives to use renewable energy resources in accordance with its 2023 strategies. Additional law amendments are also made in order to prevent the country's foreign resource dependency and to protect the environment and human health. If the bureaucratic obstacles can be eliminated and effective incentives are given to new renewable power plants, including wind power plants, there will be a dramatic increase in renewable energy use in Turkey. Another important problem for Turkey is the terrorist attacks in its eastern regions. The high wind potential in these regions cannot be utilized efficiently because of these attacks. If the problems in the region can be solved and if security can be assured, there will be a great wind energy investment in these regions, which will increase the wind power energy production capacity of Turkey. Finally, it can be said that the use of renewable energy resources will increase in Turkey in the near future, and these valuable resources of Turkey will be exploited by numerous investors of the energy sector.

References

1. Şahin, A.D. Progress and Recent Trends in Wind Energy. *Prog. Energ. Combust. Sci.* 2004, 30, 501–543.
2. Ackerman, T.; Soder, L. Wind Energy Technology and Current Status: A Review. *Renew. Sustain. Energ. Rev.* 2000, 4, 315–374.
3. U.S. Energy Information Administration. http://www.eia.gov/cfapps/ipdbproject/iedindex3.cfm?tid=2&pid=28&aid=12&cid=ww,&syid=1980&eyid=2012&unit=BKWH (accessed Oct 16, 2012).
4. Ahmed, A.S. Wind Energy as a Potential Energy Production Source at Ras Benas, Egypt. *Renew. Sustain. Energ. Rev.* 2010, 14, 2167–2173.
5. Najafi, G.; Ghobadian, B. LLK 1694—Wind Energy Resources and Development in Iran. *Renew. Sustain. Energ. Rev.* 2011, 15, 2719–2728.
6. International Energy Agency [IEA]. www.iea.org. (accessed Feb 1, 2012).
7. Güler, Ö. Wind Energy Status in Electrical Energy Production of Turkey. *Renew. Sustain. Energ. Rev.* 2009, 13, 473–478.
8. Karsli, V. M.; Geçit, C. An Investigation on Wind Power Potential of Nurdağı-Gaziantep, Turkey. *Renew. Energ.* 2003, 28, 823–830.
9. Ozerdem, B.; Turkeli, M. An Investigation of Wind Characteristics on the Campus of Izmir Institute of Technology, Turkey. *Renew. Energ.* 2003, 28, 1013–1027.
10. Hepbasli, A.; Ozgener, O. A Review on the Development of Wind Energy in Turkey. *Renew. Sustain. Energ. Rev.* 2004, 8, 257–276.
11. Köse, R. An Evaluation of Wind Energy Potential as a Power Production Source in Kütahya, Turkey. *Energ. Conv. Manag.* 2004, 45, 1631–1641.
12. Şahin, B.; Bilgili, M.; Akilli, H. The Wind Energy Potential of the Eastern Mediterranean Region of Turkey. *J. Wind Eng. Ind. Aero.* 2005, 93, 171–183.
13. Akpinar, S.; Akpinar, E.K. Wind Energy Analysis Based on Maximum Entropy Principle (MEP)-Type Distribution Function. *Energ. Conv. Manag.* 2007, 48, 1140–1149.

14. Gökçek, M.; Genç, M.S. Evaluation of Electricity Generation and Energy Cost of Wind Energy Conversion Systems (WECSs) in Central Turkey. *Appl. Energ.* 2009, 86, 2731–2739.
15. İlkılıç, C.; Türkbay, İ. Determination and Utilization of Wind Energy Potential for Turkey. *Renew. Sustain. Energ. Rev.* 2010, 14, 2202–2207.
16. Celik, A. N. Review of Turkey's Current Energy Status: A Case Study for Wind Energy Potential of Çanakkale province. *Renew. Sustain. Energ. Rev.* 2011, 15, 2743–2749.
17. Yaniktepe, B.; Koroglu, T.; Savrun, M. M. Investigation of Wind Characteristics and Wind Energy Potential in Osmaniye, Turkey. *Renew. Sustain. Energ. Rev.* 2013, 21, 703–711.
18. Akpinar A. Evaluation of Wind Energy Potential At Coastal Locations Along the North Eastern Coasts of Turkey. *Energy* 2013, 50, 395–405.
19. Karaduman, Ö. Mevcut Durum Özel Sektörün Yatırım Yapma İştahını Azaltıyor. *Enerji ve Çevre Dergisi* 2012, 86, 2012 (in Turkish).
20. Intelli News. *Turkey Energy Sector Report*. June 2010: p 4.
21. Özturk, R.; Kincay, O. Potential of Hydraulic Energy. *Energ. Sources* 2004, 26, 1141–1156.
22. Yuksel, I. Hydropower in Turkey for a Clean and Sustainable Energy Future. *Renew. Sustain. Energ. Rev.* 2008, 12, 1622–1640.
23. Invest in Turkey. http://www.invest.gov.tr/tr-tr/sectors/Pages/Energy.aspx (accessed Feb 15, 2012).
24. Türkiye Elektrik İletim A.Ş. Electricity Generation & Transmission Statistics of Turkey 2010. http://www.teias.gov.tr/T%C3%BCrkiye Elektrik%C4%B0statistikleri/istatistik2010/%C4%B0statistik%202010.htm (accessed Oct 16, 2014).
25. Türkiye İstatistik Kurumu. www.turkstat.gov.tr/PreIstatistikTablo.do?istab_id=1578. (accessed Oct 16, 2014).
26. Keskin, M.; Mert, A. Türkiye'de Enerji ve Çevre Konusunda Yapılan Hataların Laboratuvarı: Yatağan – Yeniköy - GökovaTermik Santralları (A Laboratory of Mistakes on Energy and Environment: Yatağan – Yeniköy - Gökova Power Plants). *Mühendis & Makina* 2002, 509, 24–36. (in Turkish).
27. Türk Tabipleri Birliği. http://www.ttb.org.tr/odd/index.php?option=com_content&task=view&id=25&Itemid=46 (accessed Feb 15, 2012).
28. Türkiye İstatistik Kurumu. http://www.turkstat.gov.tr/PreHaber Bultenleri.do?id=13482 (accessed Oct 16, 2012).
29. Kaygusuz, K. Renewable and Sustainable Energy Use in Turkey: A Review. *Renew. Sustain. Energ. Rev.* 2002, 6, 339–366.
30. Aras, H. Wind Energy Status and Its Assessment in Turkey. *Renew. Energ.* 2003, 28, 2213–2220.
31. Environmental Protection Agency (EPA), Clean Energy http://www.epa.gov/cleanenergy/energy-resources/refs.html (accessed Oct 14, 2014).
32. Turkish Wind Energy Association, Turkish Wind Energy Statistics Report. http://tureb.com.tr/attachments/article/168/Turkiye%20Ruzgar%20Enerjisi%20Istatistik%20Raporu%202012-Turkish%20W%C4%B1nd%20Energy%20Statistics%20Report%202012.pdf (accessed Feb 20, 2012).
33. Bubik, S. Teşvikler ve Yönlendirmeler. *Enerji ve Çevre Dergisi*. 2012 17 May. (In Turkish).
34. Altuntaşoğlu, T.Z. *Yerli Rüzgar Enerji Teknoloji Üretimi Destek Politikaları ve Türk Mevzuatı*. Proceedings of TMMOB Türkiye VI. Enerji Sempozyumu, Küresel Enerji Politikaları ve Türkiye Gerçeği (22–24 Oct 2007), 377–389; Ankara, Turkey. (in Turkish).
35. Gökçınar, R.E.; Uyumaz, A. *Rüzgar Enerjisi Maliyetleri ve Teşvikleri*. Proceedings of VII Ulusal Enerji Sempozyumu (UTES 2008) (17–19 Dec 2008), 699–706; İstanbul, Turkey (in Turkish).
36. Meteoroloji Genel Müdürlüğü. http://www.mgm.gov.tr/FILES/haberler/2010/rets-seminer/2_Mustafa_CALISKAN_RITM.pdf (accessed Oct 17, 2014).

10

Review of Wind Energy Development and Policy in India

RAVINDRA B. SHOLAPURKAR and YOGESH S. MAHAJAN*

Department of Chemical Engineering, Babasaheb Ambedkar Technological University, Lonere, Tal Mangaon, District Raigad, Maharashtra, India

Abstract: Due to growing population, economic growth, and socioeconomic development, energy is the most essential need. Worldwide, about 86.4% of energy is produced by fossil fuels. Globally, India ranks fourth among the countries that produce wind energy. The last five years' growth in wind energy in India is about 16%. This growing Indian wind energy market is compared with the world scenario along with the state of Maharashtra in India, which is at the second position in the country, backed up by a detailed database. The review includes the details of Maharashtra's wind energy progress in the last few years with predictions of the future for some years, considering factors such as wind project installations, total capacity, declared wind sites, wind power density at different altitudes, and so forth. This review also covers the potential wind sites with monitoring stations and velocity. The significance of this work is that it explores the wind potential and facilitates the reader to judge the upcoming market in this region and also for possible investment in wind power generation. It also briefly considers the scope and policies in wind energy development and investment.

Keywords: Wind energy, wind potential areas, wind power projects and operators, Maharashtra, India

1. Introduction

Day by day, the world population is rising at an exponential rate due to which the demand for energy is increasing. Generation of energy is considered an indispensable part of life and of national advance. There exists a direct relationship between development and energy consumption. Hence we need to produce more energy to be progressive.[1] To produce more energy, we depend on fossil fuels. The use of fossil fuels increases the emission of pollutants such as SO_x, NO_x, and carbon monoxide that have a detrimental effect on the environment.[2] Hence, the use of alternate energy sources such as wind, solar, and hydrogen is gaining importance.[3] Wind has proven to be a very effective source of energy due to technological richness, infrastructure, and relative cost attractiveness.[4,5] Wind can offer several advantages such as being inexhaustible, pollution free, and requiring minimal or no fuel.[6] Renewable energy and especially wind energy does not emit any CO_2 in the atmosphere—thus it protects us from global warming as well. This is the reason why many countries

are using wind energy as a source of energy.[7] In developing countries, one-third of the world's population still lives without electricity. Wind energy systems have made it possible to harvest cost-effective power generation.[8,9] Wind turbines have traditionally been used for almost 200 years to generate electricity.[10] The use of wind turbines for renewable energy has become one of the most viable substitute sources of power generation due to some compensations such as being lucrative and eco-friendly.[11,12] Many companies, institutions, organizations, and researchers have reported that wind turbines with higher productivities are needed to realize the energy mandate.[13,14] Countries that do not possess natural reservoirs of fuel stock need to be very careful and attentive to the use of alternate sources of energy, including wind.[15]

Wind energy is the most gifted and potential source of energy, even though its obtainability varies from place to place. However, the main drawback is to obtain a continual power supply. Wind turbine technology has seen a dramatic change in the last three decades. Continuous development in the technology of wind turbines such as power electronics, aerodynamics, and mechanical drive train design has made it an efficient source of energy. The disadvantage is of noise pollution, which is much less than other power plants.[16]

Recently, in the United States, many leading research organizations, namely the U.S. Department of Energy, National Renewable Energy Laboratory (NREL), Electric Power Research Institute (EPR), and the University of Colorado (CU) have explored how the contribution of wind power will make active power controls (APC) and how it could benefit the total power system economics, increase revenue, and improve the reliability and secure the national power systems. The APC will help

*Address correspondence to: Dr. Yogesh S. Mahajan, Department of Chemical Engineering, Babasaheb Ambedkar Technological University, Lonere, Tal Mangaon, District Raigad, Maharashtra 402103, India. Email: yogesh_mahajan66@yahoo.com or ysmahajan@dbatu.ac.in

to stabilize the electric grid as a generating source and control the repetitively fluctuating needs. NREL gives a complete view to approach the needs. Primary frequency control increases synchronous (spinning) machines that instantly provide energy to the grid by converting rotational wind energy. Researchers of NREL have shown that wind plants could earn additional revenue by providing APC. NREL has evaluated a timeframe that ranges from milliseconds to minutes to the lifetime of wind turbines and scales ranging from small-scale wind turbines to large-scale wind plants.[17]

2. World Wind Energy Development

The onshore wind energy potential is very large as paralleled to the total world power consumption. It is nearly $20,000 \times 10^9$ to $50,000 \times 10^9$ KWh per year against the total consumption of power of about $15,000 \times 10^9$ KWh. The impending of wind energy depends on aspects such as average wind speed, wind speed distribution, turbulence intensities, and cost of wind turbine systems. The global wind energy council is operating in the wind energy sectors of several countries. These participants operate in more than 100 countries with over 2,500 organizations involved in hardware manufacture, project development, power generation, and finance consultancy on a build, operate, and transfer (BOT) basis. These participants are also involved in research and academics.[16]

Most of the developed countries are facing challenges to meet the power demand due to the increase in population and industrialization. The U.S. Department of Energy (DOE) has given the directions toward the development of energy by wind: Wind energy technology must be cost-effective with increasing viability. The supporting research is also conducted in the area of wind development technology such as power system integration, resource information, industry support, and market acceptance.[18] The power produced by wind technology of the future needs to be cost-effective and capable of competing with other energy such as from coal and natural gas.[19] Present R&D is successful toward wind power development; many countries have adopted it as a main source of energy, and a number of growth studies have been presented.[20]

According to the report by the Intergovernmental Panel on Climatic Change (IPCC), by 2050 wind energy will play a major role in electricity generation among renewable energy. Its contribution will be about 80% of the world's energy demand. Wind energy is known for its simplicity and limited space needs. Most of the houses and housing complexes can have independent energy production facility. Offshore wind power projects have become a trend in European countries.[21] For a clean climate, renewable energy is a must, and it is going to last for a long time with less maintenance.[22] In the development of renewable energy, there are many barriers including investment. All these barriers are not solved by only technological, social, political, or economic factors. There must also be a multidimensional approach toward identifying and explaining the fundamentals of these barriers to develop a viable solution.[23] At the end of 2009, the total installed capacity of wind turbines in the world was around 160 GW. By recent guidelines, the European Union (EU) wind electricity production will be 20% of energy demand by

2020.[24] In the year 2014, the potential sites for solar and wind power generation for three economic and industrial zones were predicted in Iran. This was predicted on the basis of dividing the year in two seasons: warm season (April–September) and cold season (October–March).[25] Such study needs to be done for the entire world as well.

In the United States, between 2008 and 2010 a substantial number of wind turbines were installed. Due to this excessive installation of wind turbines, radar communication got affected, which caused the abandonment of an estimated 20,000 MW of wind energy capacity. This resulted in the reduction in the growth of wind energy and also affected the target of the U.S. Department of Energy to produce 20% of energy by wind turbines by the end of 2030.[26] Such effects, if any, should also be thoroughly investigated.

The global wind power installation capacity has been increased in 2012 to 44,799 MW, and it became a total of 282,587 MW with an 18.7% increase on the 238,050 MW installed in 2011. More than half of the wind power generation is added in the year 2010–2011; the main contribution is from China, North America, and Europe. China alone had an installation of 18,000 MW in 2011 and 75,324 MW of wind power installed in 2012.[27] Other countries include: 21% Denmark, 18% Portugal, 16% Spain, 14% Ireland, and 9% Germany.[28] Presently, the global wind power installation capacity at the end of 2014 is 364,540 MW at a percentage growth of 26.2.[29]

Several wind power installations have come up in the past few years. For example, in the United States, there is Roscoe Wind Farm in 2011 with a capacity of 781 MW. Other major installations are Thanet Wind Farm in the United Kingdom of 300 MW, followed by Homs Rev II 209 MW in Denmark. Some of the other largest wind farms under construction in different parts of the world are Bard Offshore 400 MW, Clycde Wind Farm 548 MW, Greater Gabbard wind farm 500 MW, Lincs Wind Farm 270 MW, London Array 1,000 MW, and another in the UK is 2014 with 12,440 MW.[27] At the end of 2014, UK wind energy is 65,879 MW.

China has witnessed a tremendous increase in renewable energy sources, especially after 2006. In fact, in the last 4–5 years, China had the maximum growth rate in terms of wind energy. In 2012 the total wind energy produced in China was 75,564 MW, which increased to 114,763 MW in 2014. It is forecasted that by 2020 China will become the world leader in wind energy production. China has set for itself a target of offshore wind development of 5 GW by 2015 and 30 GW by 2020. A report from the Chinese government has claimed that by 2020 China would be able to generate 440 TWh of electricity annually.[30,31]

The next largest wind energy–producing country is Germany. The 2006 production of wind energy was 20,622 MW, which in 2014 became 39,165 MW; thus Germany became the third largest wind energy–producing country. The target set by Germany itself is to produce a total of 35% of power consumption by wind alone, which they wish to slowly achieve by phasing out energy production by nuclear power.[32]

Spain is the country with the fourth largest wind energy production in the world. In 2006 the capacity was 11,630 MW, which almost doubled to 22,987 MW in 2014. Although the financial crises among other problems are there, Spain has put forth 20%

of total energy from renewable sources, out of which 35.75% GW will be of wind power.[33]

The detailed wind power installation capacity in the world is shown in Table 1.

3. Wind Energy Development in India

India is now fourth among the several countries that produce more electricity from wind power. In India, the Ministry of New and Renewable Energy (MNRE) and the Indian Renewable Energy Development Agency (IREDA) work in coordination with the state government's wind energy department. Each state has its own wind energy department as in Maharashtra (MEDA). IREDA deals with identification of wind potential sites, wind resource assessment, setting up government policy, financing/profitability, availability of equipment, service, perception of investors, constraints/barriers and suggestions, and so on. IREDA has set up Anemometry Masts (AM) all over India to measure wind power density (WPD). AMs have recorded the qualifying criteria of WPD in the country, being above 200 W/m²

at a height of 50 meters and at an 80-meter height above ground level, and the number of stations along with the potential power capacity and achievement up to 2014 is shown in Table 2.

In India, in 1985, demonstration type wind energy projects were started. In this, about 69.6 MW power projects were started at various locations in India. In 1986 five wind farms were started with a capacity of 3.3 MW power generation. The first commercial wind power generation started in 1990 at Kattadilalai, Muppandal, in Tamil Nadu. Until 1992, many wind turbines were installed in coastal areas of Tamil Nadu, Gujarat, Maharashtra, and Orissa. After 1996, India has seen steep growth in wind power production. In 2004, Asia's biggest and tallest wind turbine was built in India. IREDA under the guidance of MNRE has drawn long-term policy for the international market to invest in India for the development of wind power energy. It has also designed policy for local investors in the renewable energy development sector. In the last two decades, the growth of wind energy in India is from 41.3 MW in 1992 to 22,465 MW at the end of 2014.[29] Among the Indian states, Tamil Nadu has the highest growth in wind power energy development and is producing 5867.165 MW, which is 41.61% of total wind energy

Table 1. Installed wind power capacity of different countries in the world (MW).[29,34]

Sr. No	Nation	2006	2007	2008	2009	2010	2011	2012	2013	2014	Last yr. % growth	6 yr. % growth
1	European Union	48.122	56,614	65,255	74,919	84,278	93,957	106,454	117,289	127,429	8.6	11.8
2	China	2,599	5,912	12,210	25,104	44,733	62,733	75,564	91,424	114,763	25.5	40
3	United States	11,633	16,819	25,170	35,159	40,200	46,919	60,007	61,091	65,879	7.8	18
4	Germany	20,622	22,247	23,903	25,777	27,214	29,060	31,332	34,250	39,165	14.3	8.6
5	Spain	11,630	15,145	16,740	19,149	20,676	21,674	22,796	22,959	22,987	0.1	5.5
6	India	6,270	7,850	9,587	10,925	13,064	16,084	18,421	20,150	22,465	11.5	15.3
7	United Kingdom	1,963	2,389	3,288	4,070	5,203	6,540	8,445	10,531	12,440	18.1	24.8
8	Italy	2,123	2,726	3,537	4,850	5,797	6,747	8,144	8,552	8,663	1.3	16.7
9	France	1,589	2,477	3,426	4,410	5,660	6,800	7,196	8,254	9,285	12.5	18.3
10	Canada	1,460	1,846	2,369	3,319	4,008	5,265	6,200	7,803	9,694	24.23	26.7
11	Denmark	3,140	3,129	3,164	3,465	3,752	3,871	4,162	4,772	4,845	1.5	7.4
12	Portugal	1,716	2,130	2,862	3,535	3,702	4,083	4,525	4,724	4,914	4.0	9.6
13	Sweden	571	831	1,067	1,560	2,163	2,970	3,745	4,470	5,425	21.3	31.5
14	Brazil	237	247	339	606	932	1,509	2,508	3,456	5,939	71.8	52.7
15	Poland	153	276	472	725	1,107	1,616	2,497	3,390	3,834	13.1	42.6
16	Australia	651	824	1,306	1,712	1,991	2,176	2,584	3,239	3,806	17.5	19.7
17	Turkey	65	207	433	801	1,329	1,799	2,312	2,959	3,763	27.2	45
18	Netherlands	1,571	1,759	2,237	2,223	2,237	2,328	2,391	2,693	2,805	4.1	3.9
19	Japan	651	1,528	1,880	2,056	2,304	2,501	2,614	2,661	2,789	4.8	6.9
20	Romania	2	7	10	14.1	462	982	1,905	2,600	2,954	13.6	579
21	Ireland	746	805	1,245	1,260	1,379	1,614	1,738	2,037	2,272	11.5	10.6
22	Mexico	84	85	85	520	733	873	1,370	1,992	2,381	19.5	115.5
23	Greece	758	873	990	1,087	1,208	1,629	1,749	1,865	1,980	6.1	12.7
24	Austria	965	982	995	995	1,011	1,084	1,378	1,684	2,095	24.4	13.7
25	Belgium	194	287	384	563	911	1,078	0	0	0		35.5
26	Norway	325	333	428	431	441	512	704	811		–	0
27	New Zealand	171	322	325	497	530	623	623	623	623	0	12.8
28	Taiwan	188	280	358	436	519	564	564	614	633	3.0	10.2
29	Bulgaria	36	70	120	177	500	612	0	0	0		70
30	South Korea	176	192	278	348	379	407	483	561	609	8.5	12.5
31	Egypt	230	310	390	430	550	550	550	550	610	0	6.3
	Other countries	1,756	1,157	1,603	2,125	3,631	4,146	5,582	8,383	379	0	4.6
	Total	74,151	93,927	121,188	157,899	197,637	238,035	282,482	318,137	364,540	14.5	26.2

Table 2. Detailed estimated gross potential wind power density.[35]

| Sr. No. | State | Wind power density (WPD range W/m² and no of stations) | | | | Potential (MW) | | Achievement up to 2011 (MW) | Achievement up to 2014 (MW) |
		200–250	251–300	301–350	351 and above	@ 50 M	@ 80 M		
1	Andhra Pradesh	15	11	5	4				
2	Gujarat	27	8	2	3	10,609	35,071	2,176.76	3,405.61
3	Karnatak	7	8	3	8	8,591	13,593	1,763.92	2,331.30
4	Madhya Pradesh	3	3	–	–	920	2,931	214.040	355.85
5	Maharashtra	18	11	4	–	5,439	5,961	2,304.35	4,024.65
6	Rajasthan	5	3	–	–	5,005	5,050	1,528.39	2,809.42
7	Tamil Nadu	8	6	9	24	5,374	14,152	5,867.16	7,254.01
8	Others					7,798	11,533	35.97	33.85
Total						49,130	102,788	14,097.48	20,953.53

development in India by 2011 and 7,254 MW at the end of 2014.[35] Among other states, Maharashtra 4024.65 MW, Gujarat 3405.605 MW, Karnataka 2331.295 MW also achieved significant growth of 16.30%, 15.44%, and 12.51%, respectively, at the end of 2014. The center for wind energy technology (C-WET) has published a wind power density atlas of India; it shows the various areas of wind potential of more than 200 Watt/m² at 50 meter height.[35] The Indian wind atlas shows that India has the potential of producing wind energy of 49,130 MW at 50 m height and 102,788 MW at 80 m height above the ground level.[35]

India currently has an installed wind power generation capacity of 22,465 MW. Also, barely 12% of the total power generation is from renewable sources. The remaining 88% of power depends on other energy resources. Offshore wind power policies should be developed; it helps to increase the wind power energy. European countries, most notably the UK and Germany, have adopted effective offshore policies. Resolving the Indian power sector the renewable energy policies have to be improved. These policies are classified into five categories, namely:

1. Government support
2. Fiscal and quota-based incentives
3. Local expertise
4. Capital for investments
5. Building and enabling ecosystem.

By these policies, the economy of India and power production can be improved.[36] Most of the renewable energy systems have some involvement with climate change because fuel is combusted at the time of production and also during operation. Accurate calculation of greenhouse gas has been carried out for each project.

The carbon intensity, energy return period, and carbon reimbursement period for the system can be calculated.[37] Energy is the prime agent of wealth and the important factor in the financial development of any country.[38] India has achieved the fifth position in the world by installing 6,018 MW capacity in 2008.[39] Now by the end of 2014, India is in the fourth position. The progress of wind energy depends on government backing and fiscal quota-based incentives, along with supporting R&D.[40,41] Geographical information system (GIS) is used to map the wind energy sources, which gives the idea of average wind velocity to assess the wind potential areas. It helps to divide the windy sites according to zones for optimum extraction, wind power projects, and site selection.[42] Modern wind turbines are more efficient than previous wind turbines. Modern wind turbines produce annually about 180 times more electricity at less than half the cost per unit (KWh) than it used to 20 years ago. Modern wind turbines can produce 5 MW power at a rotor diameter of 126 m and are easy to install from a few MW to several hundred MW as wind farms.[43]

A climatic model developed for the Indian national electricity market helps to study the relative balance in renewable and non-renewable energy. It helps in the proper allocation of resources and their utilization and in the development of a country's stronger economy and security.[44] Indian major growth in renewable energy has taken place in the last few years. Wind energy is an uncontaminated, sustainable, renewable supply besides being nonpolluting.[38] In India, wind energy is determined to be of the highest level of importance compared to other sustainable energy forms. New technologies have developed to get optimum power from the minimum wind with optimum design.[39]

Presently, India is emitting CO_2 of about one ton/year-capita. To reduce this, India is keen on investing in renewable energy technologies. In the year 2011 investment in renewable energy technologies was $12.3 billion compared to the investment made in the previous year (2010: ~$7.5 billion), a hike of nearly 36% in the turnover. This quantity is to be reduced by increasing the renewable energy and decreasing the consumption of fossil fuel. This indicates the large scope there is in India for the development of renewable energy. The present extenuation of CO_2 in the Indian energy sector is about 203 million tons with fixed a capacity of 24 GW in 2012. Nevertheless, a large amount of scope is seen in the Indian market for the growth of wind energy.[45]

4. Wind Energy Policies in India

India needs to sustain an economic growth of at least 9% over the next 25 years if it is to eradicate poverty and meet its larger human development goals. The primary energy supply (including the gathered noncommercial one such as wood and dung) must increase at a rate of 5.8% annually for fueling the growth. Meeting this requirement is a challenge that needs to be addressed through an integrated energy policy. The broad vision behind the integrated energy policy is to reliably meet the demand for energy services of all sectors, including the lifeline energy needs of vulnerable households in all parts of the country with safe, clean, and convenient energy at the least cost.[46]

4.1. Generation-Based Incentive

Generation-based incentive (GBI) of 50 paisa (half an Indian rupee) per unit was launched in December 2009. The purpose of this subsidy/incentive was to shift the mechanism of payment from installation-based to generation-based methods of rewarding wind farms. Even before the GBI was introduced, tax benefits in the form of accelerated depreciation were made available to the wind farm developers. But this mechanism failed to encourage the wind projects to produce more power. GBI is a way to encourage development of more efficient wind farms.

Another reason for GBI coming into picture was that although the development of renewable energy has been significant, the achievement cannot be compared to the potential that exists. There is a potential of more than 45,000 MW wind capacity against which only 11,000 MW has been commissioned. Also, it was felt that the fiscal incentives in place were not sufficient to meet the RPO targets under the National Action Plan on Climate Change (NAPCC). The GBI was introduced to act as a booster to the capacity addition. As against a target of 10,500 MW capacity additions in the 11th plan, less than 4,000 MW had been commissioned by December 2009, and at the end of 2014 the cumulative capacity increased to 22,465 MW.[47,48]

4.2. State Wise Tariff for Wind Power

Talking about wind energy in India, all started well with wind since around 2002 until the end of 2011 as it enjoyed the benefits of accelerated depreciation (AD) till April 2012. The generation-based incentive (GBI), announced in 2011, was later discontinued. Now the government has launched its first wind energy mission to give a boost to the wind energy sector and putting it in the same league as the high-profile solar mission. The National Wind Energy Mission (NWEM) has been launched. This might provide a great boost to the wind energy sector, which is experiencing slowdown since 2011 continually. In 2011 approximately 29,536 MW of renewable power capacity was installed in India, which included about 19,933 MW from wind, 2,079 MW from solar, 3,746 MW from small hydro, and 3,776 MW from bio energy.[47]

India is the fourth largest wind power producer in the world with an installed total capacity of close to 23 GW in 2014.

4.3. Renewable Energy Certificate Scheme (REC)

Renewable energy is promoted by the Ministry of New and Renewable Energy (MNRE), the central authority for all policies, regulations, and approvals relating to renewable energy.

It is supported by the Ministry of Power and the Central and State Electricity Regulatory Commissions (CERC and SERCs). CERC deals with the national grid and interstate transfer/trading of power, while SERCs manage regional distribution and transmissions. These play a key role in the promotion of renewable energy as they have the sole authority to ascertain the feed-in

Table 3. Detailed solar and non-solar price.[46]

	Non-solar REC (/MWh)	Solar REC (/MWh)
Forbearance Price	3,300	13,400
Floor Price	1,500	9,300

tariffs and other policy matters, such as the Renewable Portfolio Standard (RPS). Energy Development Agencies (EDAs) represent the MNRE at the state level. Their main purpose is to assess and promote renewable energy frameworks for individual states and to advise the MNRE, state governments, and SERCs. IREDA promotes financial assistance for renewable energy and energy efficiency projects in India. The price of REC would be determined in power exchange. REC would be traded in power exchange within the forbearance price and floor price determined by CERC from time to time. CERC has determined the floor price and forbearance price on August 23, 2011 applicable from April, 2012 until FY 2016–17. The forbearance and floor price of solar and non-solar energy sources in 2012 are given in Table 3.[46]

4.4. National Clean Energy Fund (NCEF)

NCEF was proposed in the Union Budget 2010–11 for funding research and innovative projects in clean energy technology. In many areas of the country, the pollution level has reached alarming proportions. While it must be ensured that the principle of "polluter pays" remains the basic guiding criterion for pollution management, there should also be a positive thrust for development of clean energy. And to build on the purpose of the NCEF, the government of India proposed to levy a clean energy process on coal produced in India at a nominal rate of Rs.50 per ton, which will also be applicable to imported coal. By the end of March 2012, NCEF was worth rupees 3,864 crore. The latest economic survey reveals that the government expects to collect rupees 10,000 crore under the Clean Energy Fund by the end of 2015. An allocation of rupees 200 crore from the fund was proposed for an environmental remediation program and another rupees 200 crore for the Green India Mission in 2013–14.[48]

4.5. Land Allocation Policy

The government of India amended the Wind Power Policy 2012, with an aim of attracting more investors and giving boost to renewable energy. The government wants to ensure an easy process for allocation of land and other formalities for setting up wind power projects.[47]

5. Wind Energy Development in Maharashtra

Maharashtra is third largest state of India: it covers an area of 307,713 square kilometers. Its western side is aligned by the Sahyadri mountain; both sides of the mountain are gently steep, and the western side is surrounded by the Arabian Sea. The significant physical trait of the state is its plateau character. The state enjoys a hot tropical climate from March to June.

Table 4. Renewable energy potential in Maharashtra (MW).[48]

Source	Wind (MW)	Small Hydro (MW)	Biomass (MW)	Bagasse cogeneration (MW)	Urban Waste (MW)	Industrial Waste (MW)	Total (MW)
Potential in country	450,000	103,240	160,000	50,000	17,000	17,000	797,240
Potential in Maharashtra	45,840	6,000	7,810	12,500	2,870	3,500	78,520
Percentage of total potential	10.20%	5.80%	4.90%	25.00%	16.90%	20.60%	9.80%
Potential vs. achievement	42.50%	35.20%	12.20%	21.00%	0.00%	1.80%	32.10%

Table 5. Renewable energy capacity additions of five years.[48]

Source	2012–13	2013–14	2014–15	2015–16	2016–17
Wind (MW)	300	300	300	300	300
Biomass (MW)	25	25	25	25	25
Bagasse Cogeneration (MW)	50	50	50	50	50
Municipal Solid Waste (MW)	0.0	0.0	0.0	0.0	0.0
Industrial Waste (MW)	100	100	100	100	100
Small Hydro (MW)	10.0	10.0	10.0	10.0	10.0
Solar Thermal, Solar PV, and others (MW	50.0	50.0	50.0	50.0	50.0
Total	535.0	535.0	535.0	535.0	535.0

Table 6. Demonstration wind power projects (MEDA).[48]

SL No	Name of site	Taluka/District	Capacity of Projects in MW	Machine Capacity KW	Number of Machines	Year of Commissioning	Land Acquired by MEDA (Hectare)	Balance Land (Hectare)
1	Girye	Deogad/Sindhudurg	1.5	250	6	1994 – 95	47.6	41
2	Gudepachgani	Shirala/Sangli	1.84	230	8	1998 – 99	40.83	20.42
3	Chalkewadi	Satara/Satara	2 Under Exec. 5.0	250	1250	1996 – 97	41.08	20.5
4	Motha	Chikhaldara/Amravati	2.0	1000	2	2002 – 03	51.93	34

The average rainfall varies from 4,000 mm on the western side to 7,000 mm on its eastern plateau. Maharashtra is one of the Indian states producing the highest amounts of energy by renewable sources. It is in the second place in India. The detailed renewable energy potential in Maharashtra is shown in Table 4. The table clearly shows that among all the renewable energy, wind energy is most significant.[35]

The mean annual wind speed in Maharashtra ranges from 5.0 m/s to 7.0 m/s. The detailed projected renewable energy capacity in five years is as follows (Table 5).

5.1. Demonstration Wind Power Projects in Maharashtra

In Maharashtra, wind energy manufacturing is carried out by various agencies; among them, there are three main ones, namely, (1) Maharashtra Energy Development Agency (MEDA), (2) Maharashtra State Electricity Board (MSEB), and (3) private agencies. The various projects done by MEDA in various places, their capacities, and year of commissioning are shown in Table 6.

5.2. Wind Monitoring and Power Projects of MSEB

Another government agency, Maharashtra State Electricity Board (MSEB), has also entered into wind power generation and has executed the projects at various places and commissioned them successfully. The power generation and supply is carried out by MSEB as shown (Table 7).

5.3. Private Wind Power Projects in Maharashtra

To improve government policies, subsidies, state financial incentives, and other policies have to be developed.[49] The Maharashtra government gives subsidies for the development of private projects, which are successful in installing wind power stations. Private projects are successfully installed in nine districts of Maharashtra with 955 wind turbines and a capacity of producing 392.825 MW power (Table 8). India has one of the highest gained wind power installations in the world among the developing countries. This is achieved because of privatization in the wind power sector. India has analyzed very thoroughly

Table 7. Wind monitoring and wind power projects (MSEB).[48]

Sr. No.	Name of site	Taluka/District	Capacity of Projects in MW	Machine Capacity KW	Number of Machines	Year of Commissioning	Land Acquired by MEDA (Hect)	Balance Land
1	Deogad	Deogad/Sindhudurg	0.55	55	10	1986 − 87	0	0
			0.55	55	10	1988 − 89	0	0
2	Dahanu	Dahanu/Thane	0.09	90	1	1987 − 88	0	0

Table 8. Private wind power projects in Maharashtra.[48]

Sr. No	Site/District	No. of Wind Turbines	Subtotal	Wind Turbine Make	Capacity (kW per Turbine)	Total Cap. (MW)	Total (MW)
1	Brahmanwel/Dhule	5	9	Windia	600	3	
		4			750	3	6
2	Kavdya, Dongar/Ahmednagar	57	57	Suzlon	1,000	57	57
3	Thoseghar/Satara	36	106	Vestas	225	8.1	30.86
		52		Enercon	230	11.96	
		18		Enercon	600	10.8	
4	Vankusavade/Satara	540	638	Suzlon	350	189	242.825
		7		Suzlon	1,000	7	
		28		AWT	750	21	
		25		Enercon	230	5.75	
		21		Enercon	600	12.6	
		1		IWPL	750	0.75	
		4		IWPL	250	1	
		1		Vestas	222	0.225	
		11		Vestas	500	5.5	
5	Chalkewadi/Satara	2	65	NEPC	225	0.45	22.57
		1		REPL	320	0.32	
		16		Suzlon	350	5.6	
		8		Vestas	225	1.8	
		10		Vestas	500	5	
		20		Enercon	230	4.6	
		8		Enercon	600	4.8	
6	Gudepachgani/Sangali	14	26	Enercon	230	3.22	10.42
		12		Enercon	600	7.2	
7	Matrewadi/Satara	25	25	Enercon	230	5.75	5.75
8	Varekarwadi/Satara	15	15	Enercon	600	9	9
9	Dhalgaon/Sangali	14	14	Enercon	600	8.4	8.4
							Total: 392.825

the clean and environmentally friendly development of energy. More importance is given toward development of energy by sustainable methods for the development of the country's economy.[50]

5.4. Declared Wind Sites in Maharashtra

Wind energy generation depends on environmental factors such as wind speed, temperature, pressure, precipitation, and lightning, which will directly or indirectly affect energy produced. The salinity of the atmosphere in a particular region will affect the static and dynamic stresses on the wind turbine parts and result in cyclic thermal/mechanical/electrical environments

that produce the failure of equipment and transmission line components.[51] In India, different states have different policies for wind energy production. The data of all the states are collected and compiled in a systematic format, and the diffusion of innovation theory is used to predict growth and to rank the states. The state-level wind power data are used to develop the model called the diffusion model (Bass model). This model includes the parameters such as land availability, preferential tariffs, wheeling and banking, third-party sales (TPS), and state-specific incentives. This model can be used to predict the future growth of wind energy in different states.[52] In Maharashtra there are 26 declared windy sites in different districts at about 20–25 m elevation, and their details are tabulated (Table 9).

Table 9. Declared windy sites.[48]

Sr. No	Station	Tal/District	Elevation MASL	Annual Mean Annual Wind Speed (KMPH) measured at 20/25 M	Annual mean Wind Power density W/m² measured at 20/25 M	Extrapolated at 50 m
1	Alamprabhu Pathar	Hatkanagale/Kolhapur	790	20.5	164	224
2	Chalkewadi	Satara/Satara	1,160	20.2	206	218
3	Dhalgav	KavatheMahakal/Sangali	810	21.2	216	260
4	Matrewadi	Patan/Satara	898	20.8	211	253
5	Panchgani	Mahabaleshwar/Satara	1,372	18.4	133	205
6	Thoseghar	Satara/Satara	1,140	21.7	229	489
7	Vankusavade	Patan/Satara	1,100	21.2	231	293
8	Varekarwadi	Patan/Satara	920	21.04	257	216
9	Mander Deo	Wai/Satara	1,280	19.4	153	206
10	Kas	Javali/Satara	1,240	20.5	194	277
11	Amberi	Khatav/Satara	960	23	237	275
12	Palsi	Patan/Satara	970	18.85	137	254
13	Khandke	Ahmednagar/Ahmednagar	920	19.6	146	250
14	Kolgaon	Srigonda/Ahmednagar	800	20.5	177	238
15	Kavadya Donger	Parner/Ahmednagar	910	23.2	224	277
16	Panchpatta	Akole/Ahmednagagar	1,080	20.51	201	236
17	Gudepanchgani	Shirala/Sangali	903	19.8	178	296
18	Dongerwadi	Miraj/Sangli	820	21.4	179	284
19	Raipur	Sakri/Dhule	500	18.9	162	214
20	Takar Mouli	Sakri/Dhule	600	20.8	186	224
21	Brahmanwel	Sakri/Dhule	600	23.1	278	324
22	Lonavala	Manvel/Pune	560	15.5	122	285
23	Sautada	Patoda/Beed	800	21.2	167	223
24	Vijayadurg	Deovgad/Sindhudurg	100	19.6	207	253
25	Gawalwadi	Dindori/Nashik	740	19	140	278
26	Chakala	Nandurbar/Nandurbar	380	23.7	242	323

Table 10. Wind monitoring in progress.[48]

SL No	Site	Tal./Dist.	Ele.	Approx. Available Land (Ha)	Approx. Wind power Cap. (MW)	Mast Height (m)
1	Kasarsirshi	Nilanga/Latur	679	500	50	25
2	Murud	Latur/Latur	716	1000	100	50
3	Rohina	Chakur/Latur	676	200	20	25
4	Bhud	Khanapur/Sangali	834	500	50	24
5	Vhaspet	Jath/Sangali	681	1000	100	50
6	Dhanagarwadi	Kankawli/Sindhudurg	191	500	50	25
7	Mahalunge	Deogad/Sindhudurg	196	600	60	50
8	Nerkewadi	Rajapur/Ratnagiri	267	500	50	25
9	Ambed	Sangameshwar/Ratnagiri	240	300	30	50
10	Deoud	Ratnagiri/Ratnagiri	232	500	50	25
11	Pacheri	Guhagar/Ratnagiri	172	500	50	50
12	Aundhewadi	Sinnar/Nashik	860	150	15	25
13	Kankora	Aurangabad/Aurangabad	920	400	50	25

5.5. Wind Power Production Capacity in Maharashtra

There are 13 sites identified in various places of Maharashtra where at a certain elevation the wind turbine can be installed with appropriate wind power to be produced as detailed in Table 10.

5.6. Wind Power Stations Closed in Maharashtra

Some of the sites that are closed and not feasible for work are listed by the MEDA as shown (Table 11).

Table 11. List of non-feasible, closed wind monitoring stations.[48]

SL No	Station	Taluka/District	SL No	Station	Taluka/District
2	Motha	Chikhaldara/ Amrawati	24	Pimpalgaon	Parner/Ahmednagar
1	Ranigaon	Dharni/Amrawati	23	Arag-Bedag	Miraj/Sangali
3	Mogarale	Man/Satara	25	Mhismal	Khultabad/Aurangabad
4	Deogad	Deogad/Sindhudurg	26	Pirtanda	Udgir/Latur
5	Malvan	Malvan/Sindhudurg	27	Kuchi	K-Mahankal/Sangali
6	Shaptshringigad	Kalwan/Nashik	28	Vinchur	Nifad/Nashik
7	Vengurla	Vengurla/Sindhudurg	29	Rajapur	Yeola/Nashik
8	Roti	Daund/Pune	30	Malegaon	Nandgaon/Nashik
9	MasaiPathar	Panhala/Kolhapur	31	Kamrawad	Shahada/Nandurbar
10	Nandivade	Ratnagiri/Ratnagiri	32	Thokalmalegaon	Kannad/Aurangabad
11	Kotoli	Shahuwadi/Kolhapur	33	Vagera	Panvel/Raigad
12	Khokade	Man/Satara	34	Vedi	Palghar/Thane
13	Mahi-Jalgaon	Karjat/Ahmednagar	35	Suryamal	Mokhada/Thane
14	Dhakale	Ambegaon/Pune	36	Mal	Shahapur/Thane
15	Ambral	M-Shwar/Satara	37	Davdi	Khed/Pune
16	Jambhulmure	Satara/Satara	38	Bedarwadi	Patoda/Beed
17	Waghapur	Patan/Satara	39	Kogil	Karvir/Kolhapur
18	Rajachikurli	Khatav/Satara	40	Shirasgaon	Sangamner/Ahmednagar
19	Renavi	Khanapur/Sangali	41	Kharumbapada	Jawhar/Thane
20	Alkud	Kavathe-Mahakal/ Sangali	42	Kogda	Jawhar/Thane
21	Kavaldara	Tuljapur/Osmanabad	43	Rajewadi	Ambegaon/Pune
22	Kamathi	Khatav/Satara	44	Elephanta	Uran/Raigad

6. Comments and Discussion

The above detailed study and analysis of data at various sites in the world, in India, and in the state of Maharashtra indicates that India is facing the challenge of sustaining its rapid economic growth. The threat of climate arises from the emissions of greenhouse gases emitted from continuous generation of energy from nonrenewable sources, intensive industrial growth, and high ingestion lifestyles. While engaged with the international community to jointly and supportively deal with this hazard, India needs a national approach to, first, acclimate to climate change and, second, to further augment the ecological sustainability of India's enlargement path.

Climate change may adversely affect India's natural resources and also the livelihood of its people. This climate change will affect agriculture, water, and forestry. In charting out the development in India, the above data analysis clearly indicates that India has a wider spectrum of choices in the sustainable development of energy because it is at an early stage of development and that wind energy would be one of the viable options.

Identifying the global climate change, India is actively participating in the UN Framework convention on climate change. The main objective is to establish an accurate, compatible, and equitable sustainable development of energy based on the principles designed by the United Nations Framework convention on climate change (UNFCCC). India is not lagging behind in the development of renewable energy and protecting the climate. In parallel to this, India has developed the organization known as Indian National Action Plan for Climate Change (NAPCC). NAPCC has made the target of producing approximately 15% of the energy mix of India by 2020. To achieve this target, the Indian Ministry of Power launched the Renewable Energy Certificate

(REC) mechanism in November 2010. This REC will assess the performance of the adjusting wind turbine projects for low-cost renewable energy generation. REC has decentralized distribution in and generation of renewable energy to different states. The participation in REC as per state data is Tamil Nadu 27%, Maharashtra 23%, and Uttar Pradesh 22%. REC helps to make the best policies for the state to implement on existing energy generation projects.[53]

The Indian electric supply system consists of a centralized generation system. Nowadays a centralized system is not capable of handling all the problems related to conventional energy. It is difficult for the authorities to visit each and every site and make the decision. For this reason the work will slow down and get delayed. This centralized electricity supply system is to be decentralized with the authority of the decision on the development of conventional energy.

This electric utility has been restructured into a number of subcontrolling authorities capable of making decisions on the development of small-scale projects. Nowadays the resources are being utilized through small and modular energy systems known as distributed generation system (DGS) based on renewable energy resources. This helps to start small-scale projects.[54]

7. Conclusion

It is essential that clean energy be produced in large amounts at reasonably less cost. One way of doing this is to use nonconventional energy sources such as wind energy. This review has discussed in detail the current position of wind energy in India with a focus on the situation in the state of Maharashtra. Here stock is taken of the capacity of the state of Maharashtra for possible wind energy production. The data of present installations, their capacity, and windy sites along with wind-power density has been given in detail. A detailed literature survey has been done, and sufficient relevant information has been provided.

References

1. Bilgili, M.; Sahin, B. Electrical Power Plants and Electricity Generation in Turkey. *Energy Sources part B*, **2010**, *5*, 81–92.
2. Sedar Genc, M.; Gokcek, M. Evaluation of Wind Characteristics and Energy Potential in Kayseri, Turkey. *J. Energy Eng.* **2009**, *135*, 33–43.
3. Demirbas, A. Biomass and the Other Renewable and Sustainable Energy Options for Turkey in Twenty-First Century. *Energy Sources* **2001**, *23*, 177–187.
4. Fung, K. T.; Scheffler, R. L; Stolpe, J. Wind Energy: A Utility Perspective. *IEEE Trans. Power Appar. Sys.* **1981**, *100*, 1176–1182.
5. Ezio, S.; Claudio, C. Exploitation of Wind as an Energy Source to Meet the World's Electricity Demand. *Wind Eng.* **1998**, *74*, 375–387.
6. Mustafa, S. G. Economic Viability of Water Pumping Systems Supplied by Wind Energy Conversion and Diesel Generator Systems on North Central Anatolia, Turkey. *J. Energy Eng.* **2011**, *33*, 21–35. DOI:10.1061/(ASCE)EY.1943-7897.0000033.
7. Sedar Genc, M. Economic Analysis of Large-Scale Wind Energy Conversion Systems in Central Anatolia, Turkey. In *Clean Energy Systems and Experiences*; Eguchi, K, Ed.; Sciyo: Rijeka, Croatia, 2010; pp. 131–154.
8. Cavallo, A. J.; Grubb, M. J. Renewable Energy Sources for Fuels and Electricity; London: Earthscan Publications, 1993. DOI:ORG/10.1016/S0038-092X (96)00087-4.
9. Celik, A. N. A Simplified Model for Estimating the Monthly Performance of Autonomous Wind Energy Systems With Battery Storage. *Renew. Energy* **2003**, *28*, 561–572.
10. Gokcek, M.; Sedar Genc, M. Evaluation of Electricity and Energy Cost of Wind Energy Conversion Systems (WECSs) in Central Turkey. *Appl. Energy* **2009**, *86*, 2731–2739.
11. Hansen, A. D.; Iov, F.; Blaabjerg, F.; Hansen, L. H. Review of Contemporary Wind Turbine Concepts and Their Market Penetration. *J. Wind Eng.* **2004**, *28*, 247–263.
12. Keith, D. W. The Influence of Large-Scale Wind Power on Global Climate. *Proc. National Acad. Sci. Washington DC* **1987**, *101*, 12–56.
13. Musgrove, P. J. Wind Energy Conversion: Recent Progress and Future Prospects. *Sol. Wind Tech.* **1987**, *4*, 37–49.
14. Hansen, A.; Sorensen, D.; Iov, P.; Blaabjerg, F. Centralized Power Control of Wind Farm With Doubly Fed Induction Generators. *J. Renew. Energy* **2006**, *31*, 935–951.
15. Sahin, B.; Bilgili, M. Wind Characteristics and Energy Potential in Belen-Hatav, Turkey. *Int. J. Green Energy* **2009**, *6*, 157–172.
16. Nair Gopalkrishan, K.; Thyangarajan, K. Optimization Studies on Integrated Wind Energy Systems. *Renew. Energy* **1999**, *16*, 940–943.
17. U.S. Department of Energy Report. *2012 Strategic Sustainability Performance Plan*; United States Department of Energy: Washington, DC, 2012.
18. Calvert, S.; Thresher, R; Hock, S.; Laxson, A.; Smith, B. U.S. Department of Energy Wind Energy Research Program for Low Wind Speed Technology of the Future—Discussion. *J. Solar Energy Eng.—Trans. ASME* **2002**, *124*, 455–463.
19. Bet, F.; Grassmann, H. Upgrading Conventional Wind Turbines. *Renew. Energy* **2003**, *28*, 71–78.
20. Thor, S. E.; Taylor, P. W. Long-Term Research and Development Needs for Wind Energy for the Time Frame 2000–2020. *Wind Energy* **2002**, *5*, 73–75.
21. Sun, X.; Huang, D.; Wu, G. The Current State of Offshore Wind Energy Technology Development. *Energy* **2014**, *41*, 298–312.
22. Pechak, O.; Mavrotas, G.; Diakoulaki, D. Role and Contribution of the Clean Development Mechanisms to the Development of Wind Energy. *Renewable Sustainable Energy Rev.* **2011**, *7*, 75–85.
23. Richards, G.; Noble, B.; Beicher, K. Barriers to Renewable Energy Development: A Case Study in Saskatchewan, Canada. *Energy Policy* **2012**, *42*, 691–698.
24. Michalak, P.; Zimmy, J. Wind Energy Development in the World: Europe and Poland from 1995 to 2009: Current Status and Future Prospective. *Renew. Sustain. Energy Rev.* **2011**, *15*, 2330–2341.
25. Mohammadi, K.; Motafaeipour, A.; Sabzpooshani, M. Assessment of Solar and Wind Energy Potentials for Three Economic and Industrial Zones of Iran. *Energy* **2014**, *67*, 117–128.
26. Auld, T.; McHenry, M. P.; Whale, J. U.S. Military, Airspace, and Meteorological Radar System Impacts From Utility Class Wind Turbines: Implications for Renewable Energy Targets and the Wind Industry. *Renew. Energy* **2013**, *55*, 24–30. DOI:10.1016/j.renene.2012.12.008.
27. Global Wind Energy Council. *Global Wind Statistics 2012;* Global Wind Energy Council: Brussels, Belgium, 2013. http://www.gwec.net/wp-content/uploads/2013/02/GWEC-PRstats-2012_english.pdf (accessed October 21, 2015).
28. Global Wind Energy Council. *Global Wind Report Annual Market Update 2013;* Global Wind Energy Council: Brussels, Belgium, 2013. http://www.gwec.net/wp-content/uploads/2014/04/GWEC-Global-Wind-Report_9-April-2014.pdf (accessed October 21, 2015).
29. Global Wind Energy Council. *Global Wind Statistics 2014;* Global Wind Energy Council: Brussels, Belgium, 2015. http://www.gwec.net/wp-content/uploads/2015/02/GWEC_GlobalWindStats2014_FINAL_10.2.2015.pdf (accessed October 21, 2015).
30. Global Wind Energy Council. *Global Wind Report Annual Market Update 2014;* Global Wind Energy Council: Brussels Belgium, 2014 http://www.gwec.net/wpcontent/uploads/2015/03/GWEC_Global_Wind_2014_Report_LR.pdf (accessed October 21, 2015).
31. Junfeng, L., et al. *China Wind Power Outlook*, GWEC, Greenpeace and CREIA, Beijing, 2010 http://gwec.net/wp-content/uploads/2012/06/wind-report0919.pdf (accessed October 21, 2015).
32. Global Wind Energy Council (GWEC). *VDMA/BWE: The German Wind Industry Takes a Breather*; Global Wind Energy Council: Brussels, Belgium, 2014. http://www.gwec.net/vdmabwe-the-german-wind-industry-takes-a-breather/ (accessed October 21, 2015).
33. Global Wind Energy Council Spain (GWEC). *Spain Market Overview*; International Renewable Energy Agency: Masdar City, Abu Dhabi, United Arab Emirates. https://www.irena.org/DocumentDownloads/Publications/GWEC_Spain.pdf (accessed October 21, 2015).
34. Global Wind Energy Council. *Global Wind Report Annual Market Update 2012;* Global Wind Energy Council: Brussels, Belgium, 2012. http://www.gwec.net/publications/global-wind-report-2/global-wind-report-2012/ (accessed October 21, 2015).
35. Indian Wind Power, volume 1, issue 2, February–March 2015. http://www.indianwindpower.com/pdf/Indian%20Wind%20Power%20Magazine%20-%20Feb.-March%202015%20Issue.pdf. (accessed October 21, 2015).
36. Mani, S.; Dhingra, T. Policies to Accelerate the Growth of Offshore Wind Energy Sector in India. *Renew. Sustain. Energy Rev.* **2013**, *24*, 473–482.
37. Marimuthu, C.; Kirubakaran, V. Carbon Payback Period for Solar and Wind Energy Project Installed in India: A Critical Review Article. *Renew. Sustain. Energy Rev.* **2013**, *23*, 80–90.
38. Sharma, A.; Srivastava, J.; Kar, S. K.; Kuma, A. Wind Energy Status in India: A Short Review. *Renew. Sustain. Energy Rev.* **2012**, *16*, 1157–1164.
39. Mabel, C. M.; Fernandez, E. Growth and future trends of wind energy in India. *Renew. Sustain. Energy Rev.* **2008**, *12*, 1745–1757.
40. Mani, S.; Dhingra, T. Critique of Offshore Wind Energy Policies of the UK and Germany—What Are the Lessons for India. *Energy Policy* **2013**, *63*, 900–909.

1. Mani, S.; Dhingra, T. Offshore Wind Energy Policy for India—Key Factors to Be Considered. *Energy Policy* **2013**, *56*, 672–683.
2. Ramachandra, T.V.; Shruth, B.V. Wind Energy Potential Mapping in Karnataka, India, Using GIS. *Energy Convers. Manage.* **2005**, *46*, 1561–1578.
3. Golait, N.; Moharil, R. M.; Kulkarni, P. S. Wind Electric Power in the World and Perspectives of Its Development in India. *Renew. Sustain. Energy Rev.* **2009**, *13*, 233–247.
4. Chattopadhya, D. Modeling Renewable Energy Impact on the Electricity Market in India. *Renew. Sustain. Energy Rev.* **2012**, *3*, 9–22.
5. Mahesh, A.; Jasmin Shoba, K.S. Role of Renewable Energy Investment in India: An Alternative to CO_2 Mitigation. *Renew. Sustain. Energy Rev.* **2013**, *4*, 414–424.
6. Global Wind Energy Council. *India Wind Energy Outlook 2012;* Global Wind Energy Council: Brussels, Belgium, 2012. http://www.gwec.net/wp-content/uploads/2012/11/India-Wind-Energy-Outlook-2012.pdf (accessed October 21, 2015).
7. Global Wind Energy Council. *Global Wind Energy Outlook, 2014;* Global Wind Energy Council: Brussels, Belgium, 2014. http://www.gwec.net/wp-content/uploads/2014/10/GWEO2014_WEB.pdf (accessed October 21, 2015).
48. Renewable Energy Atlas. www.mahaurja.com/Atlas%20Sale.html (accessed September 9, 2015).
49. Sawhney, A.; Rahul, M. Examining the Regional Pattern of Renewable Energy CDM Power Projects in India. *Energy Econ.* **2014**, *42*, 240–247.
50. Rajsekhar, B.; Van Hulle, F.; Jansen, J.C. Indian Wind Energy Program: Performance and Future Directions. *Energy Policy* **1999**, *27*, 669–678.
51. Trivedi, M. P. Environmental Factors Affecting Wind Energy Generation in Western Coastal Region of India. *Renew. Energy* **1999**, *16*, 894–898.
52. Rao, U. K.; Kishore, V.V.N. Wind Power Technology Diffusion Analysis in Selected States of India. *Renew. Energy* **2009**, *34*, 983–988.
53. Gupta, S. K.; Purohit, P. Renewable Energy Certificate Mechanism in India: A Preliminary Assessment. *Renew. Sustain. Energy Rev.* **2013**, *22*, 380–39.
54. Singh, A. K.; Parida, S. K. National Electricity Planner and Use of Distributed Energy Sources in India. *Sustain. Energy Tech. Assess.* **2013**, *2*, 42–54.

11

IEC 61850 Substation Communication Network Architecture for Efficient Energy System Automation

IKBAL ALI, MINI S. THOMAS, SUNIL GUPTA*, and S. M. SUHAIL HUSSAIN

Department of Electrical Engineering, Faculty of Engineering & Technology, Jamia Millia Islamia, New Delhi, India

Abstract: High-speed peer-to-peer IEC 61850-8-1 GOOSE and IEC 61850-9-2 sampled values based information-exchange among IEDs in modern IEC 61850 substations have opened the opportunity for designing and developing innovative all-digital protection applications. The transmission reliability and real-time performance of these SVs and GOOSE messages, over the process-bus network, are critical to realize these all-digital IEC 61850 substation automation systems (SASs) protection applications. To address the reliability, availability, and deterministic delay performance needs of SAS, a novel IEC 61850-9-2 process-bus based substation communication network (SCN) architecture is proposed in this article. Reliability of the proposed as well as the traditional process-bus based SCN architectures is evaluated using the reliability block diagram (RBD) approach. Network components are modeled, and end-to-end (ETE) time-delay performance is also evaluated for all-digital protection applications running on the SCN architectures simulated in the OPNET modeler platform. The reliability and performance results of the proposed architecture compared to the traditional architectures confirmed its highly reliable, fast, and deterministic nature.

Keywords: All-digital protection system, IEC 61850-9-2 process-bus architecture, OPNET modeler, reliability, SAS, switched Ethernet networks

Nomenclature

Abbreviations

CB	Circuit Breaker
CB_IED	Circuit Breaker IED
DHP	Dual Homing Protocol
EHV	Extra High Voltage
ES	Ethernet Switch
ETE	End To End
FTP	File Transfer Protocol
GOOSE	Generic Object Oriented Substation Event
GPS	Global Positioning System
HSR	High-availability Seamless Redundancy
IEC	International Electrotechnical Commission
IED	Intelligent Electronic Device
MTTF	Mean Time To Failure
MU	Merging Unit
NCIT	Non Conventional Instrument Transformer
OPNET	Optimized Network Engineering Tool
PPS	Pulse Per Second
PRP	Parallel Redundancy Protocol
PTP	Precision Time Protocol
QoS	Quality of Service
RBD	Reliability Block Diagram
RSTP	Rapid Spanning Tree Protocol
SAS	Substation Automation System
SCN	Substation Communication Network
SV	Sampled Values
VLAN	Virtual Local Area Network

*Address correspondence to: Sunil Gupta, Research Scholar, Department of Electrical Engineering, Faculty of Engineering & Technology, Jamia Millia Islamia, New Delhi 110025, India. Email: sun16delhi1@gmail.com

1. Introduction

To realize the concept of the smart grid, the recent trend is to utilize advanced grid assets, state-of-the-art communication techniques, and information technologies in electric utility. Modern substation automation systems (SAS) are implementing switched Ethernet-based international communication standard IEC 61850 for achieving the smart grid goals. The standard IEC 61850, *Communication Networks and Systems for Power Utility Automation,* permits communication interoperability among substation IEDs from different vendors at lower integration cost. Moreover, it allows the time-critical information low of IEC 61850-8-1 GOOSE and IEC 61850-9-2 SVs messages over the process-bus network in IEC 61850 SASs.[1,2] Thus, his type of information exchange between the process-level primary and the bay-level secondary devices provides the opportunity for designing and developing an innovative all-digital protection application.

The reliability of the process-bus network has a strong impact on the reliability of these all-digital protection applications n IEC 61850 SAS. Also, there are some real-time performance requirements for IEC 61850 SVs/GOOSE messages on he process-bus for implementing SAS applications. The most critical information exchange is related to the protection function, i.e., the transmission of the SVs from the conventional or NCITs/merging units (MUs) at the process level to the protection IEDs on the bay level. It also involves the transmission of GOOSE trip commands from the protection IEDs to the circuit breaker IEDs or the transmission of interlocking data between IEDs. Thus, the reliability and performance of a process-bus network is critical and presents one of the most challenging issues to the substation communication network (SCN) design engineer.[3-7]

To reduce the transmission delay, SVs from MUs and GOOSE messages among IEDs are mapped directly to the link layer of the Ethernet in IEC 61850 communication stack.[8] This feature accelerates the transmission of time-critical GOOSE and SVs messages but adversely affects their transmission reliability. Moreover, the use of switched Ethernet technology with quality of service features allows the efficient use of available network bandwidth and minimizes the delays by segregating and prioritizing the network traffic.[9] However, these features do not ensure the deterministic delivery of these real-time messages over the process-bus network during worst-case scenarios, i.e., the arrival of high-priority SVs/GOOSE messages during the transmission of the lower-priority client-server traffic with large packet size. In this situation, the higher-priority packets will have to wait in a queue until the lower-priority packets are transmitted. The worst-case scenario also depends on the packet size and traffic on SCN. Unlike conventional hardwired schemes, the performance of IEC 61850 communications-based protection applications are influenced primarily by the SCN topology along with communication network parameters, network load conditions, and the processing capabilities of the devices used. For this, the transmission time performance requirements of SVs/GOOSE messages as per IEC 61850-5 standard must be ensured under any network operating conditions.[10] Thus it is crucial to focus on the reliability and the real-time performance of process-bus SCN network under different network parameters and load scenario.

A significant work is reported in the literature to evaluate the reliability and ETE delay performance of time-critical messages in IEC 61850 substations.[11-13] T. Skeie et al.[14] demonstrated the feasibility of designing switched-Ethernet based SCN that fulfills the real-time demands of protection functions. T. S. Sidhu et al.[15] introduced the designing of IEC 61850 IED models in OPNET and then analyzed the dynamic performance of SCNs in traditional Ethernet network topologies. M. G. Kanabar et al.[16] have focused on the importance of analyzing the reliability and availability of traditional Ethernet network configurations for designing IEC 61850 SCN architectures. Kanabar et al.[17,18] have investigated the feasibility of implementing

ring-type process-bus SCN, based on SVs packet ETE delay and drop performance under different network parameters. They have also highlighted the importance of some corrective measures to be taken to ensure the real-time performance of time-critical messages under worst-case scenarios. D. M. E. Ingram et al.[19-21] critically examined the performance of the process-bus network in terms of Ethernet frame latency in switches under varying SVs data loading conditions. They have shown the adverse impact of increasing SVs data on the switching performance. L. Yang et al.[22] have evaluated the performance of process-bus based protection scheme and highlighted the need of efficient SAS design and configuration. From the above literature survey, it is found that none of the traditional Ethernet network configurations, i.e., star, ring, or star-ring fulfills the reliability and real-time needs of IEC 61850 SCNs.

Different SCN architectures, to evaluate their applicability in terms of reliability and ETE delay performance, for substation protection functions are presented in the following literature. M. S. Thomas et al.[23] proposed and analyzed the dynamic performance of SCN architecture, without process-bus, simulated in OPNET modeler. X. Liu et al.[24] modified ring Ethernet topology by incorporating communication path redundancy and presented a cobweb topology but did not include the critical component redundancy. Thus an optimized process-bus SCN is required to provide high reliability and transmission time performance as per IEC 61850 standard even under critical components/communication path failure and worst-case scenario.

This article presents a novel IEC 61850-9-2 process-bus SCN architecture that has exploited the advantages of both the structural changes in the network and that of the components/communication path redundancy. Reliability of the proposed as well as the traditional SCN architectures is evaluated using the reliability block diagram (RBD) approach. Network components are modeled and SCN architecture is simulated using OPNET modeler.[25] ETE time delay performance is also evaluated for all-digital protection applications running on the SCN architecture. The reliability and ETE performance results of the proposed architecture compared to the traditional SCN architectures confirmed its highly reliable, fast, and deterministic nature.

2. Proposed Process-Bus Based SCN Architecture

Any IEC 61850-9-2 process-bus based all-digital protection system consists of the electronic components such as NCITs, MUs, protection IEDs, and circuit breaker IEDs (CB_IEDs), where these components are connected to an Ethernet switch (ES) to form an SCN. MU IED at the process level collects process data from NCITs and transmits these data in an SV data packet format, as per communication mechanisms described in IEC 61850-9-2, to the protection IEDs on the bay level in IEC 61850 SAS.[26] Protection IED performs substation protection functions by sending GOOSE event triggered data to another bay and process level IEDs. CB_IED, representing the circuit breaker controlling/monitoring device, receives the GOOSE/interlocking commands data from protection IEDs and monitors the status and condition of the breaker. Time synchronization (TS) source provides the reference timing signals for the time stamping of sampled values generated from MUs so that the subscribers such as protection IEDs can use them to align the samples for further processing. Network-based IEEE 1588 PTP, as per *IEC 61850-9-2 LE* process-bus implementation guidelines, is used to synchronize MUs by supplying 1-PPS timing signals with less than 1 microsecond accuracy.[27] The PTP timing network allows synchronization directly over the Ethernet and includes a specialized networking hardware such as Grandmaster clock, generally a GPS receiver, acting as a primary time reference for PTP. The end users of PTP, e.g., merging units, have a slave clock that regenerates 1-PPS timing signals. Time-stamped and time-synchronized data from MU IEDs in SCN enables the measurement of several delays affecting the performance of substation operations.

Figure 1 shows traditional IEC 61850 intra-bay SCN architecture, which is prone to single point of failures, first, from a communication point of view as ES provides the only link for connecting the process level and the bay level equipment, and second, from the communication path point of view as there exists only one path for accessing primary equipment by protection IED through MU IED. Moreover, this architecture consists of only single critically important protection IED whose non-availability directly affects the performance of the protection function.

The proposed SCN architecture for intra-bay communication is shown in Figure 2. Here the protection system consists of two redundant and independent protection IEDs (main1 and main2), i.e., protection IED1 and protection IED2 per bay, from different manufacturers operating with the different protection principles. Moreover, only one protection IED, i.e., primary, out of redundant protection IEDs, i.e., protection IED1 and protection IED2, works at a time to clear the fault. Each dual-port protection IED, MU IED and CB_IED, are connected to two different Ethernet switches, i.e., with its own bay Ethernet switch and to the adjacent bay Ethernet switch. In Figure 2, it is illustrated that the Ethernet switches ES0, ES1, and ES2 correspond to bay 0, bay 1, and bay 2, respectively. Here the protection IED1 is connected to ES1 and ES2, i.e., Ethernet switch of its bay (ES1) and adjacent bay (ES2). Similarly, protection IED2 is connected to ES1 and ES0, i.e., Ethernet switch of its bay (ES1) and adjacent bay (ES0). In case of failure in the communication network of a protection system, e.g., main1, i.e., protection IED1, transfers the control to the redundant port through dual homing protocol (DHP) port switchover mechanism and uses the alternate communication path for further communication. Thus, only single Ethernet switch and protection IED per bay is utilized effectively in the protection function implementation at a time.

It can be observed that a protection scheme implemented through this component redundant proposed SCN architecture can survive the failure of anyone protection IED as well as an Ethernet switch. The above SCN intra-bay architecture is extended to SAS for a typical 220/132 kV D2-1 type substation, and the resultant architecture is shown in Figure 3. This EHV substation is of type D2-1 that consists of six feeder bays (F1–F6), two transformer bays (TI and T2), and one bus section (S) bay.[15]

The proposed SCN architecture for the whole substation is constructed by forming a ring network of bay Ethernet switches, i.e., F1_ES ... F6_ES, T1_ES, T2_ES, and S_ES. The formation of an Ethernet ring provides an alternate data path to the message flow in case of a link failure. The proposed SCN architecture is designed on the basis that modern substation IEDs have dual Ethernet communication ports, which can automatically switch communication to back up port using the DHP (in case the primary port fails) and hence ensures system operation continuity. Considering F1 bay, F2 bay, and S bay, the protection IED1, CB_IED and MU IED of F1 bay are connected to redundant F1 and F2 bay switches, i.e., F1_ES and F2_ES, in redundant star configuration. Similarly, the protection IED2 of F1 bay is connected to the Ethernet switch of its bay, i.e., F1_ES, and to the adjacent bay Ethernet switch, i.e., S_ES, and so on. In this way, the proposed architecture exhibits communication path redundancy for improved reliability and performance of the protection function.

Modern managed Ethernet switches with rapid spanning tree protocol (RSTP) manage the flow of mission-critical messages in redundant paths of a network and, at the same time, identify any communication failure in ring network topology with a reconfiguration time of only a few milliseconds. Thus, it improves the robustness of a proposed SCN against any failed communication path that maintains the stability of substation protection applications. IEC 61850 Ed.2, however, recommends the PRP/HSR protocols for achieving bump less redundancy and zero recovery time against failure in a substation operations.[28] However, the PRP/HSR technology needs specialized IEDs, which is not only expensive but also the major challenge encountered in the successful configuration of devices and the system. Therefore, the usage of specialized IEDs with PRP/HSR technology makes the overall system architecture expensive and complicated.

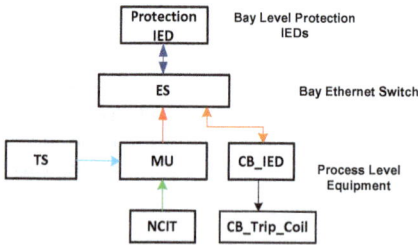

Fig. 1. Traditional IEC 61850 intra-bay SCN architecture.

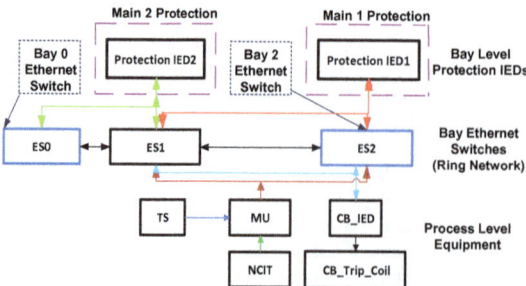

Fig. 2. Proposed IEC 61850 intra-bay SCN architecture.

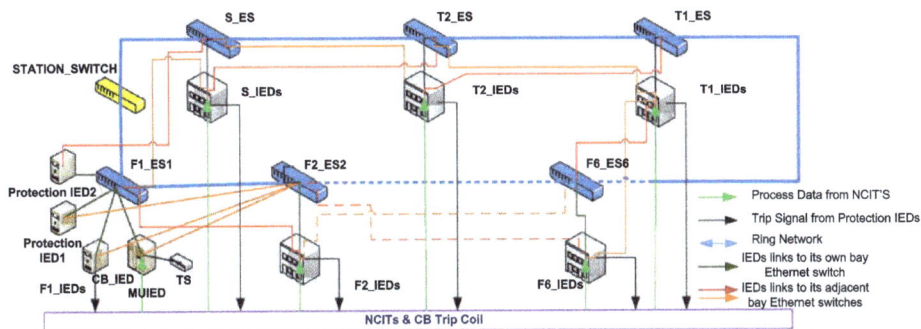

Fig. 3. Proposed process-bus SCN architecture.

Thus this article proposes an SCN architecture that incorporates the component as well as the communication path redundancy using DHP IEDs and managed Ethernet switches running RSTP. The proposed solution presents lesser packet processing complexities with lower cost compared to PRP/HSR-based solutions. Further, to handle bulk substation data from several IEDs integrated in substations and to reduce network congestion under severe fault conditions, virtual LANs (IEEE 802. 1Q) in the proposed SCN segregate and prioritize the transmission of time-critical information in a secure manner.[29] The proposed IEC 61850-9-2 process-bus based SCN architecture uses dual-port IEDs and multiple communication links that show significant improvements in the reliability and ETE delay performance required for time-critical protection functions in SAS.

3. System Reliability Analysis

3.1 System Reliability Based on the Reliability Block Diagram

The RBD method among the various available techniques such as Markov model, fault tree, minimal cut set, and minimal tie-set methods can effectively be used to determine the relative reliability of SCNs in different configurations and hence is used in this research. The article evaluates the reliability of process-bus based all-digital protection systems for traditional and proposed architectures by considering a substation layout as discussed in section 2. According to RBD method, the reliability calculation of IEC 61850 SAS application involves the construction and analysis of an RBD that shows the logical relationship among the substation components in terms of a successful SCN.[30] The SAS components are arranged in series and parallel arrangements between the system input and output nodes needed to realize a protection function successfully. The vital components required to perform the protection function effectively are put in series, while the redundant components are put in parallel, where at least one component must function for the protection system to perform.

Modern SCN architectural components such as NCIT, MU, protection IED, CB_IED, ES, and TS are electronic devices, and hence their reliability distribution is taken as exponential. Since the failure rate of components i is constant, the reliability function of these components is expressed as in Eq. (1)

$$R_i(t) = \exp(-\lambda_i t) \tag{1}$$

where t is the mission time and λ is the component failure rate.

The reliability of a series system, $R_s(t)$, is given by Eq. (2). Here it is assumed that the reliability of individual components is independent of each other.

$$R_s(t) = \prod_{i=1}^{n} R_i(t) = \exp\left[-\left(\sum_{i=1}^{n} \lambda_i\right) t\right] \tag{2}$$

Similarly, the reliability of a parallel system, $R_p(t)$, is given by Eq. (3).

$$R_p(t) = 1 - \prod_{i=1}^{n} Q_i(t) \tag{3}$$

where $Q_i(t) = 1 - \exp[-\lambda_i(t)]$, and represents the unreliability of ith component. The system unreliability is thus given by Eq. (4):

$$Q_{sys}(t) = 1 - R_{sys}(t) \tag{4}$$

Also, the mean-time-to-failure (MTTF) of a system is given by Eq. (5):

$$MTTF_{sys} = \int_0^{\infty} R_{sys}(t)\,dt \tag{5}$$

The MTTF and failure rate of various SCN architecture components for reliability calculations are presented in Table 1.[16,31]

3.2 Reliability Block Diagram (RBD)

Figures 4(a)–4(e) show the RBDs drawn for process-bus based traditional and proposed SCN architectures for D2-1 type substation as described in section 2. In traditional architectures, each bay has ES, protection IED (PR_IED), MU, CB_IED, CB components, and these bay components must work together to realize protection function successfully in IEC 61850 SAS. In cascade architecture, all the Ethernet switches are connected

Table 1. MTTF and failure rate of SAS components.

IEC 61850 SAS components	MTTF (yr)	Component failure rate (λ) (Yr^{-1})
Protection IED	100	0.01000
Circuit breaker IED (CB_IED)	150	0.00667
Merging unit (MU)	150	0.00667
Ethernet switch (ES)	50	0.02000
Time synchronization (TS)	150	0.00667
Nonconventional instrument transformer (NCIT)	150	0.00667
Circuit breaker trip coil (CB)	150	0.00667

in a cascade manner without forming any loop. Each switch is connected to the previous switch or next switch in the cascade via one of its ports. Hence, in RBD of cascade architecture, as shown in Figure 4(a), all Ethernet switches are connected in series. Star architecture offers the least amount of latency because of direct point-to-point connection between IEDs. But if a central switch fails, all bay switches are isolated, which reduces its transmission reliability. Hence, in RBD for star network, as shown in Figure 4(b), the central switch is connected in series with other critical components. Ring architecture is similar to cascade, but a loop is formed by connecting the last switch to the first switch. It offers N-1 redundancy in the network against any communication link failure. Hence, in RBD of a ring network, as shown in Figure 4(c), only 6 out of 7 managed Ethernet switches,

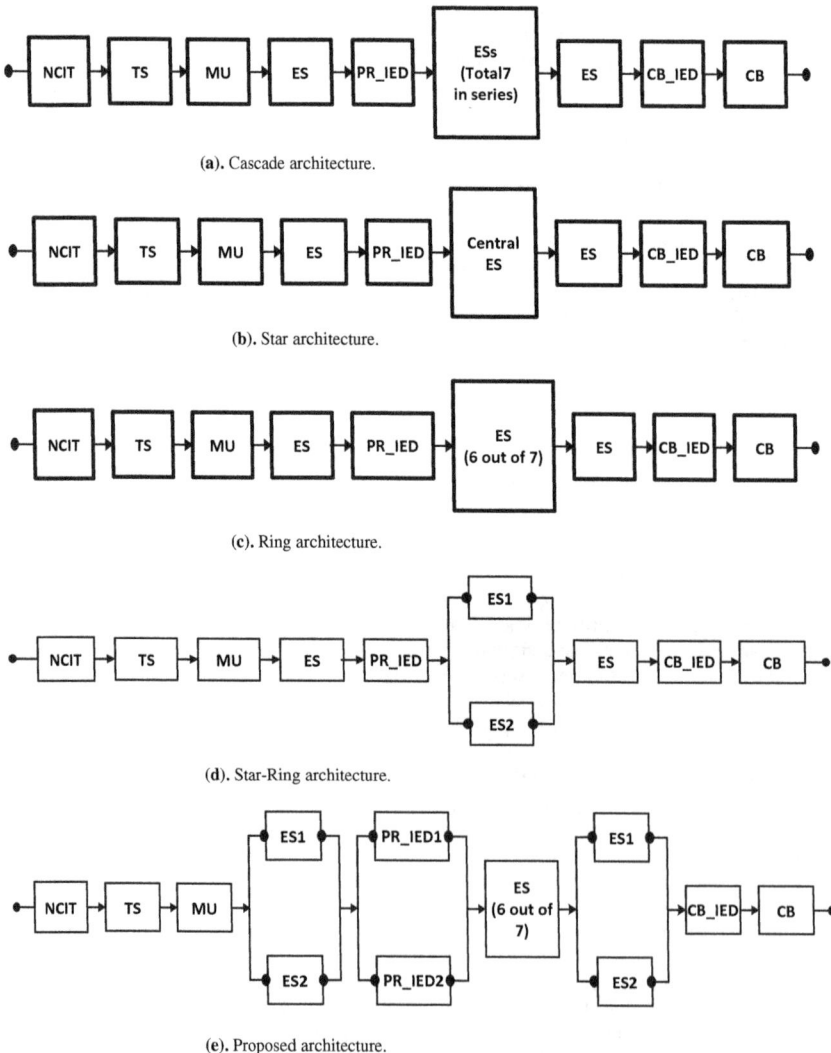

(a). Cascade architecture.

(b). Star architecture.

(c). Ring architecture.

(d). Star-Ring architecture.

(e). Proposed architecture.

Fig. 4. RBDs for various process-bus based SCN architectures for D2-1 type substation.[15]

considering the worst-case scenario, are required for inter-bay communication. In star-ring architecture, each bay-level switch is connected directly to two main Ethernet switches, which are connected in the ring to provide higher redundancy as well as low latency. Hence, in RBD for star-ring network, as shown in Figure 4(d), redundant ESs are shown to be connected in parallel with other critically important components within the protection system.

In RBD for proposed architecture, as shown in Figure 4(e), each IED has two redundant Ethernet ports, and all IEDs in each bay are connected to two redundant bay switches in star configuration. Also, all bay switches are connected in a ring configuration. Thus redundant ESs per bay would be in parallel, and also the ring configuration is shown to be separately connected in series with other essential components. Since each ring offers N-1 redundancy, only 6 out of 7 Ethernet switches are required for inter-bay communication. The reliability analysis of traditional and proposed SCN architectures, based on the RBD approach, is discussed in the following subsection.

3.3 System Reliability Equations and Calculations

Here the reliability calculation of an all-digital protection system is based on the assumptions that the failure modes are independent of each other and the components of the same type possess the same reliability. The reliability is decided by both the reliability of the IEDs/ESs and the communication links. However, the failure probability of links is quite low compared to ESs, and hence is neglected in the reliability calculations. Thus only the IEDs and Ethernet switches are considered in reliability calculations. The reliability of NCITs, MUs, TSs, ESs, protection IEDs, CB_IEDs, and CB trip coils are designated as R_{NCIT}, R_{MU}, R_{TS}, R_{ES}, R_{PRIED}, R_{CBIED}, and R_{CB}, respectively. To consider the worst-case reliability scenario, the idea is to compute the reliability of a protective function that involves inter-bay communication between IEDs placed at extreme ends in the SCN. The reliability of the protection function using cascade architecture, based on its RBD as shown in Figure 4(a), is given by Eq. (6).

$$R_{sys}^{Cascade} = R_{NCIT}.R_{TS}.R_{MU}.R_{PRIED}.R_{ES}^9.R_{CBIED}.R_{CB} \quad (6)$$

Similarly, the reliability of the protection function using star and ring architectures, as shown in Figure 4(b) and 4(c), are given by Eqs. 7 and 8, respectively.

$$R_{sys}^{Star} = R_{NCIT}.R_{TS}.R_{MU}.R_{PRIED}.R_{ES}^3.R_{CBIED}.R_{CB} \quad (7)$$

$$R_{sys}^{Ring} = R_{NCIT}.R_{TS}.R_{MU}.R_{PRIED}.R'_{ES_{6/7}}.R_{ES}^2.R_{CBIED}.R_{CB} \quad (8)$$

where

$$R'_{ES_{6/7}} = 7.R_{ES}^6.(1 - R_{ES}) + R_{ES}^7 \quad (9)$$

The reliability in Eq. (9) is computed using binomial distribution assuming the condition that a minimum of 6 or more ESs are required for inter-bay communication under the worst-case scenario. The reliability of the protection function for star-ring architecture, using its RBD as shown in Figure 4(d), is given by Eq. (10).

$$R_{sys}^{Star-Ring} = R_{NCIT}.R_{TS}.R_{MU}.R_{PRIED}.R'_{ES}.R_{ES}^2.R_{CBIED}.R_{CB} \quad (10)$$

where $R_{ES1} = R_{ES2} = R_{ES}$ and $R'_{ES} = 1 - (1 - R_{ES1}).(1 - R_{ES2}) = 2R_{ES} - R_{ES}^2$.

Now the reliability of the protection function for proposed SCN architecture, based on its RBDs as shown in Figure 4(e), is given by Eq. (11).

$$R_{sys}^{Proposed} = R_{NCIT}.R_{TS}.R_{MU}.R'_{ES}.R'_{PRIED}.R_{ES_{6/7}}.R'_{ES}.R_{CBIED}.R_{CB} \quad (11)$$

where $R_{PRIED1} = R_{PRIED2} = R_{PRIED}$ and $R'_{PRIED} = 1 - (1 - R_{PRIED1}).(1 - R_{PRIED2}) = 2R_{PRIED} - R_{PRIED}^2$.

It is shown in Figure 5 that the protection system unreliability increases with an increase in mission time for all process-bus architectures. However, the proposed architecture possesses the highest reliability, as the system failure increase rate is slowest in comparison to traditional architectures. Cascading architecture has the lowest reliability, as no-redundant Ethernet switches are used. Star architecture is more reliable than the cascade architecture, but the availability of star architecture is considerably less than the other network topologies. Both the ring and star-ring architectures are more reliable than star and cascade architectures, but the reliability of these architectures is lower when compared to the proposed architecture.

The protection system reliability for all SCN architectures, using the reliability equations, at a mission time of 1000 hours is calculated and presented in Table 2.

The proposed architecture possesses the highest reliability of 99.61% compared to traditional architectures. The MTTF of the

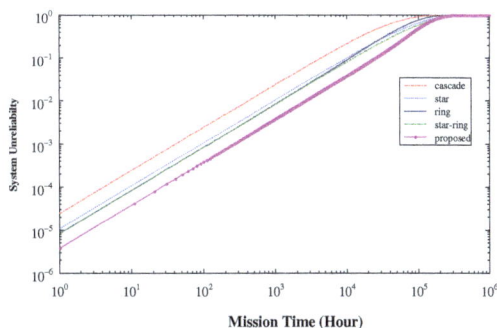

Fig. 5. Protection system unreliability versus mission-time of various process-bus architectures.

Table 2. Protection system reliability at mission time of 1000 h.

SCN architecture	Reliability, R_{sys} (%)
Cascade	97.48%
Star	98.83%
Ring	99.04%
Star-Ring	99.28%
Proposed	99.61%

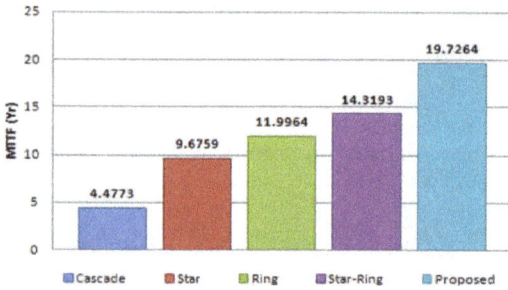

Fig. 6. MTTF comparison of various process-bus based SCN architectures.

proposed architecture, as shown in Figure 6, has a significantly higher value of 19.72 years, and concludes that it has the most reliable long operating life in SAS compared to the other existing traditional SCN architectures.

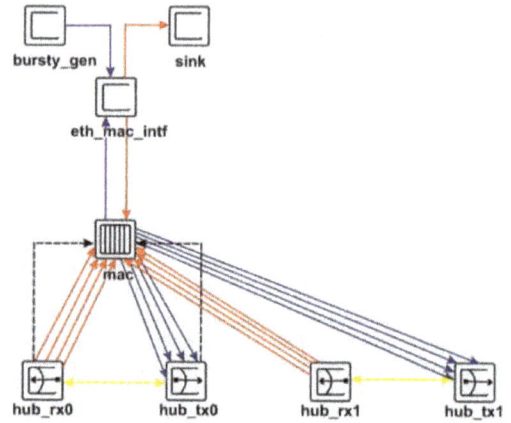

Fig. 8. Dual communication port node model for MU IED.

4. Modeling and Performance Analysis

Figure 7 shows the OPNET's simulated proposed process-bus, SCN architecture for a 220/132 kV D2-1 type substation as described in section 2, where the feeder bays, transformer bays, and the bus section bay are modeled into subnets. Each subnet represents "bay" and carries its corresponding protection IEDs, MU IEDs, CB_IEDs, and ESs. The node model editor and process model editor available with the OPNET modeler facilitate the construction of the dynamic models of various bay components, i.e., protection IEDs, MU IEDs, CB_IEDs, ESs, and communication links, based on the object-oriented modeling approach.

The ethernet_station_adv node model in OPNET is customized to support dual Ethernet communication ports (as shown in Figure 8) to function as MU IEDs as its communication stack has only application, Ethernet, and physical layers. The MU

IED transmits the SVs packets of process value data, through the process-bus, to the protection IEDs. The protection IED and CB_IED node model, as shown in Figure 9, is designed to support both the client-server stack and GOOSE stack. With this, the protection IED is able to communicate with other IEDs such as MU IED, station PC, server, and CB_IED in the network. Also, the protection IED and CB_IED node models feature two communication ports for connecting to redundant bay Ethernet switches in proposed SCN.

The ethernet16_switch node model with 16 port interfaces is selected to function as an Ethernet switch model. The100BaseT link node model that supports communication at the data rate of 100 Mbps is selected to function as communication links in the SCN. Finally, the standard ethernet_server node and ethernet_wkstn node models are simulated as server and station PC, respectively. Further, for network traffic management, port-based VLANs are configured in standard Ethernet switch node

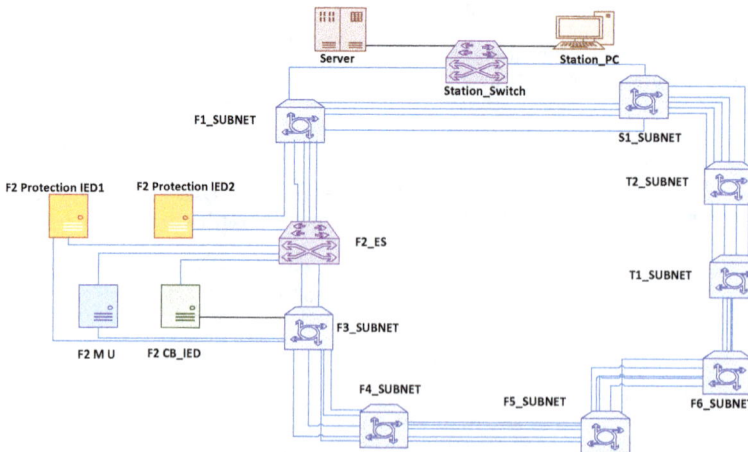

Fig. 7. Proposed IEC 61850-9-2 process-bus based SCN architecture simulated in OPNET Modeler.

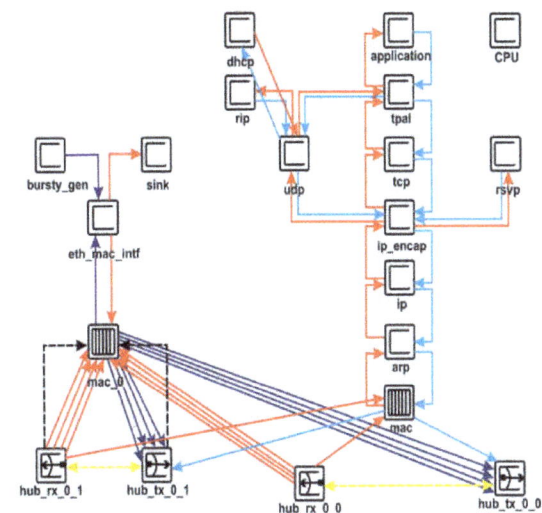

Fig. 9. Dual communication port node model for protection IED and CB_IED.

models that limit the flow of multicast SVs traffic to only its subscribed protection IEDs in an SCN.

The Application Configuration window in OPNET allows different IEC 61850 messages such as SVs, GOOSE trip, file transfer, status updates, and interlock data to be defined for establishing the communication among IEDs in the simulated SCN. Each MU IED in SCN transmits SVs packets of 126 B to its corresponding bay protection IEDs at a sampling rate of 4800 Hz when the simulation starts. These SV data packets are sent once to the network. Protection IEDs under fault condition send trip commands to the corresponding bay CB_IEDs. The GOOSE trip message size consists of 50 bytes. Each trip message is sent four times to ensure the correct delivery of the message; hence it possesses the higher transmission reliability.

Also, all protection IEDs are simulated to support file transfer, acting as background traffic with a message size of 300 KB, with the station server using heavy FTP functions available in OPNET. Further, all the protection IEDs and CB_IEDs send updated status values to a station server at a rate of 20 Hz. These are type 2 messages with a message size of 200 bytes. Each message is

sent once to the network. These messages are configured either using standard applications available in OPNET or by designing customized applications such as GOOSE, interlock, etc., by using the Task Configuration window. The Profile Configuration window in OPNET allows the creation of various profiles such as MU IED, protection IED, CB_IED, station server, and station PC that set the traffic flow among IEDs in the simulated SCNs. Each profile supports different applications defined in the Application Configuration node. Table 3 enlists the detail of network traffic configured in the simulated SCNs using the Application Configuration window, Profile Configuration window, and Task Configuration window of OPNET.

The ETE delay performance of the protection function is taken as the sum of the time spent for SVs to reach the protection IED from MU IED and the delay of GOOSE trip command to reach CB IED from the protection IED. The performance of the traditional and the proposed process-bus network has been analyzed for a feeder F2 fault where protection IEDs have corresponding MU IEDs connected to the F1 bay Ethernet switch. Assume that the fault causes MU IEDs to send fault data to corresponding protection IEDs in F1 bay, which, on detecting a fault, further sends a GOOSE trip message to the F1 CB_IEDs.

Figure 10 shows the comparison of the ETE delay characteristics of proposed and traditional process-bus based SCN architectures analyzed under normal network conditions. It can be observed that the ETE delay in each architecture is less than the upper limit of 3 ms. But it results the least in the proposed architecture. However, the reliability of traditional architectures, due to the presence of a single point of failures and limited components/communication path redundancy, is considerably less than that of the proposed architecture as is calculated in section 3. Moreover, the proposed architecture survives even under critical component and link failure situations.

In the worst scenario of a component and/or communication link failure situation, the ETE delay performance of the traditional architectures deteriorated either because of nil (star, cascade) or limited (ring, star-ring) communication path/component redundancy, whereas the worst-case performance of the proposed architecture, due to critical component redundancy/communication path redundancy with DHP support is not much affected. Because on detecting primary communication links or an Ethernet switch failure, communication is shifted to another Ethernet interface instantaneously to maintain the system operation continuity.

Table 3. Message type and size among IEDs in the SCN.

Applications	Source IED	IEC 61850 Message Type	SCN Traffic Type	Destination IED	Sampling Frequency (Hz)	Packet Size (Bytes)
Sampled value data	MU IED	4	Raw data message	Protection IEDs	4800 Hz	126
Protection	Protection IED	1, 1A	GOOSE trip signal	CB_IEDs	–	50
Controls		3	Control signals	Protection IED, CB_IED	10 Hz	200
File transfer		5	Background traffic	Station server	1 Hz	300 KB
Status updates	Protection IED CB_IED	2	Status signals	Station server	20 Hz	200
Interlocks	Protection IED	1, 1A	GOOSE signal	CB_IEDs	–	200

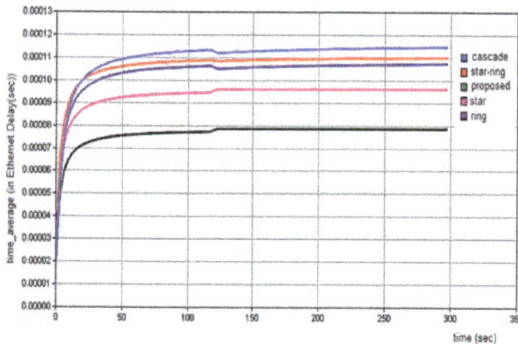

Fig. 10. ETE delays of process-bus based SCN architectures.

5. Conclusion

This article presents a novel IEC 61850-9-2 process-bus based SCN architecture that fulfills the transmission reliability and the real-time performance requirements of time-critical SVs and GOOSE messages for all-digital protection functions of the substation. Reliability block diagrams have been demonstrated for the proposed and traditional Ethernet SCN architectures, considering inter-bay communication among IEDs in substations, to quantitatively evaluate the reliability of these all-digital protection systems. The feasibility of the proposed architecture in implementing critical applications of substations has been analyzed by calculating ETE delay using the OPNET simulation tool. It has been demonstrated that the traditional architectures fulfill the IEC 61850 communication needs under normal network traffic. However, none of them achieves the strict performance requirements of the standard under critical components/communication path failure and worst network traffic scenario. Thus it is discovered that the proposed architecture achieves the highest reliability and performance, among all other process-bus based SCN architectures. It signifies the importance of utilizing the redundant critical components/communication paths in achieving the crucial SAS operations success. However, it is recommended that the proposed architecture should be selected considering the required reliability and performance (cost constraint) for real-time substation applications.

References

1. *Communication Networks and Systems for Power Utility Automation—Part 8-1: Specific Communication Service Mapping(SCSM) – Mappings to MMS (ISO 9506-1 and ISO 9506-2) and to ISO/IEC 8802-3*, IEC 61850-8-1 ed2.0. 2011.
2. *Communication Networks and Systems for Power Utility Automation—Part 9-2: Specific Communication Service Mapping(SCSM) – Sampled Values Over ISO/IEC 8802-3*, IEC 61850-9-2 ed2.0. 2011.
3. Wen-Long, W.; Ming-Hui, L.; Xi-Cai, Z. Research on the Shared-network of SMV and GOOSE in Smart Substation. *J. Int. Council Elec. Eng.* **2014**, *4*, 136–140.
4. Lee, N. H.; Jang, B. T. Development of the Model-driven Test Design System for IEC 61850 Based Substation Automation System. *J. Int. Council Elec. Eng.* **2013**, *3*, 20–24.
5. Skendzic, V.; Ender, I.; Zweigle, G. *IEC 61850-9-2 Process-bus and Its Impact on Power System Protection and Control Reliability*. Schweitzer Engineering Laboratories. Technical Report. 2007.
6. Andersson, L.; Brunner, C.; Engler, F. Substation Automation Based on IEC 61850 with New Process-Close Technologies. *IEEE Power Tech Conference Proceedings*, Bologna, 2006.
7. Ren, Y.; Xiao Y.; Jin, Y.; Peng, S. Impact of IEC61850 on Substation Design. *J. Int. Council Elec. Eng.* **2013**, *3*, 210–214.
8. Sidhu, T. S.; Gangadharan, P. K. Control and Automation of Power System Substation Using IEC61850 Communication. *Proceedings IEEE Conference on Control Applications*. Toronto, Canada. 28–31 August 2005, 1331–1336.
9. Decotignie, J. D. Ethernet-based real-time and industrial communications. *Proc. IEEE* **2005**, *93*, 1102–1117.
10. *Communication Networks and Systems in substations—Part5: Communication Requirements for Functions and Device Models, IEC International Standard*. July 2003.
11. Andersson, L.; Brand, K. P.; Brunner, C.; Wimmer, W. Reliability investigations for SA communication architectures based on IEC 61850. *Proceedings IEEE Power Technology*. August 2005.
12. Andersson, L.; Brand, K. P.; Fuechsle, D. Optimized Architectures for Process Bus with IEC 61850-9-2. Presented at the CIGRE Session Paris, France. Aug. 2008, paper B5-101.
13. Yang, L.; Crossley, P. A.; Zhao, J.; Li, H.; An, W. Impact Evaluation of IEC 61850 Process Bus Architecture on Numerical Protection Systems. *International Conference on Sustainable Power Generation and Supply*, 2009.
14. Skeie, T.; Johannessen, S.; Brunner, C. Ethernet in Substation Automation. *IEEE Control Sys.* **2002**, *22*, 43–51.
15. Sidhu, T. S.; Yin, Y. Modelling and Simulation for Performance Evaluation of IEC61850-based Subsation Communication Systems. *IEEE Trans. Power Deliv.* **2007**, *22*, 1482–1489.
16. Kanabar, M. G.; Sidhu, T. S. Reliability and Availability Analysis of IEC 61850 Based Substation Communication Architectures. *presented at the IEEE Power Eng. Soc. General Meeting Calgary*. Alberta, Canada, July 2009.
17. Kanabar, M. G.; Sidhu, T. S. Performance of IEC 61850-9-2 Process Bus and Corrective Measures for Digital Relaying. *IEEE Trans. Power Deliv.* **2011**, *26*, 725–735.
18. Kanabar, M. G.; Sidhu, T. S.; Zadeh, M. R .D. Laboratory Investigation of IEC 618509-2 Based Busbar and Distance Relaying With Corrective Measure for Sampled Value Loss/Delay. *IEEE Trans. Power Deliv.* **2011**, *26*, 2587–2595.
19. Ingram, D. M. E.; Schaub, P.; Taylor, R. R.; Campbell, D. A. Direct Evaluation of IEC 61850-9-2 Process Bus Network Perfromance. *IEEE Trans. Smart Grid* **2012**, *3*, 1853–1854.
20. Ingram, D. M. E.; Schaub, P.; Taylor, R. R.; Campbell, D. A. Performance Analysis of IEC 61850 Sampled Value Process Bus Networks. *IEEE Trans. Indust. Inform.* **2013**, *9*, 1445–1454.
21. Ingram, D. M. E.; Schaub, P.; Taylor, R. R.; Campbell, D. A. System-level Tests of Transformer Diffrential Protection Using an IEC 61850 Process Bus. *IEEE Trans. Power Deliv.* **2014**, *29*, 1382–1389.
22. Yang, L.; Crossley, P. A.; Wen, A.; Chatfield, R.; Wright, J. Design and Performance Testing of a Multivendor IEC 61850-9-2 Process Bus Based Protection Scheme. *IEEE Trans. Power Deliv.* **2014**, *5*, 1382–1389.
23. Thomas, M. S.; Ali, I. Reliable, Fast, and Deterministic Substation Communication Network Architecture and Its Performance Simulation. *IEEE Trans. Power Deliv.* **2010**, *25*, 2364–2370.
24. Liu, X.; Pang, J.; Zhang, L.; Xu, D. A High Reliability and Determinacy Architecture for Smart Substation Process-Level Network Based on Cobweb Topology. *IEEE Trans. Power Deliv.* **2014**, *24*, 842–850.
25. OPNET Modeler – OPNET Technologies. Version 17.5. 2014.
26. IEC 61850-9-2 LE: *Implementation Guideline for Digital Interface to Instrument Transformers Using IEC 61850-9-2*. UCA International Users Group.

27. Ingram, D. M. E.; Schaub, P.; Campbell, D. A. Use of Precision Time Protocol to Synchronize Sampled-Value Process Buses. *IEEE Trans. Instrum. Meas.* **2012**, *61*, 1173–1180.

28. Kirrmann, H.; Rietmann, P.; Kunsman, S. *Network Redundancy Using IEC 62439*. PAC World, ABB, Switzerland, Autumn 2008.

29. Ingram, D. M. E.; Schaub, P.; Campbell, D. A. Multicast Traffic Filtering for Sampled Value Process Bus Neytworks. *Proc. 37th Annu.*

Conf. IEEE Indust. Electron. Soc. Melbourne, Australia, Nov. 7–10. 2011, 4710–4715.

30. Billinton, R.; Allan, R. N. *Reliability Evaluation of Engineering Systems: Concepts and Techniques*. Springer: USA, 1992.

31. Brand, K. P.; Lohmann, V.; Wimmer, W. *Substation Automation Handbook*. Utility Automation Consulting Lohmann: Bremgarten, Switzerland, 2003.

12

Use of Jatropha Biodiesel as a Future Sustainable Fuel

AMBARISH DATTA* and BIJAN KUMAR MANDAL

Department of Mechanical Engineering, Bengal Engineering and Science University, Howrah, West Bengal, India

Abstract: This article briefly discusses the present status and future scope for use of jatropha biodiesel as an alternative to diesel (fossil derived) in India. The big gap between the production and the use of petroleum fuels is presently met by imports from other countries. Therefore, it is obligatory on the part of India to go in for some alternative, renewable, and eco-friendly fuels that can be cultivated in the otherwise barren land available in the country. In this respect, it may be mentioned that the government of India has identified jatropha as a possible and promising alternative to diesel. However, the bio-fuel policy adopted by the government through its bio-fuel mission launched in 2003 and 2007 in two phases did not evoke much success because of various challenges faced by the commercial production of jatropha, in spite of its many advantages. Therefore, in this era of energy crisis and fast degradation of the environment, the government must devise an appropriate plan of action to overcome these challenges and to implement the bio-fuel policy to promote the use of jatropha biodiesel as a partial substitute to mineral diesel fuel.

Keywords: Alternative energy, jatropha, biodiesel, policy

1. Introduction

Fast depletion of fossil fuels demands an immediate and urgent need for extensive research so that some viable alternative is obtained and sustainable energy demand with less environmental impact is met. The major environmental concern, as expressed in an IPCC report is "Most of the observed increase in globally averaged temperatures since the mid-20th century is due to the observed increase in anthropogenic greenhouse gas concentrations."[1] Since the combustion of fossil fuels is known to increase greenhouse gas concentrations in the atmosphere, these fuels are likely sources of global warming. Another concern is advocated by the peak oil theory which predicts a rising cost of fossil fuels caused by a severe shortage of petroleum reserves underground during an era of growing energy consumption. According to the peak oil theory, the demand for oil will exceed the supply, and the gap between the demand and supply will continue to grow. This could cause a growing energy crisis starting between 2010 and 2020. However, the crisis has not yet come, and it may be delayed further. The reason is that the peak oil theory did not take into consideration the growing technological developments in the energy sector.[2] Lastly, since the majority of the known petroleum reserves are located in the Middle East Asia, there is a general concern that the fuel shortage worldwide could intensify the unrest in this region. This may even lead to further conflict and war. An alternative fuel, also known as a nonconventional fuel, is any material or substance that can be used as a fuel other than conventional fuels. Conventional fuels include fossil fuels (petroleum oil, coal, propane, and natural gas) and also nuclear fuels such as uranium in some instances. A host of alternative fuels have already been identified, and these include biodiesel, bio-alcohol (methanol, ethanol, and butanol), hydrogen, non-fossil methane, non-fossil natural gas, vegetable oils, and other fuels derived from biomass sources. Among all those alternative fuels, biodiesel is the most promising and popular in the transport sector and other CI engine applications.

2. Necessity of Alternate Fuels in India

In 2005, India consumed 30 million tons of oil in the transport sector, of which 29% was gasoline and 71% was diesel. The Indian energy demand is expected to grow at an annual rate of 4.8% over the next couple of decades. It has been projected that India will double its oil consumption, at least, by 2030, when India will become the third largest oil consumer in the world. Bio-fuel production could, therefore, potentially play a major role in this respect. A number of developmental and statutory activities are being introduced in the country for the production of biofuels, and one of these is to include a 5% mandatory blend of ethanol in gasoline. The trials with such blends are ongoing in various states of India. The government of India has set a target to increase the blend of biofuels with gasoline and diesel to 20% by 2017.[3] In India, the domestic production of crude oil in the year 2003–2004 was 33.38 million tons, whereas a quantity

*Address correspondence to: Ambarish Datta, Department of Mechanical Engineering, Bengal Engineering and Science University, Shibpur, Howrah 711103, West Bengal, India. Email: ambarish.datta84@gmail.com

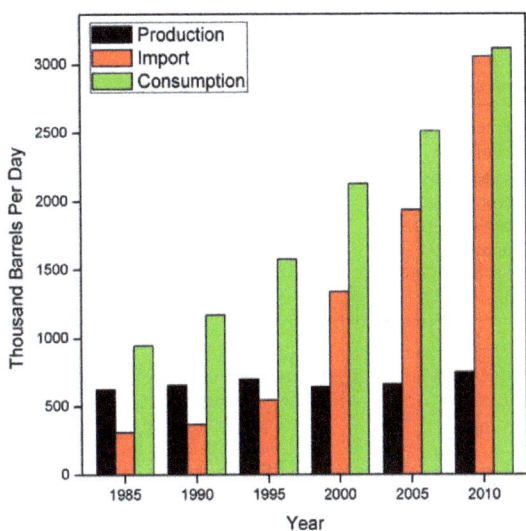

Fig. 1. Production, importation, and consumption of crude oil in India.[4]

of 90.43 million tons of oil (which amounts to 73% of total oil consumed) was imported. Figure 1 illustrates the status of production, consumption, and importation of crude oil until 2010, and it demonstrates that the consumption of crude oil in India has been increasing rapidly.[4] On the contrary, the production rate has remained almost static. As the fuel importation rate is increasing at a rapid pace, the future of fuel economy of the country is becoming heavily dependent on the major fuel-producing countries. The ever-increasing price of fuel may also affect the domestic economy of such a fuel-importing country as India in an adverse manner. So it is time that a viable alternative fuel was developed to decrease crude oil import. It will be wise to explore the possibility of the use of various alternative fuels that can be produced indigenously. This will certainly reduce the oil import bill and improve the domestic economy of the country.

3. Biodiesel: The Popular Bio-Fuel in India

The consumption of diesel fuel in India is approximately six times that of gasoline fuel, as shown in Table 1.[3] The table predicts a rising cost of oil-derived fuels caused by severe shortages

Table 1. Demand of gasoline and diesel in India.[3]

Year	Gasoline demand (MMT)	Diesel demand (MMT)
2001–2002	7.07	39.81
2002–2003	7.62	42.15
2003–2004	8.20	44.51
2004–2005	8.81	46.97
2005–2006	9.42	49.56
2006–2007	10.07	52.33
2011–2012	12.85	66.90

of oil because of growing energy demand. So it is necessary that an appropriate policy decision be made in the country so that the future demand of diesel fuel is fulfilled in compliance with stringent emission norms. Renewable fuels, particularly biodiesel, should get more attention in India because of the many promises it offers. Researchers are trying to find several ways to make biodiesel from different feedstock such as edible oil, nonedible oil, waste vegetable oil, algae, animal tallow and fats, etc. The transesterification process, employed to manufacture biodiesel from raw feedstock, yields a byproduct, glycerol, which has many applications in the pharmaceutical, cosmetic, and food industries. Sharma and Singh[5] produced biodiesel from nonedible feedstock such as karanja, mahua, and a hybrid mixture (50:50 v/v) of the two. Saka and Kusdiana[6] produced biodiesel from rapeseed vegetable oil. Venkanna and Reddy[7] produced biodiesel from honne oil. Ghadge and Raheman[8,9] produced biodiesel from mahua oil. But all of the above production processes were carried out in the laboratory, and the products were used only for research purposes.

Biodiesels produced from different sources, as of now, may be considered supplementary fuels to the diesel fuel in CI engine applications. In addition, it also promises employment to rural people through the opportunities of cultivation of vegetable oil–bearing plants, and this may help improve the domestic economy. Biodiesel is a clean burning fuel, and it can be produced from 100% renewable resources.[10] Several experimental studies[11–14] available in the literature have shown that biodiesel and biodiesel–diesel blends reduce smoke opacity, emission of particulate matter, unburnt hydrocarbon and carbon monoxide as well as life-cycle carbon dioxide emissions. However, the emission of nitrogen oxides increases to some extent with the use of biodiesel as fuel. However, the NO_x emissions can be reduced by the use of some post-combustion techniques such as exhaust gas recirculation and catalytic conversion. As biodiesel contains about 11% excess oxygen, the calorific value is lower than diesel, but it enhances the combustion process. The peak pressure rise after TDC with biodiesel makes the combustion process safer and more efficient. Another advantage is the shorter ignition delay of biodiesel, which results in a decrease in maximum heat release rate. Biodiesel does not contain petroleum, but it can be mixed with petroleum and a biodiesel blend—thus made it can be used in a number of vehicles. Biodiesel is biodegradable and nontoxic. It has even been claimed to be less toxic than common salt. Biodiesel is made from vegetable oil through a process called transesterification. This process involves removing the glycerine from the vegetable oil or fat. Biodiesel is free from some environmentally harmful substances such as sulphur and aromatics that are found in the traditional diesel fuels. The overall reaction of the transesterification process is shown below in Figure 2.

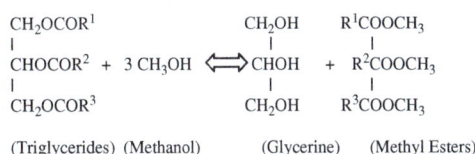

$$\begin{array}{lll} CH_2OCOR^1 & CH_2OH & R^1COOCH_3 \\ | & | & | \\ CHOCOR^2 + 3\ CH_3OH \Longleftrightarrow CHOH & + & R^2COOCH_3 \\ | & | & | \\ CH_2OCOR^3 & CH_2OH & R^3COOCH_3 \end{array}$$

(Triglycerides) (Methanol)   (Glycerine)   (Methyl Esters)

Fig. 2. Transesterification reaction for biodiesel production.[6]

4. GHG Emission and the Vehicular Norms in India

The carbon cycle of biodiesel consists of the release and absorption of carbon dioxide. The combustion and respiration processes release carbon dioxide into the atmosphere, and crops absorb carbon dioxide during the photosynthesis process. Thus the accumulation of carbon dioxide in the atmosphere is reduced, and a balance is maintained. The carbon-cycle time for fixation of CO_2 and its release after combustion of biodiesel is quite small (a few years) compared to the carbon-cycle time of petroleum oils (a few million years). The transportation sector contributes a significant amount of carbon dioxide, one of the principal greenhouse gasses, to the atmosphere. CO_2 from combustion of fossil fuels is the predominant GHG (greenhouse gases) produced by transportation, accounting to over 95% of the annual global warming potential produced by the sector.[15] Vehicles equipped with catalytic converters and internal combustion engines emitting methane account for nearly all the remainder. The total share of CO_2 emission by the transportation sector alone is about 21% globally, and with the rising number of vehicles every day, this share is expected to go up to 23% by the year 2030. Consumption in the transportation sector is also a key factor in the overall increase in oil demand in the world. Almost two-thirds of the worldwide increase in oil demand between the years 2004 to 2030 will emanate from the transportation sector only.[16] The U.S. Department of Energy has predicted in their most recent report (July 2013) that the energy consumption in the transportation sector will increase by an average of 2.2% per year up to 2040.[17]

The Bharat Stage II emission norms were enforced in the entire country from 1 April 2005, and Euro IV equivalent emission norms were enforced in 2010. In addition to four metro cities where Bharat Stage II norms had already been in place, Bangalore, Hyderabad, Ahmedabad, Pune, Surat, Kanpur, and Agra had also came under the per view of this norm from April 1, 2003. The four metros and the other seven cities had already complied with Euro III since 2005 and Euro IV since 2010. The two- and three-wheelers were conforming to Bharat Stage II norms from April 1, 2005 all over the country and Bharat Stage III norms from April 1, 2008. For new vehicles, a drastic reduction in sulphur content (<350 ppm) and higher cetane numbers (>51) will be required in the petroleum diesel produced by Indian refineries.[18] Biodiesel meets the specifications of these two emission norms and helps in improving the lubricity of low sulphur diesel when blended with it. Also, Section 52 of the Indian Motor Vehicles Act already allows conversion of an existing engine of a vehicle for the use of biofuels. Engine

Table 2. Estimated yield of nonedible oil from different plants.[20]

Scientific name (Indian name)	Plant type	Plant part	Oil yield (kg/ha)
Azadirachta indica (Neem)	Tree	Seed	2670
Jatropha curcas (Jatropha, Ratanjyot)	Tree/shrub	Seed	1900–2500
Pongamia pinnata (Karanja)	Tree	Seed	225–2250
Ricinus communis (Castor)	Tree/shrub	Seed	450

manufacturers would need to address the issue of compatibility with biofuels wherever necessary[19] with suitable modifications of the engines.

5. Oil Yield and Lands for Cultivation in India

The government of India has adopted several policies in order to bolster the biofuel sector. It starts with the selection of appropriate feedstock having a relatively higher oil yield and higher availability. The selection of the feedstock should be made in such a manner so that the cultivation throughout the country with minimum cost is possible. It also includes the identification of lands and seeds and the task of motivating farmers for cultivating the biodiesel feedstock, mainly the jatropha. So there is a long chain of activities that starts with the identification of oil seeds, extraction of oil from the seed, transesterification of the raw vegetable oil for the production of biodiesel, transportation of biodiesel to the oil depot, making blends with mineral diesel, and, finally, distribution to retailers. Table 2[20] and Table 3 show the estimated yield of oil seed and the potential of nonedible oil seed, respectively, in the Indian scenario. It can be seen from Table 2 that the oil yield of jatropha per hectare of land is highest among the prospective nonedible oil seeds. It is quite interesting to note from the data in Table 3 that although jatropha has lower potential, it is still found to be very attractive in the sense that it can grow in adverse agro-climatic conditions throughout India. Also it has a much shorter gestation period compared to other nonedible oil plants. Being a shrub, jatropha offers an easier harvest than most of the other nonedible sources. Other identified plants, as shown in Table 3, have various other useful applications, other than as fuels, in the medical and industrial sectors. For the

Table 3. Potential nonedible oil seed plants in India.[21]

Common name	Distribution	Potential (metric tons/annum)	Oil (%)	Different uses
Neem	All over India	5,00,000	35–45	Ayurvedic, unani, and homeopathic medicine, preventing stored grains
Jatropha	All over India	15,000	40–60	Biodiesel, erosion control, to reclaim land, grown as live fence
Karanja	Maharastra, Karnataka, Assam	2,00,000	30–40	Tanning leather, soap, illuminating oil, lubricant, water-paint binder
Castor	All over India	7,90,000	46–55	Adhesives, coatings, soaps, lubricants, paints, and dyes

above-mentioned reasons, jatropha has been identified as one of the most acceptable biodiesel-producing feedstock in India.

In a country such as India, rice and jute are cultivated mainly in the rainy season. The former is the main food grain while the other is one of the main revenue-generating crops. The jatropha seed collection season does not coincide with the rainy season; rather, it can be collected at any time of the year. Consequently, its cultivation makes an attractive proposition for people for an extra income in the slack agricultural season.

In order to develop a successful biodiesel industry, it is necessary that appropriate pricing policies for jatropha seeds and biodiesel along with financing policies for organized cultivation be evolved. To achieve the target, the government of India launched Biofuel Mission in two phases in the year of 2003. The first phase was initiated in 2003 and was scheduled to be completed in 2007. It was supposed to cover experimentation, demonstration, and communication to farmers, but it failed to make sufficient publicity. The second phase was started in 2007 and was scheduled to be completed in 2012. The weakness in the second phase is the omission of the sustainable issues and corrective measures for them.

According to the planning commission report, 13.4 million hectares (Mha) of land are required for the cultivation of jatropha, and this is sufficient for a 20% blending target. In Table 4, the details of the lands are described. The land estimated to be available for cultivation of jatropha by the end of 2012 is shown in Table 5.[3] However, the target of 20% biodiesel blending with diesel is not yet fulfilled in India for several reasons.

Table 4. Land available for jatropha curcas plantations.[3]

Type of land	Area (Mha)	Potential for jatropha plantation (Mha)
Forest areas (understocked forests)	31.0	3.0
Agriculture (protective hedge around agricultural field)	142.0	3.0
Agriculture (agro-forestry)	2.0	
Cultivable fallow lands	24.0	2.4
Wastelands under integrated/watershed development program of Ministry of Rural Development	2.0	
Strip-lands such as roads, railways, canal banks, etc.	1.0	
Total	197.0	13.4
Additional wastelands	4.0	

6. Features of Jatropha Curcas

Jatropha curcas, a shrub of 3–4 m in height, belongs to the family of Euphorbiaceae.[14] The harvesting of jatropha is easier than a tree. Jatropha has also a much shorter gestation period (about 2–3 years) than most of the second-generation biofuel feedstock.[21] Jatropha can be cultivated in tropical parts of the world with an annual rainfall of 300–1000 mm. It occurs mainly at lower altitudes (0–500 meters above sea level) with an average annual temperature well above 20°C, but it can also be cultivated at higher altitudes and adverse climatic conditions with an inferior quality of soil. Also, it requires very few nutrients to survive and, therefore, can be grown on less fertile land. The oil content of jatropha varies from 35–40% of the seed mass. Other parts of jatropha curcus have various other uses. The root is used as an antidote for snakebite, and the latex obtained from jatropha has anti-cancerous properties. Also, its cultivation, seed collection, oil extraction, and biodiesel production can generate large-scale employment.[14]

The properties of the triglyceride and the biodiesel fuel are determined by the amounts of different fatty acids present in the molecules. Chain length and the number of double bonds determine the physical characteristics of both fatty acids and triglycerides.[22] The fatty acid compositions of jatropha oil compiled from the previous work of Bamgboye and Hansen,[23] Sarin et al.,[24] Foidl and colleagues,[25] and Ma and Hanna[26] are presented in Table 6.

The byproducts obtained during the production of biodiesel are also quite useful as biofertilizer and glycerine. India has a large wasteland of about 17.45% of the total geographical area[27] suitable for jatropha cultivation. This can generate large volumes of biodiesel, provided jatropha's cost of production becomes compatible with the price of petroleum diesel.[28]

7. Properties of Jatropha Biodiesel

Jatropha oil can be converted to its methyl esters via the transesterification process in the presence of a catalyst. The

Table 6. Fatty acid compositions of jatropha oil.[22–26]

Fatty acid	Formula	Structure	Wt%
Palmitic	$C_{16}H_{32}O_2$	16:0	11.3
Stearic	$C_{18}H_{36}O_2$	18:0	17.0
Arachidic	$C_{20}H_{40}O_2$	20:0	4.7
Oleic	$C_{18}H_{34}O_2$	18:1	12.8
Lenoleic	$C_{18}H_{32}O_2$	18:2	47.3

Table 5. Biodiesel demand and required land for Jatropha curcas plantations.[3]

Year	Diesel demand (Mt)	5% Biodiesel blend (Mt)	Area for 5% biodiesel blend (Mha)	10% Biodiesel blend (Mt)	Area for 10% biodiesel blend (Mha)	20% Biodiesel blend (Mt)	Area for 20% biodiesel blend (Mha)
2006–2007	52.33	2.62	2.19	5.23	4.38	10.47	8.76
2011–2012	66.90	3.35	2.79	6.69	5.58	13.38	11.19
2016–2017	83.58	4.18	3.48	8.36	6.97	16.72	13.98

purpose of the transesterification process is to lower the viscosity of the oil. Ideally, transesterification is potentially a less expensive way of transforming the large, branched molecular structure of the bio-oils into a smaller, straight chain structure so that the smaller molecules conform to the type required in regular diesel combustion engines. Biodiesel from jatropha oil is free from sulphur and still exhibits excellent lubricity, which is an indication of the amount of wear that occurs between two metal parts covered with the fuel as they come in contact with each other. It contains very small amounts of phosphorus and sulphur; therefore, emission of oxides of sulphur (SO_x) is almost negligible. In comparison with the commercial petro-diesel, jatropha biodiesel has higher density and cetane number. In addition, the higher flash point (more than 100°C) of jatropha biodiesel makes the storage and transportation issues much simpler. It is a much safer fuel than diesel because of its higher flash point and fire point. The cloud filter pour point is generally higher than that of diesel, and this may create some complications for the operation in cold weather. The amount of carbon residue from the hot decomposition of vegetable compounds with higher molecular weight is greater than that of commercial diesel oil. Some of the important fuel properties of jatropha oil, jatropha biodiesel, and conventional petro-diesel are compiled from the previous work of Ramos and colleagues,[22] Pradeep and Sharma,[28] Graboski and McCormick,[29] and Barnwal and Sharma.[30] These properties are presented in Table 7 for comparison.

8. Economics of Biodiesel Utilization

The major hindrance to the use of biodiesel as a fuel is its higher cost of production than that of mineral diesel. The estimated cost for biodiesel production was much lower in 2003, as reported by the Planning Commission of India. The cost of biodiesel was estimated as Rs. 16.59–14.98 per liter. The details of the cost components have been provided in Table 8. But nowadays the cost of production is much more. Actually, the cost of production depends on the scale of production. In some cases, it also depends on government policies. Sustained commitment from the government regarding budget allocation, coordination between different sectors, and incentives for the producers, processors, and consumers are the need of the hour for successful implementation

Table 8. Cost of biodiesel production in 2003.[3]

Cost components	Rate (Rs./kg)	Quantity (kg)	Cost (Rs.)
Seed	5.00	3.280	Rs. 16.40
Cost of collection and oil extraction	2.36	1.050	Rs. 2.48
Less cake produced	1.00	2.230	− Rs. 2.23
Transesterification cost	6.67	1.000	Rs. 6.67
Less cost of glycerol produced	40–60	0.095	− Rs. 3.8 to −5.70
Cost of biodiesel per kg	−	−	Rs. 19.52 – 17.62
Cost of biodiesel per liter (Sp. gravity 0.85 at 15°C)	−	−	Rs. 16.59 – 14.98

of biofuel programs.[14] From another point of view, the biofuel industry increases employment opportunities, indigenous energy sufficiency, and savings of foreign exchange.

To date, the production of biofuel crops is hit by a lack of research and development work and price support for farmers. As a protection measure for domestic producers, the Biodiesel Association of India has advocated for an increase in the import duty of biodiesel by 100%. On the basis of the current market survey, the cost analysis of biodiesel production based on the latest data of Kumar and colleagues[31] has also been shown in Table 9. They estimated the unit price on a per liter basis in 2012 as Rs. 46.45.

9. Indian Scenario

As discussed earlier, India will be the third largest oil consumer in the world by increasing its oil demand by 2030. Therefore, the implementation of biofuel policy in commercial sectors has been carried out by different government and nongovernment organizations. This includes planting of jatropha by the National Oilseed and Vegetable Oil Development Board (NOVOD).[32] There are some pilot plants for biodiesel production at different Indian Institutes of Technology (IIT), Indian Institute of

Table 7. Properties of jatropha oil, jatropha biodiesel, and diesel.[22,28–30]

Property	Jatropha oil	Jatropha biodiesel	Diesel	Biodiesel standards ASTM D 6751-02	DIN EN 14214
Density at 15°C (kgm^{-3})	940.00	880.000	850.00	−	860–900
Viscosity (mm^2s^{-1})	24.50	4.800	2.60	1.9–6.0	3.5–5.0
Flash point (°C)	225.00	135.000	68.00	>130	>120
Pour point (°C)	4.00	2.000	−20.00	−	−
Water content (%)	1.40	0.025	0.02	<0.03	<0.05
Ash content (%)	0.80	0.012	0.01	<0.02	<0.02
Carbon residue (%)	1.00	0.200	0.17	−	<0.30
Acid value ($mgKOHg^{-1}$)	28.00	0.400	−	<0.80	<0.50
Calorific value ($MJkg^{-1}$)	38.65	39.230	42.00	−	—

Table 9. Cost of biodiesel production in 2012.[31]

Cost components	Rate (Rs./kg)	Quantity (kg)	Cost (Rs.)
Seed	16.00	3.280	Rs. 52.48
Cost of collection and oil extraction	2.36	1.050	Rs. 2.478
Less cake produced	1.00	2.230	– Rs. 2.23
Transesterification cost	6.67	1.000	Rs. 6.67
Less cost of glycerol produced	50.00	0.095	– Rs. 4.75
Cost of biodiesel per kg	–	–	Rs. 54.65
Cost of biodiesel per liter (Sp. gravity 0.85 at 15 °C)	–	–	Rs. 46.45

Petroleum, Dehradun and Council for Scientific and Industrial Research (CSIR) laboratories, mainly for research and development purposes. Field trial runs, by automobile giants such as Mahindra & Mahindra, Daimler Chrysler, have been started with jatropha as the base feedstock in their tractors and cars, respectively. Indian Railways have already started to use biodiesel made from nonedible oils such as jatropha and karanja in short-term usage.[33] In Kolkata, the Calcutta Tramways Company and Calcutta State Transport Corporation used the blend of biodiesel with diesel at their buses on a regular basis. CSIR encouraged Tata Motors and Indian Oil Corporation Limited to bring biodiesel as a way out to the world's fuel crisis.

10. Sustainability Analysis of Jatropha

For an assessment of the future prospect of jatropha biodiesel in the Indian subcontinent, a sustainability analysis of the above-mentioned fuel is needed in terms of its strength, weakness, opportunities, and challenges. In Indian perspective, we need to put our attention to the regular use of biodiesel for commercial purpose. This analysis may give a clear vision of the future of jatropha biodiesel as a fuel in the Indian subcontinent and other countries, from the standpoint of both economic and environmental aspects.

10.1 Strength and Weakness

The jatropha biodiesel in India has definitely some strengths and advantages. Jatropha is a renewable biological crop, and it maintains a closed carbon cycle. This explains its eco-friendly nature as a fuel. A plantation of jatropha also promotes the use of otherwise barren lands and controls soil erosion. It has high yield potential of more than 2 tons of oil per hectare per year.[34] Straight vegetable oil (SVO) of jatropha can also be used directly in small-scale diesel generators, oil lamps, and stoves. As India imports a huge amount of fossil-based crude oil to fulfil its normal demand, jatropha-based biodiesel may decrease its dependence on imported petroleum. But this fact should not be overemphasized, as the percentage of the country's fuel supply that can be replaced with biodiesel will not be too high.[11] However, an additional

source of fuel can have a surprising impact on the stability of fuel prices under fluctuating global petroleum market conditions.

On the other hand, jatropha oil and its biodiesel suffer from some inherent drawbacks. First, jatropha is still a wild species, and its cultivation as an oil crop has not yet been standardized. Its payback period is more than two years. The byproducts produced during oil extraction in the form of cake is unsuitable as animal food due to its toxic nature. India is yet to form a standard protocol for biodiesel use. Biodiesel production technologies are not firmly established for commercial purposes in India. The production cost of jatropha biodiesel is still prohibitively high due to small-scale production.

10.2 Opportunities and Challenges

In spite of some of its weaknesses, as pointed out by different researchers, jatropha biodiesel still has opportunities to be used as a partial replacement for diesel fuel. The cultivation of oil seeds, production of oil, conversion to biodiesel, and its marketing can generate employment and improve the economic condition of a country such as India. Jatropha biodiesel also reduces health hazards as the toxic pollutant emission is lower compared to that for existing fossil fuels. It is expected that jatropha biodiesel blending (10–20%) with petro-diesel may be made mandatory in the future, and this will definitely promote the production and use of biodiesel in India. Biodiesel production may also increase due to the expected incentives and subsidies from the government of India.

Jatropha biodiesel is facing some challenges. This includes the deficiency in domestic knowledge, as the localized technology is not yet standardized. The lack of actual field results is also responsible for the non-popularization of jatropha oil among farmers and users. Considering the "food vs. fuel" debate, people are showing more interest for crops that can either yield fuel along with food annually or can be cultivated in rotation with food crops. Another challenge jatropha biodiesel may face is the tough competition with algae-based biodiesel (third-generation biofuels), since the oil content per unit mass of algae is much higher than that of jatropha. The minimum selling cost of biodiesel should be fixed at a certain level by the appropriate authority so that it can compete with the cost of fossil-derived fuels.

11. Conclusion

India as well as other oil-importing countries needs alternative fuels to replace petroleum-based fossil fuels for several reasons. The ever-increasing energy demand in the automobile sector, alarming oil import bills, and stringent emission norms are considered to be the main reasons behind this. Biodiesel from jatropha is popular in India because the wasteland can be used for its cultivation. Jatropha biodiesel, obtained from jatropha oil through the transesterification process, has similar properties of petro-diesel and meets the ASTM (US) and DIN EN (European) standards. Although jatropha biodiesel has enormous scope, it does not yet show popularity in India from a commercial point of view. The experimentation with jatropha and the implementation of the biofuel policy related to jatropha are limited to research and development in laboratories and some major automobile companies only. The high cost of production and lack of firm government policy regarding the use of biodiesel are the two key barriers for its successful implementation in the fuel sector. Third-generation biofuels, particularly algal biodiesel, may put a challenge to jatropha-based biodiesel. Since India is an agro-economic country, if the cultivation of jatropha is carried out more scientifically, jatropha may turn out to be more economical than algae. Formation of a definite biodiesel policy and its proper implementation by the government of India may improve the situation in the near future.

References

1. A Report of Working Group of the Intergovernmental Panel on Climate Change. http://www.ipcc.ch/pdf/assessment-report/ar4/wg1/ar4-wg1-spm.pdf (accessed March 12, 2009).
2. Fusco, L. Peak Oil Theory. http://www.ucs.mun.ca/~oilpower/documents/Peakoil2-1.doc.pdf (accessed July 14, 2013).
3. Report of the Committee on Development of Bio-Fuel, Planning Commission, Government of India, 2003. http://planningcommission.nic.in/reports/genrep/cmtt_bio.pdf (accessed June 15, 2013).
4. Report of United States Energy Information Administration. http://www.eia.doe.gov/ (accessed March 8, 2013).
5. Sharma, Y. C.; Singh, B. A Hybrid Feedstock for a Very Efficient Preparation of Biodiesel. Fuel Process. Technol. **2010**, *91*(10), 1267–1273.
6. Saka, S.; Kusdiana, D. Biodiesel Fuel From Rapeseed Oil as Prepared in Supercritical Methanol. Fuel **2011**, *80*, 225–231.
7. Venkanna, B. K.; Reddy, C. V. Biodiesel Production and Optimization from Calophyllum Inophyllum Linn Oil (Honne Oil)—A Three Stage Method. Bioresour. Technol. **2009**, *100*, 5122–5125.
8. Ghadge, S. V.; Raheman, H. Biodiesel Production From Mahua (Madhuca indica) Oil Having High Free Fatty Acids. Biomass Bioenergy **2005**, *28*, 601–605.
9. Ghadge, S. V.; Raheman, H. Process Optimization for Biodiesel Production From Mahua (Madhuca indica) Oil Using Response Surface Methodology. Bioresour. Technol. **2006**, *97*, 379–384.
10. Minnesota Department of Agriculture. A Biodiesel Blend Handling Guide. http://www.biodiesel.org/docs/using-hotline/a-biodiesel-blend-handling-guide.pdf?sfvrsn=4 (accessed May 23, 2013).
11. Palit, S.; Chowdhuri, A. K.; Mandal, B. K. Environmental Impact of Using Biodiesel as Fuel in Transportation: A review. Int. J. of Global Warming **2011**, *3*(3), 232–256.
12. Agarwal, A. K.; Das, L. M. Biodiesel Development and Characterization for Use as a Fuel in Compression Ignition Engines. J. Eng. Gas Turbines Power. **2000**, *123*(2), 440–447.
13. Shaoo, P. K.; Das, L. M. Combustion Analysis of Jatropha, Karanja and Polanga Based Biodiesel as Fuel in a Diesel Engine. Fuel **2009**, *88*, 994–999.
14. Misra, R. D.; Murthy, M. S. Jatropa—The Future Fuel of India. Renewable Sustainable Energy Rev. **2011**, *15*, 1350–1359.
15. A Report of Working Group of the Intergovernmental Panel on Climate Change. http://www.ipcc.ch/ipccreports/tar/wg3/pdf/3.pdf
16. Reducing Transport Greenhouse Gas Emissions: Trends & Data 2010. http://www.internationaltransportforum.org/Pub/pdf/10GHGTrends.pdf (accessed Nov 21, 2013).
17. International Energy Outlook 2013. www.eia.gov/ieo/ (accessed Aug 15, 2013).
18. Society of Indian Automobile Manufacturers. Emission Norms. http://www.siamindia.com/scripts/emission-standards.aspx (accessed Jan 23, 2011).
19. National Policy on Biofuels, a Report by Government of India, Ministry of New & Renewable Energy. http://mnre.gov.in/file-manager/UserFiles/DIREC_2010_Report.pdf.
20. Kumar, A.; Sharma, S. Potential Non-edible Oil Resources as Biodiesel Feedstock: An Indian Perspective. Renewable Sustainable Energy Rev. **2011**, *15*, 1791–1800.
21. Borugadda, V. B.; Goud, V. V. Biodiesel Production From Renewable Feedstocks: Status and Opportunities. Renewable Sustainable Energy Rev. **2012**, *16*, 4763–4784.
22. Ramos, M. J.; Fernandez, C. M.; Casas, A.; Rodrigez, L.; Perez, A. Influence of Fatty Acid Composition of Raw Materials on Biodiesel Properties. Bioresour. Technol. **2009**, *100*, 261–268.
23. Bamgboye, A. I.; Hansen, A. C. Prediction of Cetane Number of Biodiesel Fuel From the Fatty Acid Methyl Ester (FAME) Composition. Int. Agrophys. **2008**, *22*, 21–29.
24. Sarin, R.; Sharma, M.; Sinharay, S.; Malhotra, R. K. Jatropha-Palm Biodiesel Blends: An Optimum Mix for Asia. Fuel. **2007**, *86*, 1365–1371.
25. Foidl, N.; Foidl, G.; Sanchez, M.; Mittelbach, M.; Hackel, S. Jatropha curcas L. as a Resource for the Production of Biofuel in Nicaragua. Bioresour. Technol. **1996**, *58*, 77–82.
26. Ma, F.; Hanna, M. A. Biodiesel Production: A Review. Bioresour. Technol. **1999**, *70*, 1–15.
27. Biswas, P. K.; Pohit, S.; Kumar, R. Biodiesel from Jatropa: Can India Meet the 20% Blending Target? Energy Policy **2010**, *38*, 1477–1484.
28. Pradeep, V.; Sharma, R. P. Use of HOT EGR for NOₓ Control in a Compression Ignition Engine Fuelled with Bio-Diesel From Jatropha Oil. Renewable Energy **2007**, *32*, 1136–1154.
29. Graboski, M. S.; McCornic, R. L. Combustion of Fat and Vegetable-oil Revived Fuels in Diesel Engines. Prog. Energy Combust. Sci. **1998**, *24*, 125–164.
30. Barnwal, B. K.; Sharma, M. P. Prospects of Biodiesel Production From Vegetable Oils in India. Renewable Sustainable Energy Rev. **2005**, *9*, 363–378.
31. Kumar, S.; Chaube, A.; Jain, S. K. Critical Review of Jatropha Biodiesel Promotion Policies in India. Energy Policy. **2012**, *41*, 775–781.
32. National Oilseeds and Vegetable Oils Development Board, Ministry of Agriculture, Department of Agriculture and Cooperation, Government of India. Jatropha: An Alternate Source for Biodiesel. http://www.novodboard.com/Jatropha-english.pdf (accessed Sept 25, 2013).
33. Government of India, Ministry of Railways. Report on Testing of Biodiesel as an Alternate Fuel. 2004. http://irsme.nic.in/files/MP-Misc-158.pdf (accessed Aug 3, 2013)
34. Comprehensive Jatropha Report. http://www.biozio.com/ref/report/jat/Comprehensive_Jatropha_Report_Preview.pdf (accessed May 14, 2013).

13

Two Methods for Converting a Heavy-Water Research Reactor to Use Low-Enriched-Uranium Fuel to Improve Proliferation Resistance After Startup

R.S. KEMP*

Department of Nuclear Science and Engineering, Massachusetts Institute of Technology

Abstract: This article demonstrates the feasibility of converting a heavy-water research reactor from natural to low-enriched uranium in order to slow the production of weapon-usable plutonium, even if the core cannot be physically reconfigured. The analysis was performed for Iran's IR-40 reactor at Arak in support of negotiations with Iran, but the methods have application to future reactors that present similar nonproliferation challenges. Two methods are considered, and both retain identical power, thermal-hydraulic, and safety profiles as the original reactor design. The conversion options can be implemented at any time during the reactor's life. The two methods have competing effects on achievable burnup, and they can be combined to produce an optimized core that matches both the fresh-core reactivity and maximum burnup of the original reactor. For the IR-40 example, the optimized design produces weapon-grade plutonium at only about 19% of the rate of the unmodified reactor for the same power level. Additionally, a reactor so converted could not be readily converted back to natural-uranium fuel without replacement heavy water, and it would retain the ability to produce medical isotopes at rates that exceed the original design through the use of LEU targets.

Keywords: Iran, Arak, nonproliferation, plutonium, LEU

1. Policy Context

Natural-uranium fueled, heavy-water moderated reactors produce plutonium quickly and have been used to support nuclear-weapon programs throughout history. Converting such a reactor to use low-enriched uranium (LEU) fuel reduces the rate of plutonium production. This happens because enriched fuels have a lower concentration of U-238, the isotope that is transmuted into plutonium by the reactor's neutrons. In February 2014, Iran's IR-40 heavy-water research reactor at Arak became the focus of nonproliferation concern. Iran initially rejected a proposal for converting the reactor to LEU on the basis that the design was committed and conversion to a new design was no longer possible. This article, initially a white paper produced in support of negotiations with Iran, demonstrated that it would be possible to convert the reactor even if the reactor could not be physically reconfigured. Iran subsequently agreed to conversion and has produced its own modified design.

*Address correspondence to: R.S. Kemp, 77 Massachusetts Ave., Cambridge, MA 02139, USA. Email: rsk@mit.edu

This study follows ground-breaking work by Thomas Mo Willig, Cecilia Futsaether, and Halvor Kippe, who performed the first study on converting the IR-40 reactor.[1] The authors suggested increasing the enrichment and power density of the core, but maintaining identical reactor power at 40 MWt, by reducing the number of active fuel tubes from 150 to 60. This modification increases the power density in the fuel region by a factor of 2.5. To maintain an unchanged temperature profile in the fuel, it would require the mass-flow rate of the coolant to increase. If the fuel geometry remained unchanged, the increased cooling requirement would require increasing the flow velocity by the same factor of 2.5. The pressure drop, under normal turbulent flow, is proportional to the velocity squared. The result is that the cooling system for the "Willig" modification would require cooling pumps with more than 6 times the pumping power planned for the original IR-40 design. It might also require larger pipework, heavier control systems, different pressurizes, and different safety strategies.

An alternative approach would be to use the Willig core design, or a similar such design, but reduce the power to no more than 16 MWt so as not to increase the demand on the cooling system. In many ways, this approach might be regarded as the best conversion strategy, as it avoids major changes of infrastructure but provides benefits in the way of increased neutron flux in target regions. However, it still requires modifications to the infrastructure in the core—specifically, the plugging of unused fuel tubes and possibly the introduction of new channels. A study of options along these lines was undertaken in parallel to this work.[2] Such a strategy is most viable if implemented before the reactor became critical; once the reactor operates, the core infrastructure becomes radioactive and modifications become difficult. This risks creating a rationale for refusing to convert and an incentive to bring the reactor into operation early in order to quiet political demands for conversion. This study showed that a strategy of avoidance would be futile, as the methods proposed herein could be implemented at any time without any changes to the physical infrastructure.

2. Constraints on the Conversion of the IR-40 Reactor

The conversion of the IR-40 reactor is subject to both political and technical constraints. Politically, conversion is more likely to be accepted by Iran if it requires few design changes and maximally uses existing investments. Additionally, the international community has made clear that it prefers Iran to keep its domestic enrichment levels to less than 5% U-235. For this reason, this study will assume that the LEU fuel to be used in a modified IR-40 reactor will be enriched to 5 percent. It is worth noting that this might not be the optimal level. There is, in principle, a balance to be struck between the proliferation potential of the plutonium and uranium aspects of Iran's program, and the optimal enrichment level may be something else. Given the difficulty of trying to optimize multiple technical constraints through diplomatic talks, however, 5% is a reasonable working presumption.

The primary technical constraint is that the conversion process must result in a safely operating reactor. The use of LEU will increase reactivity, and it must be safely controlled. This study assumes that the fresh core reactivity should equal that of the unmodified core. A second consideration—albeit more of an economic factor—is that the core life should not be significantly compromised. Again, conversion to LEU is deemed possible if the modified design can provide the same core life as the original design. It is this principle of equivalency that establishes feasibility, not the quantitative parameters of any particular design.

A further constraint is that the reactor must be capable of fulfilling whatever scientific requirements it has been designed to execute. For example, the reactor may need to produce medical isotopes at a certain rate, and that will necessitate some minimum neutron flux in the channels in which isotope-producing targets are to be irradiated. One of the consequences of converting from natural to low-enriched uranium fuel is, for the same geometric buckling, the neutron flux will go down at any given power density. However, this can be compensated for by raising the enrichment of the isotope-producing targets. If the original reactor was designed for natural-uranium targets, then the use of LEU targets can compensate for the lower neutron flux, leaving medical-isotope production rates unchanged or improved.[3,*]

Finally, the reactor is subject to thermal-hydraulic constraints. These include the performance of the primary cooling loop during full-power operation as well as the safety systems designed to

remove decay heat in the event the cooling system fails. Changes to the power distribution within the core can have a significant impact on these systems. This study considers only conversion options that preserve or reduce the cooling requirements of the reactor.

3. Methods of Conversion

The purpose of conversion is to reduce plutonium production, which requires replacing natural uranium with LEU. The higher fraction of fissile U-235 in LEU must be compensated for to keep reactivity within manageable bounds. There are two basic methods for doing this that leave the core geometry unchanged.

1. *Reduce the density of fissile material* in the core to approximately what would have been present if the reactor were fueled with natural uranium. Such a reduction can be effected by using a dispersion fuel, in which uranium dioxide is dispersed within a neutronically inactive filler, such as pure aluminum. Dispersion fuels are commonplace in research reactors, and Iran's Tehran Research Reactor already uses dispersion fuel. The method of compensating for changes in enrichment by changing the fissile-material density is also common, and it has been the mainstay of the RERTR (Reduced Enrichment for Research and Test Reactors) program, which has converted over 40 reactors to use LEU fuel worldwide.**
2. *Absorb neutrons to control excess reactivity* arising from the increased enrichment. Because LEU introduces extra fissile material relative to natural uranium, the extra reactivity could instead be absorbed by neutron poisons, bringing the fresh-core reactivity in line with the maximum allowed value. This is routinely done by adding burnable poisons such as boron or gadolinium. In the case of the IR-40 reactor, it is interesting to consider an unusual but effective non-burnable poison: a small amount of chemically identical (and therefore perfectly compatible) light water added to the heavy water used as coolant and moderator. Light water acts as a poison because the protium-hydrogen absorbs neutrons. Unlike burnable poisons, however, light water will not extend the core life—generally considered to be a negative. However, light water has the advantage of making it difficult to convert the reactor back to natural-uranium fuel without first performing the time-consuming and highly visible step of flushing the moderator and replacing it with a fresh batch of high-purity heavy water. Furthermore, if heavy-water production is terminated, converting the reactor back may be infeasible. These factors could be significant nonproliferation advantages.

Both of the above methods will enable the reactor to use LEU fuel. Both can be implemented at any time during the reactor's life, without any changes to hardware, and for very low cost. It remains for this study to show that the two conversion methods

---

*Iran had no recourse to enriched uranium when the reactor was originally designed an previously published a study on using natural-uranium targets for medical isotope production.

**Normally, the RERTR program is reducing enrichment from high-enriched uranium or HEU (defined as uranium containing more than 20% ^{235}U) to low-enriched uranium or LEU (less than 20% ^{235}U) by increasing the density; in the case of IR-40 reactor, the conversion goes the other way.

will produce reasonable technical outcomes, specifically (1) that the reactivity can be brought within manageable bounds through practicable aluminum-dispersion or light-water dilution levels; (2) that the impact of conversion on core lifetime can be made comparable to the core life of the original reactor; and (3) that the result of conversion will reduce plutonium production rates under a proliferation scenario.

4. Design of the IR-40 Reactor

This analysis seeks to understand the impact of the proposed conversion strategies in general terms. Nonetheless, parts of the analysis will require certain design-specific assumptions to be made, specifically with regard to the fuel design and the size of the core. For this purpose, the parameters or Iran's IR-40 reactor are used.

According to data compiled by Willig and an Iranian reactor-safety analysis,[4] the IR-40 is a tank-type reactor (see Fig. 1) with pressure tubes, using heavy-water coolant and moderator, and 150 RBMK-1000-style 19-pin fuel bundles in a hexagonal lattice with a bundle-to-bundle pitch of 26.5 cm.[***] The reactor's design parameters are given in Table 1. In addition to these provided data, this analysis assumes that the pressure tubes and major structural elements of the RBMK fuel bundles are made of Zircaloy-2. It is also assumed that the center tube, which forms a key structural element of the fuel bundle, has the same inside and outside dimensions as the fuel-cladding tubes and is filled with coolant.

5. Methods of Analysis

To demonstrate the feasibility and utility of the two conversion concepts, the following analyses are performed.

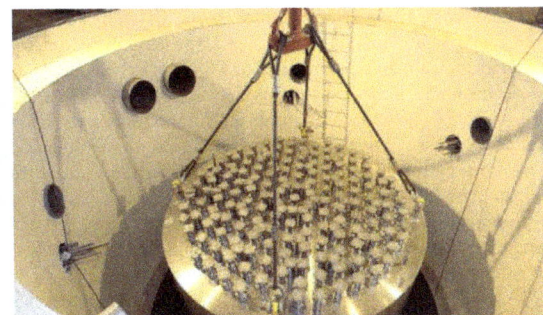

Fig. 1. Photograph of tank of the IR-40 reactor showing fixed pressure tubes and dimensional ratios consistent with data collected by Willig. Photo credit: Yasaman Hashemi, PressTV, Iran.

[***]RBMK or *Reaktor Bolshoy Moshchnosti Kanalnyy* is a class of Russian power reactors.

Table 1. Key parameters of the IR-40 reactor.

Power	40 MWt
Coolant and moderator composition	D_2O (99.75%)
Coolant and moderator inlet pressure	3.5 bar
Coolant and moderator inlet pressure	3.5 bar
Coolant average inlet temperature	50°C
Coolant average exit temperature	70°C
Pressure-tubes with fuel	150
Pressure-tube lattice pitch	26.5 cm
Pressure-tube inner radius	4.0 cm
Pressure-tube wall thickness	0.4 cm
Fuel composition and density	10.4 g (UO_2)/cm^3
Fuel outer radius	0.5740 cm
Clad inner radius	0.5965 cm
Clad outer radius	0.6815 cm
Clad and assembly material	Zircaloy
Fueled rods per bundle	18
Empty (water filled) rods per bundle	1 (center)
Active fuel length	340 cm
Approximate core radius	166 cm
Power density in fuel (calculated)	4.58 W/g(U)
Temperature of fuel (estimated)	100°C

5.1 Excess Reactivity Control in the Fresh Core

The fresh core, without neutron poisons, has the greatest level of reactivity of all possible core configurations for a given fuel type. While it is likely that higher levels of fresh-core reactivity might be desired and safely managed through the use of better control rods or burnable poisons, estimating just how much acceptable reactivity is allowed is a matter subject to debate. Therefore, to establish feasibility in general, the conservative assumption is made that if the reactivity can be made to match the original design, then reactivity cannot be considered an impediment to conversion. A search of the parameter spaces for the two conversion options (adjusting fissile-material density for method 1; and the H_2O:D_2O ratio for method 2) until solutions that gave the same fresh-core k_{eff} as the original IR-40 design was undertaken, and results compared to the original design until a configuration that yielded identical k_{eff} was found (at 0.1 MWd/kg, equivalent to a fresh core with xenon poison equilibrium).

The fuel effects for all configurations, both the converted and original design, were computed using a lattice physics code, which gives k_∞ and the migration length M^2 in an infinite lattice. The k_{eff} for a finite reactor can then be estimated by computing the probability of non-leakage for a finite core with diffusion theory, and the resulting values compared to the value for the original core. Any recourse to diffusion theory necessitates that the medium should be weakly absorbing (scattering dominates the interactions), that the flux should depend only weakly on the angular direction of the neutrons, and that the reactor be reasonably homogenous from the perspective of neutrons. These conditions hold for IR-40-like reactors. Specifically, heavy water is less neutron absorbing than light water; scattering off of deuterium in heavy water is more isotropic than scattering off light hydrogen; and the neutron mean free paths are large compared to the fuel geometry but small enough that leakage at the boundary is small. Although small errors in the computed probability of non-leakage will still result, primarily from geometry considerations, those errors will be nearly the same for both the modified and unmodified core. Since feasibility depends only on showing that the two fresh-core k_{eff} can be made to match, the effect of these errors will tend to cancel. Finally, it is important to bear in mind that both the fuel composition and the finite-core geometry are essentially approximations. Ultimately, greater flexibility is possible with LEU, and reactivity and cycle length can be better managed using fuel shuffling and burnable poisons. The purpose, however, is not to find a precise or optimal solution but to establish feasibility even under stringent limitations.

5.2 Effect of Conversion on Core Life

The core life, or maximum achievable burnup and the intimately related cycle length, are ultimately design considerations driven by external factors such as fuel economics, the reactor's operating profile, and indigenous technology limitations. Nevertheless, claims about core life requirements can be laid to rest if is possible to establish that similar burnup can be obtained even after conversion. The challenge arises because the different materials used for the modified core will alter the rate at which the reactor loses reactivity. The impact of this rate of reactivity loss can be readily observed in the infinite lattice. The true effect, however, requires a finite core analysis. Conveniently, however, the lattice results show that the two conversion strategies have opposite effects on achievable burnup. This implies the existence of a hybrid conversion strategy—a blend of fuel density reduction and moderator dilution—that will provide the same burnup as the original core. As such, feasibility can be established without solving a finite core burnup model, which would normally need to incorporate a larger set of assumptions about external constraints than this general analysis is able to make.

5.3 Effect of Conversion on Plutonium Production

Conversion is interesting only if the converted reactor, in some configuration approximating the performance objectives of the original reactor, produces considerably less plutonium. Plutonium production is a matter of per-fuel bundle exposures. This analysis makes the assumption that all fuel bundles can be burned to the same extent through reshuffling, which is reasonable for a tank-type research reactor. It then becomes possible to study the plutonium vector as a function of bundle burnup by using a lattice physics code. Both the plutonium production rate and the isotopics of the plutonium are relevant considerations and jointly evaluated.

6. Computation of k_∞, M^2, and k_{eff}

To compute k_∞ and M^2, it is necessary to model accurately the geometry of the lattice described in Table 1, and to take into account effects such as parasitic neutron capture in fission products and resonance self-shielding—effects that go beyond what is possible with diffusion theory. We do this with CASMO-4E, a lattice physics code that solves the transport equation using the method of characteristics technique. The code was developed for power-reactor designers by Studsvik Scandpower, a division

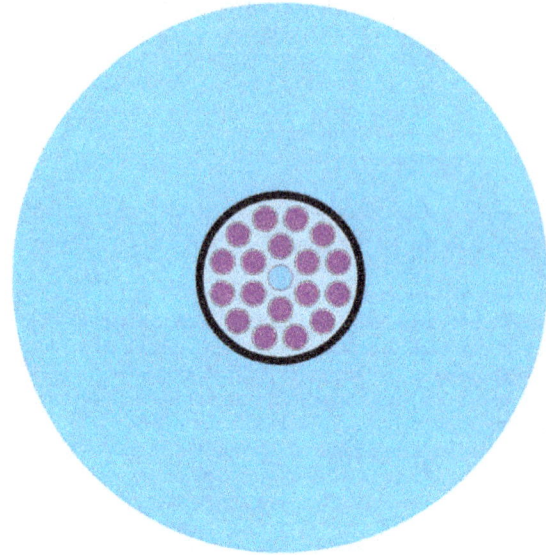

Fig. 2. Tracks for a single fuel bundle and associated moderator used for the characteristics ray-tracing solution produced by CASMO-4E. Colors relate to different materials. The 18-pin fuel bundle (purple), cladding (gray), and pressure tube (black) are clearly visible at the center of coolant and moderator (blue). The track density provide an indication of how the transport solution sees the geometry. That the tracks appear homogeneously distributed and dense (the white speckles are the residual areas not covered by tracks) indicates that the geometry well sampled.

of Studsvik AB, Nykoping, Sweden. CASMO version 4E is designed to model hexagonal lattices like that of the IR-40 by using a "white" boundary condition in which neutrons crossing the boundary are assumed to be isotropically instead of spectrally reflected.[5] The model used here is therefore a single bundle with associated coolant and moderator, as shown in Figure 2, surrounded by a white boundary, which is why the model geometry is circular rather than hexagonal. CASMO breaks the geometry into a fine mesh, shown in Figure 3, and solves the neutron-transport equations across the mesh for 70 energy groups using ENDFB-VI cross-section libraries under the assumption that the neutron source is spatially flat and isotropic inside each mesh element. CASMO also computes burn-up isotopics, tallies the corresponding plutonium production and isotope vector, and can model extensive core physics such as a Doppler broadening and thermal expansion in materials.

The result of this model is that the IR-40 infinite lattice has a $k_\infty = 1.128$ and $M^2 = 409\,cm^2$ at the fresh-core condition after equilibrium xenon poisoning (determined at 0.1 MWd/kg burnup). Detailed results at different burners are shown in Table 2.

6.1 Calculation of k_{eff} by Diffusion Theory

We use diffusion theory to convert k_∞ of the infinite lattice to k_{eff} for a finite core. Diffusion theory can be thought as the

Fig. 3. The same tracks shown in Figure 2, but with the color of each track changing at the boundary between computational mesh cells. Within each mesh cell, the neutron sources are assumed to be spatially flat and isotropic.

Table 2. CASMO output for the unmodified IR-40 infinite lattice.

Burnup MWd/kg	k_∞	M^2 cm^2	U-235 w%	Fissile Pu w%	Total Pu w%
0.00	1.144	404	0.720	0.000	0.000
0.10	1.113	397	0.708	0.007	0.007
0.50	1.113	340	0.664	0.038	0.039
1.00	1.112	391	0.612	0.072	0.076
2.00	1.101	387	0.521	0.126	0.139
3.00	1.084	384	0.443	0.167	0.192
4.00	1.065	382	0.375	0.199	0.237
5.00	1.044	382	0.316	0.223	0.278
6.00	1.022	381	0.266	0.242	0.313
7.00	1.001	381	0.222	0.257	0.345
8.00	0.980	381	0.185	0.268	0.374
9.00	0.961	380	0.153	0.277	0.401
10.0	0.943	380	0.126	0.285	0.425

steady-state balance between production, consumption, and leakage of neutrons for a fixed-size reactor of a homogenous material. For a given energy group, the continuity equation for an arbitrary volume is:

$$\partial N(r,t)/\partial t = -\nabla \cdot \mathbf{J}(r,t) + S(r,t) - s(r,t) \qquad (1)$$

where N is the neutron density in an infinitesimal volume, $\nabla \cdot \mathbf{J}$ the net neutron current across the surface enclosing that volume, S is the sum of all sources internal to that volume, and s is the sum of all internal sinks in that volume. Fick's law gives the relationship between the net neutron current (vector flux) $\mathbf{J}$ and the scalar flux-density ϕ:

$$\mathbf{J}(r,t) = -D(r)\nabla\phi(r,t) \qquad (2)$$

Where D is the diffusion coefficient. In the steady-state condition, $\partial N/\partial t = 0$, so equation 1 can be rewritten as simply the balance of sources, sinks, and leakage:

$$0 = \nabla \cdot [D(r)\nabla\phi(r)] + S(r) - s(r) \qquad (3)$$

Where the sources are S, the sinks s, arise from nuclear reactions in the volume in proportion to the flux density ϕ and the macroscopic interaction cross-sections Σ, which is typically weighted over a group of energies represented by ϕ and all the materials within the reactor. Writing the macroscopic absorption cross-section as Σ_a and the macroscopic fission cross-section as Σ_f, and assuming that neutrons are produced only by fission, gives

$$S(r) = \frac{1}{k_{\mathrm{eff}}}\nu\Sigma_f(r)\phi(r) \qquad (4)$$

$$s(r) = \Sigma_a(r)\phi(r) \qquad (5)$$

where ν is average number of neutrons (prompt and delayed) produced by a fission event, and k_{eff} is the effective neutron multiplication between generations. Substituting S and s into equation 3, assuming core materials are spatially uniform, making use of the geometrical buckling $B^2 = [-\nabla^2\phi(r)]/\phi(r)$, canceling $\phi(r)$, and solving for k_{eff} gives

$$k_{\mathrm{eff}} = \frac{\nu\Sigma_f}{DB^2 + \Sigma_a} \qquad (6)$$

Dividing through by Σ_a lets k_{eff} be written in terms of $k_\infty = \nu\Sigma_f/\Sigma_a$ and the migration area $M^2 = D/\Sigma_a$, both of which were found numerically using CASMO in section 6.

$$k_{\mathrm{eff}} = \frac{k_\infty}{1 + M^2B^2} \qquad (7)$$

Equation 7 gives k_{eff} from k_∞ by multiplication with the term $1/(1 + M^2B^2)$, which is called the probability of non-leakage, and it is the same for one or multiple energy groups. The calculation of M^2 differs for different energy groups, but since M^2 is obtained from CASMO's numerical simulation, the approximation here is primarily that of the geometric buckling, B^2, which remains unchanged across all cases.****

****The geometrical buckling B^2 for a right cylinder radius R and height H is:

$$B_{\mathrm{cyl}}^2 = \left(\frac{J_{0,1}}{R}\right)^2 + \left(\frac{\pi}{H}\right)^2$$

where $J_{0,1} = 2.4048$ is the first root of the zeroth Bessel function of the first kind. To the radius and height are usually added a small extrapolation length that helps the flux profile better fit the real distribution inside the reactor.

7. Results

7.1 Method 1: Dispersion of Fissile Material in an Aluminum Matrix

The process described in section 6 was iterated until the aluminum-LEUO$_2$ ratio was found that produced the same k_{eff} as the unmodified natural-uranium core. The solution occurs at 0.526 g/cm^3 uranium (heavy metal at 5% U-235) dispersed in 2.31 g/cm^3 aluminum.

The lattice burnup at this mix ratio is given in Table 3. Compared to the results for the unmodified lattice in shown in Table 2, the data show the core lifetime would be improved as k_∞ remains higher for any given burnup.

7.2 Method 2: Dilution of Heavy-water Moderator with Light Water

The process described in section 6 was iterated until the amount of light water that, when added to heavy water, produces the same k_{eff} as the unmodified natural-uranium-reactor was found. The solution occurs at 81.5 atom-percent heavy water.[†]

The lattice burnup at this dilution ratio is given in Table 4. Compared to the results for the unmodified lattice in shown in Table 2, that data show that cycle length would be compromised as k_∞ remains drops more rapidly for a given burnup.

7.3 Optimization: A Combined Dispersion-Dilution Conversion Strategy

The above results demonstrate that it is technically feasible to convert the IR-40 reactor to LEU fuel by either of the two methods proposed. It is notable that the use of an LEU dispersion fuel extends the core life relative to the unmodified reactor (for the same initial k_{eff}), whereas the light-water dilution method severely shortens core life but offers additional nonproliferation advantages of making conversion more difficult to reverse. The two methods can be combined to achieve the same initial reactivity and same core life as the original design, while producing less plutonium and achieving the extra nonproliferation advantages afforded by the use of light water. For the infinite lattice, this occurs at approximately 95.8 atom-percent heavy water and 1.01 g/cm^3 uranium (heavy metal at 5% U-235) dispersed in 2.18 g/cm^3 aluminum. Detailed results are given in Table 5.

8. Non-Proliferation Consequences of Conversion

Using the plutonium tallies in the CASMO outputs given in Tables 2–5, one can estimate the plutonium production rates and plutonium-isotope vectors of the various core designs. Two immediate effects are observable: the plutonium production rates

---

[†] The addition of approximately 20% light water will reduce coolant viscosity and density by less than 2%, a change that is likely too small to have a detrimental effect on any cooling system or safety profile and leaves the thermal hydraulics of the reactor effectively unchanged.

Table 3. CASMO output for the 5% enriched aluminum-dispersion fuel infinite lattice.

Burnup MWd/kg	k_∞	M^2 cm^2	U-235 w%	Fissile Pu w%	Total Pu w%
0.00	1.246	706	4.94	0.000	0.000
0.10	1.210	693	4.93	0.000	0.000
0.50	1.200	693	4.88	0.004	0.004
1.00	1.193	694	4.82	0.009	0.010
2.00	1.181	699	4.70	0.022	0.022
3.00	1.170	704	4.58	0.034	0.035
4.00	1.158	709	4.46	0.046	0.047
5.00	1.147	715	4.34	0.057	0.059
6.00	1.135	720	4.22	0.068	0.071
7.00	1.124	726	4.10	0.079	0.083
8.00	1.111	732	3.98	0.090	0.095
9.00	1.099	738	3.86	0.100	0.106
10.0	1.086	745	3.75	0.109	0.118

Table 4. CASMO output for the light-water diluted IR-40 infinite lattice.

Burnup MWd/kg	k_∞	M^2 cm^2	U-235 w%	Fissile Pu w%	Total Pu w%
0.00	1.047	121	5.00	0.000	0.000
0.10	1.033	121	4.99	0.002	0.002
0.50	1.027	121	4.94	0.010	0.010
1.00	1.022	121	4.88	0.019	0.019
2.00	1.016	121	4.76	0.038	0.039
3.00	1.010	121	4.65	0.056	0.057
4.00	1.004	121	4.53	0.073	0.075
5.00	0.998	121	4.41	0.089	0.093
6.00	0.992	121	4.30	0.105	0.110
7.00	0.985	121	4.19	0.120	0.127
8.00	0.979	121	4.07	0.134	0.144
9.00	0.972	121	3.96	0.148	0.160
10.0	0.966	121	3.85	0.161	0.175

Table 5. CASMO output for reactivity and core-life matched IR-40 infinite lattice with extra proliferation resistance.

Burnup MWd/kg	k_∞	M^2 cm^2	U-235 w%	Fissile Pu w%	Total Pu w%
0.00	1.114	367	4.94	0.000	0.000
0.10	1.086	363	4.92	0.000	0.000
0.50	1.078	363	4.87	0.005	0.005
1.00	1.071	363	4.81	0.012	0.012
2.00	1.061	365	4.69	0.025	0.025
3.00	1.051	366	4.57	0.038	0.039
4.00	1.041	368	4.45	0.050	0.052
5.00	1.030	370	4.33	0.062	0.064
6.00	1.020	372	4.21	0.074	0.077
7.00	1.009	374	4.10	0.085	0.089
8.00	0.998	376	3.98	0.095	0.101
9.00	0.987	378	3.86	0.106	0.113
10.0	0.975	380	3.75	0.116	0.125

Table 6. Weapon-grade plutonium production in the IR-40 infinite lattice.

Configuration	Burnup Target for 7% Pu-240 [MWd/kg]	Total Pu at Target Burnup [w%]	WgPu Rate at 40MWt [gPu/day]	Relative Production Rate
Unmodified	1.55	0.103	26.6	1
Optimized	9.86	0.124	5.04	0.19
Moderator Dilution Only	8.64	0.155	7.20	0.27
Dispersion Fuel Only	10.3	0.121	4.72	0.18

in the LEU cores are much lower, but the isotopic vectors in the LEU reactors retain a more weapon-favorable fraction of fissile isotopes at higher burnups. This means there is a tension between overall plutonium production rates and the weapon usability of the plutonium.[††]

Weapon-grade plutonium is defined as plutonium having less than 7 weight-percent Pu-240.[‡] Plutonium production rates and isotopic vector trends were fitted to CASMO data to deduce the effective rate at which each configuration would produce weapon-grade plutonium. The results show the reduced plutonium-production rate arising from the use of LEU fuel substantially outweighs the negative effect of the improved isotope vector. The maximum rates at which weapon-grade plutonium can be produced by each configuration are shown in Table 6.

Table 6 shows the optimized and dispersion-fuel cores have similar properties in terms of weapon-grade plutonium production potential; however, the combined dispersion-dilution core, being moderated by 95.8 weight-percent heavy water, will not sustain a chain reaction if refueled with natural-uranium fuel ($k_{eff} = 0.95$ for the infinite lattice without xenon). This small amount of light water in the moderator prevents Iran from converting the reactor back to a more proliferation-capable reactor without first producing a fresh batch of moderator and coolant. Because heavy-water production is visible, slow, and easily monitored, this would not go unnoticed.[‡‡] Better still, Iran's existing heavy-water production facility could be dismantled as a confidence-building measure with no negative impact on Iran's nuclear program (a small reserve of 95.8 weight-percent heavy water could be kept on hand for topping up the IR-40 reactor

as needed). The combined dilution-dispersion strategy provides a meaningful additional barrier to proliferation.

Finally, it is important to note that although plutonium production has been slowed by roughly a factor of five, eventually the accumulated spent fuel will contain a weapon-quantity of plutonium that could be used to make a weapon quickly. It will, therefore, be necessary to export spent fuel on an occasional basis. The frequency of export depends on the power and duty cycle at which the reactor is operated, but the modifications suggested here will significantly ease the export requirement. For example, if Iran used its 20% enriched uranium as targets for medical isotope production, the reactor could be operated at a continuous average power below 10 MWt and still produce medical isotopes at a rate equal to or above that of the original IR-40 design with natural-uranium targets. Under this arrangement, the reactor would produce plutonium at less than 5% the rate of the original design, and fuel would need to be exported no more often than once every 4 years (assuming a maximum allowed plutonium inventory in spent fuel of 4 kg). In practice, with additional isotope production ongoing at Iran's TRR reactor, the IR-40 reactor might operate at a lower duty cycle, and plutonium production and exports would be proportionally lower and less frequent.

9. Conclusions

This article demonstrates the viability of converting the IR-40 reactor at Arak to use low-enriched uranium by two methods that can be effected after the reactor becomes critical and is therefore not sensitive to the timing of the negotiations. By combining the two methods, it is possible to produce a 5% enriched LEU core with the same fresh-core reactivity and core life as the original, unmodified reactor, but which produces weapon-grade plutonium at only 19% the rate of the original reactor design. This optimized core has the additional advantage of preventing Iran from easily converting the reactor back to natural-uranium fuel.

The foregoing analysis was constrained by the assumption that the LEU fuel had to be enriched to 5% U-235. Higher enrichments, if tolerable within the context of the enrichment program, would result in even less plutonium. Conversely, lower enrichments could be used to reduce the threat from the enrichment program at the cost of somewhat greater plutonium production rates. The optimal configuration depends on the size of the enrichment program and policies regarding the accumulated LEU it produces.

This analysis did not consider other reactor of fuel-cycle optimization strategies. For example, core life could be extended by the use of burnable poisons and greater initial fissile loading. Neutron flux across the reactor could be flattened or tailored using different enrichments or U:Al dispersion ratios. To model such effects would require from Iran a clear definition of the reactor's purpose and performance requirements, as well as a finite-reflected core model. These analyses are best performed jointly by Iran and its negotiating partners and beyond the scope of this analysis. Nevertheless, this paper demonstrates the feasibility and tremendous flexibility that exists in converting Iran's existing IR-40 reactor to a more peaceful configuration.

[††]Modern weapon designs are not neutronically sensitive to plutonium isotopics, although heat from the decay of the higher plutonium isotopes may present a more practical, but not insurmountable, design challenge. The sophistication of a posited Iranian weapon design is not known to the author, but some evidence in the public domain suggests Iran may possess design knowledge that exceeds historical first-generation weapons.[6,7]

[‡]This analysis assumes CASMO's non-fissile plutonium tally (which is the sum of all even numbered isotopes) is effectively equivalent to a Pu-240 tally. The two will be very nearly equal at the low burnups considered here.

[‡‡]The Heavy Water Plant in Iran is rated at 16 tons/year. The Arak reactor needs about 100 tons of heavy water.

Acknowledgment

The author thanks Kord Smith for assistance in modeling the IR-40 reactor with CASMO.

References

1. Willig, T. M.; Futsaether, C.; Kippe, H. Converting the Iranian Heavy Water Reactor IR-40 to a More Proliferation-Resistant Reactor. *Sci. Global Secur.* **2012**, *20*, 97–116.

2. Ahmad, A.; von Hippel, F.; Glaser, A.; Mian, Z. A Win–Win Solution for Iran's Arak Reactor? *Arms Control Today* **2014**, April.

3. Sayareh, R.; Gannadi Maragheh, M.; Shamsaiea, M. Theoretical Calculations for the Production of 99Mo Using Natural Uranium in Iran. *Annals Nuclear Energ.* **2003**, *30*, 883–895.

4. Faghihi, F.; Ramezani, E.; Yousefpour, F.; Mirvakili, S. M. Level-1 Probability Safety Assessment of the Iranian Heavy Water Reactor Using SAPPHIRE Software. *Reliability Eng. Sys. Safety* **2008**, *93*, 1377–1409.

5. Knott, D.; Edenius, M.; Peltonen, J.; Anttila, M. *Results of Modeling Hexagonal and Circular Cluster Fuel Assembly Design Using CASMO-4*. American Nuclear Society Topical Meeting—Advances in Nuclear Fuel Management II, EPRI TR-107728-VI, 1997.

6. U.S. Department of Energy, *Nonproliferation and Arms Control Assessment of Weapons-Usable Fissile Material Storage and Excess Plutonium Disposition Alternatives;* DOE/NN-0007, January 1997, Box 3-1, pp. 37–39.

7. Harrington, M. Evidence Emerges of Iran's Continued Nuclear Weapons Research. *Jane's International Defence Review*, March 13, 2008.

8. International Atomic Energy Agency, *Implementation of the NPT Safeguards Agreement and Relevant Provisions of Security Council Resolutions in the Islamic Republic of Iran*, 28 August 2013 (GOV/2013/27), footnote 38.

14

Techno-Economic Study of Hydrogen Production via Steam Reforming of Methanol, Ethanol, and Diesel

SEYYED MOHSEN MOUSAVI EHTESHAMI* and SIEW HWA CHAN

Nanyang Technological University, Singapore, Singapore

Abstract: A large portion of industrial hydrogen is generated from the steam reforming (SR) of hydrocarbons.[1–7] A rational choice of fuel for hydrogen production from hydrocarbons is controversial due to the disadvantages of the fuels, including the cost, infrastructure development, and energy efficiency of the corresponding reforming process.[8–10] The optimum selection should be made considering all the above factors. A techno-economic analysis of the steam reforming of strategic fuels, including methanol, ethanol, and diesel, is carried out. The produced gas molecules, equilibrium composition of the products, appropriate operating conditions, and energy efficiency of the system operating on corresponding fuels are studied applying the minimization of Gibbs free-energy technique. It is concluded that steam reforming of methanol yields the most facile conversion. The appropriate steam reforming operating temperature for the studied fuels vary from low to high temperatures, being 180–220°C for methanol and up to 650–700°C for other fuels. Furthermore, the economical evaluations of the steam reforming process of the mentioned fuels provide a guideline for a rational selection of fuels for an integrated proton-exchange membrane fuel cell and reformer system.

Keywords: Steam reforming, methanol, ethanol, diesel

1. Introduction

A large portion of industrial hydrogen is generated from the steam reforming (SR) of hydrocarbons.[1–7] A rational choice of fuel for hydrogen production from hydrocarbons is controversial due to the disadvantages of the fuels, including the cost, infrastructure development, and energy efficiency of the corresponding reforming process.[8–10] Production of hydrogen, as an ideal energy carrier, through an economically and environmentally sustainable path is an essential goal in the development of alternative energy sources for future supply. This ideality has produced an intensified interest in the exploration of hydrogen generation from various sources including hydrocarbons, biomass, wind, solar, and nuclear energy.[11–18] The current status of the hydrogen production, storage, and distribution is not acceptable technically and economically.[19,20] To develop a hydrogen supply system fulfilling the industry requirements, considering the size and cost of the options, a wide range of fuels and processing technologies, specifically steam reforming, have been investigated extensively.[2,5,6,21–23]

*Address correspondence to: Seyyed Mohsen Mousavi Ehteshami, Nanyang Technological University, Singapore, Singapore. Email: m080054@e.ntu.edu.sg

In this study, a techno-economic analysis is carried out to evaluate the energy efficiency and economical viability of methanol, ethanol, and diesel steam reforming. The objective is to provide a guideline for a rational selection of fuels for an integrated proton-exchange membrane fuel cell and reformer system.

Natural gas, as the essential source of industrial hydrogen generation via steam reforming, includes mainly methane (Figure 1).[24] However, steam reforming of methane requires a high operating temperature for complete conversion.[3,23,25,26] Methanol as a liquid with high energy density has been well studied because of its well-known facile reforming and availability.[27,28] Moreover, the infrastructure required for production and distribution of methanol is not available.[16,29] Therefore, other high-energy-density liquid fuels that are abundant and obtained more easily have been considered, such as ethanol, glycerol, and dimethyl ether.[2,4,5,30–35] These fuels are environmentally sustainable, ensuring the renewable energy nature of the process. However, this group of fuels is not well developed and economically and technically not viable yet. Oil-derived fuels such as diesel, naphtha, and gasoline constitute another group of hydrocarbons, generally called heavy hydrocarbons.[16,36,37] These fuels are abundantly available and cheap. In addition, having high energy densities, their required infrastructure is commercially well developed. It should be noted that although there might be issues regarding the low energy efficiency and nonrenewable nature of these fuels, they have the advantage of the existing infrastructures, avoiding the economic burden of extra capital cost required for hydrogen production. Based on this discussion, methanol, ethanol, and diesel are considered for this study. Some physicochemical properties of the fuels are

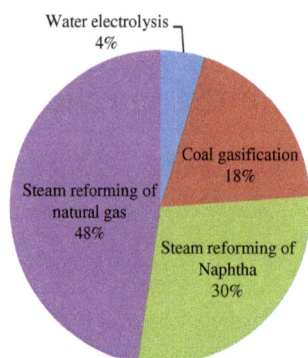

Fig. 1. The main processes for producing industrial hydrogen.

Table 2. Chemical reactions of steam reforming of the fuels.

Fuel	Chemical reaction
Methanol	$CH_3OH + H_2O \rightarrow 3H_2 + CO_2$
Ethanol	$C_2H_5OH + H_2O \rightarrow 4H_2 + 2CO$
Diesel	$C_nH_m + 2nH_2O \rightarrow (m/2 + 2n)H_2 + n\,CO_2$

presented in Table 1. In fact, a rational selection of the fuel is debatable due to the disadvantages of the fuels, including the cost, infrastructure development, and the energy efficiency of the corresponding reforming process. The optimum selection should be made considering all the effective factors.

The economic study carried out includes the evaluation of the life cycle cost of each fuel considering several key parameters. The goal of the economic study is to determine the feasibility of utilizing the studied fuels. The cost of raw materials and the cost related to the greenhouse gases emission are taken into account. The capital costs of hydrogen generation equipment and hydrogen pumping stations are excluded from the comparison as they will have a similar share for different fuel options. The end-user instrument also plays a significant role in the economical viability of the hydrogen economy. The cost of the fuel cell–driven power train in transportation applications is an essential parameter for a convincing hydrogen energy justification. It is necessary for governmental institutions to facilitate the application of fuel cell–powered systems by subsidizing the price of these systems. For example, the price of the fuel cell power must be below US $30/kW by 2015 so that utilizing hydrogen as the alternative fuel becomes justifiable.[38] Taking into account the effect of greenhouse gases, hydrogen and fuel cell–driven transportation would become more defendable.

2. Methodology

The thermodynamic simulations are carried out under chemical equilibrium conditions using the commercial package Aspen Hysys, which facilitates the thermodynamic analysis of chemical processes. Peng-Robinson equation of state is used to calculate

the physicochemical properties of the fuels.[39] The chemical reactions of the fuels reforming are summarized in Table 2. Diesel is considered to consist of an equi-volumetric amount of hydrocarbons including 14 and 16 carbon atoms. The fuels under investigation are considered to be fed into the system at room temperature and atmospheric pressure with a constant flow rate. The feedstock is premixed with water, and the aqueous solution with different water-to-fuel molar ratios is pumped into the fuel processor. The configuration of the system is depicted in Figure 2 and explained in the fuel processing section (section 2.1). Aspen Hysys is used to calculate the composition of the products and the energy requirements for each unit of the plant. The computational method and energy efficiency calculation is presented in section 2.2.

2.1. Fuel Processing Unit

Figure 2 presents the configuration of the fuel steam reforming unit integrated with a PEM fuel cell. The fuel processor includes a reformer and a CO cleanup unit. The pre-heater section heats up the feedstock to a higher temperature before entering to the reformer. The reformer is considered an isothermal reactor that is modeled by a RGIBBS module, although in the real reformer, there are non-identical temperature profiles in the axial and radial directions due to the endothermic nature of the reactions. Minimization of Gibbs free energy is the basis of the calculation of products compositions. Heat integration is essential in designing of a reforming system.[40] The reactants usually have to be heated to high temperatures and cooled down for the subsequent process unit. When taking into account the heat recovery/integration, the disadvantage of high reforming temperatures will be greatly reduced and might change the outcome of the investigation. However, since the objective of this report is to provide a comparative evaluation of the fuels concentrating on the steam reforming reaction, it is assumed that the heat requirements are supplied from external heating units, for example, electric heaters powered by the PEM fuel cell integrated with the fuel processing unit. Obviously, the energy efficiency of the system would be lower compared to others utilizing heat integration, such as systems that couple catalytic combustors with the reformer.

Table 1. Physicochemical properties of the fuels.

Fuel	Formula	Density (kg.m^{-3})	Boiling point (°C)	Heat of vaporization (kJ.mol^{-1})	Lower heating value (kJ.mol^{-1})	Energy density (MJ.L^{-1})	Heat capacity (J.mol^{-1}.K^{-1})	Flammability limits (Vol. %)
Hydrogen	H_2	0.090	−252.70	0.92	240	−0.89	28.6	4.1, 74
Methanol	CH_3OH	794.000	64.60	35.20	643	−17.00	49.0	7.3, 36
Ethanol	C_2H_5OH	790.000	78.30	38.90	1240	−22.00	77.3	4.3, 19
Diesel	$C_{16}H_{34}$	0.856	120–430	47.00	8080	−34.20	340.0	1.0, 6

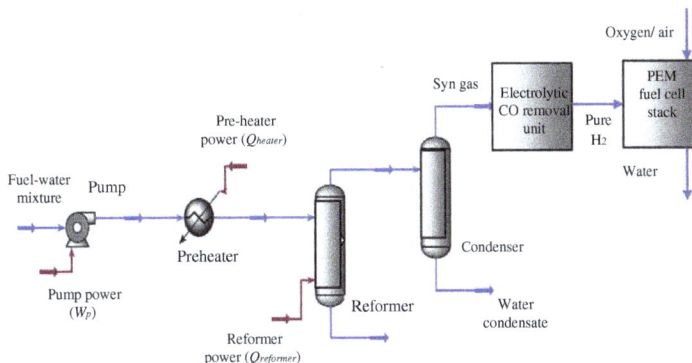

Fig. 2. The system configuration of fuel processor-PEM fuel cell system.

The hydrogen-rich gas produced in the reformer can be purified by several approaches, such as water gas shift reaction, methanation and preferential CO oxidation, pressure swing adsorption, and Pd membranes.[41-45] These processes are greatly dependent on the catalytic performance and operate at high temperatures. In the current study, it is assumed that an electrolytic CO oxidation system investigated in our group is used, which operates at the outlet temperature of the reformer and atmospheric pressure.[46] Such a system generates pure hydrogen by removing CO and other impurities available in reformate produced in the reformer. In addition, it offsets the required energy by production of extra hydrogen. The power requirements of such a unit are negligible. Therefore, the energy analysis of the reforming system is not affected. The PEM fuel cell is considered an adiabatic reactor (modeled by RSTOIC in Aspen Hysys), which operates at 80°C and 1 atm. The efficiency of the fuel cell is assumed to be 65%.

2.2. Computational Method

As mentioned, Gibbs free energy minimization method is used to perform the thermodynamic analysis. The system total Gibbs free energy should reach a minimum at the operating temperature when the system reaches thermodynamic equilibrium. The total Gibbs free energy of the system is calculated using equation (1):

$$nG = \sum_{i=1}^{N} n_i \overline{G}_i \tag{1}$$

Where the molar Gibbs free energy of each component $\overline{G}_i$ is calculated from equation (2):

$$\overline{G}_i = G_i^o + RT ln\frac{\hat{f}_i}{f_i^o} \tag{2}$$

Where f_i is the fugacity of the components. Therefore, the total Gibbs free energy of the system is rewritten to

$$nG = \sum (n_i G_i^o) + RT \sum \left(n_i ln\frac{\hat{f}_i}{f_i^o}\right) \tag{3}$$

Lagrange multiplier is then used to calculate the composition of the components at which the Gibbs free energy of the system is minimized:

$$\sum_{N}^{i=1} n_i [G_{f_i}^o + RT \, lnP + RT \, lny_i + RT \, ln\widehat{\theta}_i + \sum_k (\lambda_i a_{ik})] = 0 \tag{4}$$

According to the experimental and theoretical studies in the literature, hydrogen, carbon monoxide, carbon dioxide, methane, and acetaldehyde are considered to be present in the products.[27,36,47] In the case of methanol, only hydrogen, carbon monoxide, and carbon dioxide are considered. The presence of carbon is excluded as the core comparisons are made at sufficiently high water-to-fuel ratios at which the carbon formation is avoided. In addition, the focus of this study is to provide a broad picture of the reforming process and to present guidelines for choosing the optimum fuel for hydrogen production considering both the technical and economical aspects. However, for a more detailed study, the kinetics of the reactions and catalyst deactivation due to the presence of solid carbon should be taken into account.

2.3. Energy Efficiency Calculation

The global energy efficiency of the system is calculated using equation (5):

$$\eta = \frac{P_s - P_{aux}}{n_f . LHV_f} \tag{5}$$

Where P_s, P_{aux}, n_f, and LHV_f are the power generated by fuel cell stack (kW), power required by all of the auxiliary units (including the pump, heater, CO removal electrolyzer) (kW), the fuel molar flow rate (mol.s^{-1}) and low heating value of the fuels (kJ.mol^{-1}), respectively. The fuel cell stack power is calculated considering the hydrogen flow rate produced from the fuel processor and the fuel cell electrochemical efficiency, which is assumed to be 0.6.

3. Economic Analysis

The goal of the economic study is to determine the feasibility of utilizing the studied fuels for hydrogen generation through the SR process. The cost of raw materials and the cost related to the greenhouse gases emission and fuel efficiency of the reforming systems are taken into account. It is desired to propose the fuel that presents the optimum efficiency of the SR process to generate hydrogen. As the fuel processing price including the capital, operation, and maintenance costs have a similar share for different fuel options, they will have negligible effect on the comparative study. To calculate the greenhouse gases emission (CO and CO_2), the specific values of CO and CO_2 production from steam reforming of the fuels under the study are used.

4. Results and Discussion

A parametric study based on the developed integrated process model is used to determine the optimal operating conditions of the system. The effect of reforming temperature and water-to-fuel ratio are considered for the parametric study. The following procedure is used to perform the thermodynamics analysis: required heat input is applied to the pre-heater to maintain the reforming temperature. Having set the water-to-fuel ratio, the feed flow rate, and the reforming temperature, the thermodynamics model calculates the reformate temperature, composition, and flow rate of the products.

Figure 3 presents the equilibrium conversion efficiency of atmospheric steam reforming of different fuels at a water-to-fuel ratio of 1.5. Conversion efficiency of a fuel determines the percentage of the fuel that converts into products within the reformer. It can be seen that methanol is converted almost fully at temperatures around 200–210°C, while other fuels require higher reforming temperatures to yield high conversion efficiencies. Ethanol also has high conversion efficiencies even at low temperatures (350–450°C), but it should be noted that ethanol conversion products include acetaldehyde and methane besides hydrogen and CO/CO_2. In other words, the high conversion efficiency necessarily does not imply the high hydrogen production rate. To make a meaningful comparison among the fuels from the hydrogen generation perspective, the composition of hydrogen along with other products should be considered. In the case of diesel, high reforming temperatures (up to 650°C) are required to yield high conversion efficiencies.

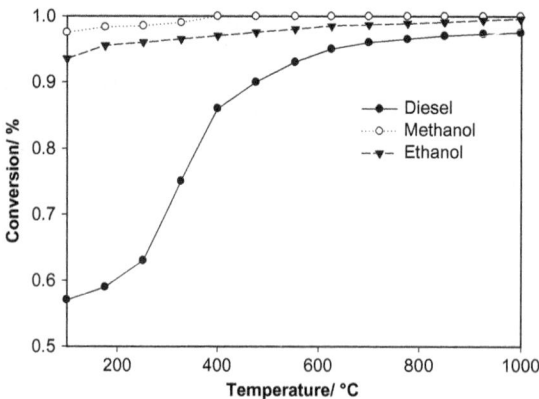

5. Effect of Water to Fuel Ratio (R)

In this section, the effect of the water-to-fuel ratio on the equilibrium composition of the products and net energy efficiency of the system is analyzed. Figure 4 presents the equilibrium composition of the steam reforming of the fuels under study. The compositions of the components are presented in molar

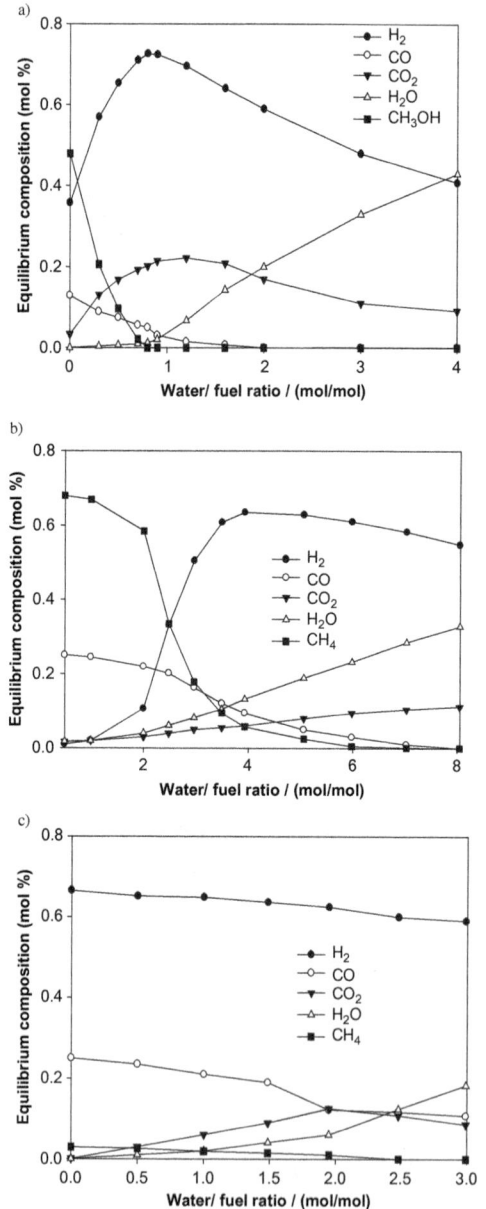

Fig. 3. Equilibrium conversion of atmospheric steam reforming of different fuels at water to fuel ratio of 1.5.

Fig. 4. Equilibrium composition of atmospheric steam reforming of (a) methanol at 250°C, (b) ethanol at 650°C, and (c) diesel at 650°C.

percentage. Maximum values for hydrogen generation molar percentage are observed as R increases to 1, 4, and 2 in the case of methanol, ethanol, and diesel, respectively.

The hydrogen equilibrium composition is observed to increase at low R values and decrease at higher R values in the case of methanol and ethanol. The increase is because of adding more steam to the water-gas shift equilibrium, which leads to generation of more hydrogen. The decrease at higher R values might be related to the lack of fuel to be steam reformed. Another observation is the decrease of CO equilibrium concentration for all the fuels as the R value increases. It is also related to water-gas shift reaction, which extends more at higher R values, leading to the decrease in the CO equilibrium concentration. However, it should be noted that high R value means the higher energy requirement to vaporize the water.

Figure 5 presents the energy efficiency of the system operating on methanol, ethanol, and diesel at 650°C. The energy efficiency of the system operating on different fuels increases in the order: diesel < ethanol< methanol. The maximum energy efficiency of the system happens at R = 1, 4, and 3 in the case of methanol, ethanol, and diesel, respectively. Methanol with an energy efficiency of 35% appears to be the fuel providing the most energy-efficient system. It should be noted that the efficiency values presented in Figure 5 are calculated for the specific system configuration depicted in Figure 2. Higher efficiency values would be achieved if heat integration is implemented.

6. Effect of the Reforming Temperature

In this section, the effect of reforming temperature on the equilibrium concentration of the products and net system efficiency are investigated. Figure 6 presents the equilibrium composition of the steam reforming of the fuels under study. The compositions of the components are presented in molar percentage. A similar incremental trend is observed for the CO equilibrium concentration with the increase in reforming temperature, in the case of all the fuels under the study. This is related to exothermic nature of the water-gas shift reaction, which increases the CO

concentration at higher temperatures. The formation of remarkable amounts of methane reduces the hydrogen generation at reforming temperatures lower than 600°C in the case of ethanol. Similarly, the same trend is observed in the case of diesel steam reforming. This indicates the decomposition of the fuel at lower temperatures. With the increase in reforming temperature, the

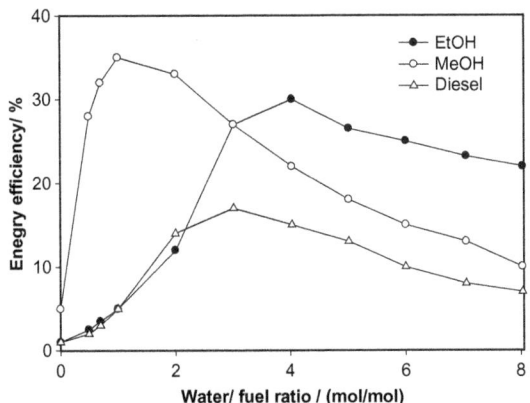

Fig. 5. Energy efficiency of the fuel processor: fuel cell system operating on different fuels at 650°C.

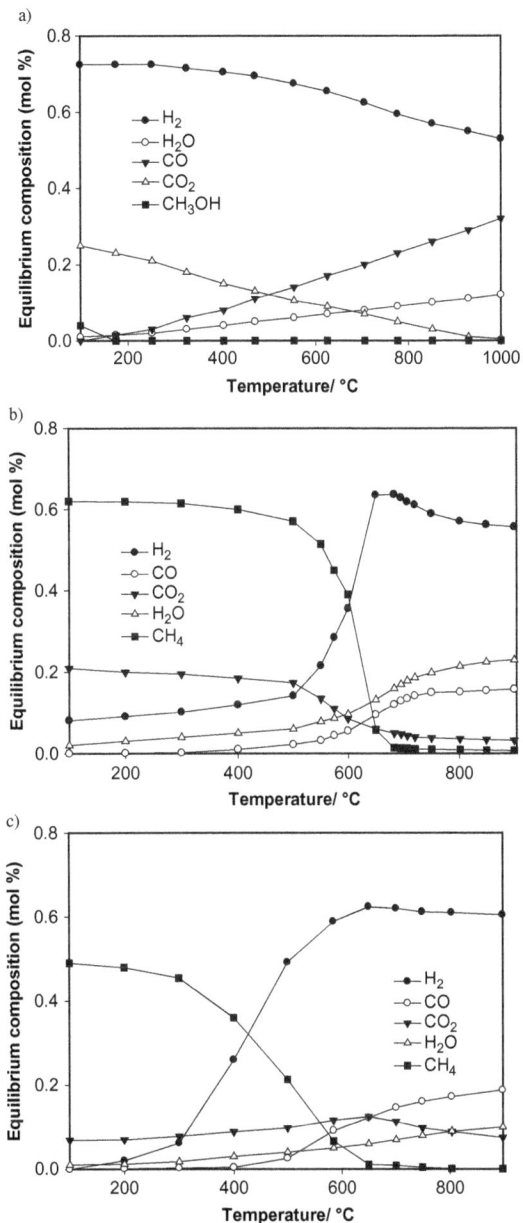

Fig. 6. Equilibrium composition of atmospheric steam reforming of (a) methanol at R = 1, (b) ethanol at R = 4, and (c) diesel at R = 2.

reduction of methane and increase of hydrogen production is observed. Therefore, the temperature range required for ethanol and diesel steam reforming would be higher than 600°C.

Figure 7 presents the energy efficiency of the system operating on methanol, ethanol, and diesel at R = 2.5. The energy efficiency of the system operating on different fuels increases in the order: diesel < ethanol < methanol. According to the model, the maximum energy efficiency of the system happens at reforming temperatures of 200°C, 700°C, and 600°C in the case of methanol, ethanol, and diesel, respectively. Methanol with energy efficiency of 36.5% provides the fuel for the most energy-efficient system.

It is essential to consider the environmental effect of the fuel to select the best option for hydrogen production through the SR process. Therefore, the corresponding specific CO and CO_2 production of the fuels under the study are calculated at the best operating conditions of the steam reforming process. These conditions are $R = 1$ and $T = 250°C$ for methanol, $R = 4$ and $T = 650°C$ for ethanol, and $R = 3$ and $T = 650°C$ for diesel. The results are presented in Figure 8, where the fuels are compared from both the energy content and greenhouse emission point of views. A negligible amount of CO is produced from steam reforming of methanol compared to ethanol and diesel. The CO content of the synthetic gas is a very important factor for selection of a fuel since PEM fuel cells, as the end-users, are much susceptible to CO concentration available in the hydrogen stream.[48] The CO concentration in the synthetic gas increases in the order: methanol< ethanol< diesel. On the other hand, CO_2 concentration follows a different trend, increasing in the order: ethanol< methanol< diesel.

Table 3 presents the cost parameters of steam reforming process of the fuels under the study. The price components of raw material cost and CO_2 and CO emission cost are considered in the analysis. The former components are obtained from the International Energy Agency (IEA) databases, while the latter ones are calculated based on the equilibrium optimum operating conditions as points at which the hydrogen and CO/ CO_2 productions are calculated per kg of fuel. There is not a big difference

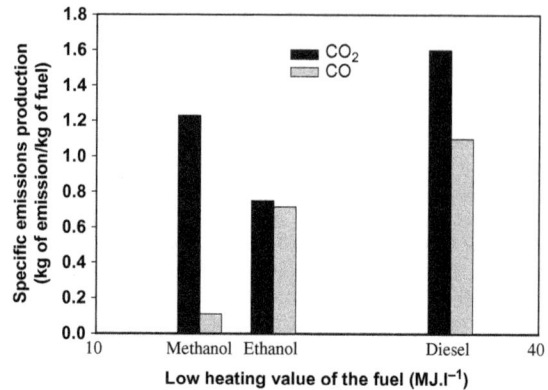

Fig. 8. Specific emission production due to steam reforming of different fuels (R = 1, T = 250°C for methanol, R = 4 and T = 650°C for ethanol, and R = 3 and T = 650°C for diesel).

Table 3. The economic analysis of the steam reforming process of the fuels.

Fuel	Raw material US$/ Gasoline equiv. (l)	Cost parameters	
		CO_2 social cost US$/ kg H_2 produced from SR	CO social cost US$/ kg H_2 produced from SR
Methanol	0.95	0.55	1.5
Ethanol	1.08	0.14	4.7
Diesel	0.95	0.37	4.3

among the raw materials costs. Significantly different values of the social costs of greenhouse gases emission are found in different studies as different approaches have been taken by other researchers to calculate this cost parameter.[49–51] Therefore, this study considers the average social cost as observed in different studies. Based on the current analysis, methanol appears to be a more promising candidate for hydrogen generation through SR. The focus of this study is to compare the key cost parameters of hydrogen generation through SR. Therefore, the raw material and the social costs induced by the hydrogen generation through SR of the corresponding fuels are taken into account. Also, it should be noted that because all the studied fuels are in liquid form, other cost parameters such as the transportation cost, infrastructure cost, and operation/maintenance costs would not be so different. Having said the above, to have a broader view of the cost parameters, the whole life-cycle cost of the hydrogen generation should be considered.

7. Conclusion

A techno-economic analysis of the steam reforming of strategic fuels, including methanol, ethanol, and diesel, is carried out. The produced gas molecules, composition of the products, appropriate operating conditions, and energy efficiency of the system

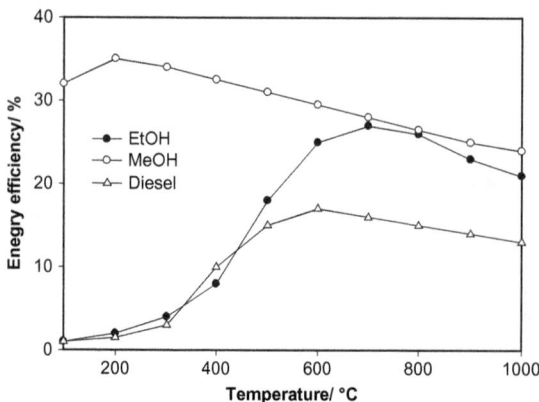

Fig. 7. Energy efficiency of the fuel processor-fuel cell system operating on different fuels at R = 2.5.

operating on corresponding fuels are studied applying the minimization of Gibbs free-energy technique. It is concluded that steam reforming of methanol yields the most facile conversion exhibiting the highest energy efficiency among the studied fuels. Water/fuel ratio and operating temperature affect the hydrogen and carbon oxides yield significantly. The appropriate steam reforming operating temperature for the studied fuels vary from low to high temperatures, being 180–220°C for methanol and up to 650–700°C for other fuels. Furthermore, an economical evaluation of the steam reforming process of the mentioned fuels is performed. Raw material and social cost of greenhouse gases emission are considered in the study. Based on the analysis, methanol appears to be the optimum choice for production of hydrogen via the SR process. The results obtained from the techno-economic study would provide guidelines for a rational selection of the optimum fuel for an integrated proton-exchange membrane fuel cell and reformer system.

References

1. Xu, X.; Li, P.; Shen, Y. Small-Scale Reforming of Diesel and Jet Fuels to Make Hydrogen and Syngas for Fuel Cells: A Review. Appl. Ener. 2013, 108, 202–217.

2. Nahar, G.; V. Dupont. Hydrogen via Steam Reforming of Liquid Biofeedstock. Biofuels 2012, 3(2), 167–191.

3. De Abreu, A.J.; Lucrédio, A.F.; Assaf, E.M. Ni Catalyst on Mixed Support of CeO 2-ZrO 2 and Al 2O 3: Effect of Composition of CeO 2-ZrO 2 Solid Solution on the Methane Steam Reforming Reaction. Fuel Proc. Tech. 2012, 102, 140–145.

4. Snytnikov, P.V. et al. Catalysts for Hydrogen Production in a Multifuel Processor by Methanol, Dimethyl Ether and Bioethanol Steam Reforming for Fuel Cell Applications. Int. J. Hydrogen Energy 2012, 37(21), 16388–16396.

5. El Doukkali, M. et al. Bioethanol/Glycerol Mixture Steam Reforming Over Pt and PtNi Supported on Lanthana or Ceria Doped Alumina Catalysts. Int. J. Hydrogen Energy 2012, 37(10), 8298–8309.

6. Mei, D. et al. A Micro-Reactor With Micro-Pin-Fin Arrays for Hydrogen Production via Methanol Steam Reforming. J. Power Sources 2012, 205, 367–376.

7. Trane, R. et al. Catalytic Steam Reforming of Bio-oil. Int. J. Hydrogen Energy 2012, 37(8), 6447–6472.

8. Dufour, J. et al. Hydrogen Production From Fossil Fuels: Life Cycle Assessment of Technologies With Low Greenhouse Gas Emissions. Energy Fuels 2011, 25(5), 2194–2202.

9. Ogden, J.M. Developing an Infrastructure for Hydrogen Vehicles: A Southern California Case Study. Int. J. Hydrogen Energy 1999, 24(8), 709–730.

10. Ogden, J.M.; Steinbugler, M.M.; Kreutz, T.G. Comparison of Hydrogen, Methanol and Gasoline as Fuels for Fuel Cell Vehicles: Implications for Vehicle Design and Infrastructure Development. J. Power Sources 1999, 79(2), 143–168.

11. Hwang, J.J. Sustainability Study of Hydrogen Pathways for Fuel Cell Vehicle Applications. Renewable Sustainable Energy Rev. 2013, 19, 220–229.

12. Kothari, R. et al. Fermentative Hydrogen Production: An Alternative Clean Energy Source. Renewable Sustainable Energy Rev. 2012, 16(4), 2337–2346.

13. Koumi Ngoh, S.; Njomo, D. An Overview of Hydrogen Gas Production From Solar Energy. Renewable Sustainable Energy Rev. 2012, 16(9), 6782–6792.

14. Alves, M. Hydrogen Energy: Terceira Island Demonstration Facility. Chemical Industry Chem. Eng. Q. 2008, 14(2), 77–95.

15. Marcus, G.H. An International Overview of Nuclear Hydrogen Production Programs. Nuclear Tech. 2009, 166(1), 27–31.

16. Navarro Yerga, R.M. et al. Catalysts for Hydrogen Production From Heavy Hydrocarbons. Chem. Cat. Chem. 2011, 3(3), 440–457.

17. Tributsch, H. Photovoltaic Hydrogen Generation. Int. J. Hydrogen Energy 2008, 33(21), 5911–5930.

18. Navarro, R.M.; Peña, M.A.; Fierro, J.L.G. Hydrogen Production Reactions From Carbon Feedstocks: Fossil Fuels and Biomass. Chemical Rev. 2007, 107(10), 3952–3991.

19. Armaroli, N.; Balzani, V. The Hydrogen Issue. Chem. Sus. Chem. 2011, 4(1), 21–36.

20. Mazloomi, K.; Gomes, C. Hydrogen as an Energy Carrier: Prospects and Challenges. Renewable Sustainable Energy Rev. 2012, 16(5), 3024–3033.

21. Authayanun, S. et al. Theoretical Analysis of a Glycerol Reforming and High-temperature PEMFC Integrated System: Hydrogen Production and System Efficiency. Fuel 2013, 105, 345–352.

22. Chen, C.C. et al. Low-Level CO in Hydrogen-Rich Gas Supplied by a Methanol Processor for PEMFCs. Chem. Eng. Sci. 2011, 66(21), 5095–5106.

23. Cipitì, F. et al. Experimental Investigation on a Methane Fuel Processor for Polymer Electrolyte Fuel Cells. Int. J. Hydrogen Energy 2013, 38(5), 2387–2397.

24. International Energy Agency (IEA). IEA Energy Technology Essentials: Hydrogen Production & Distribution. In IEA report.

25. Özkara-Aydınolu, Ş. Thermodynamic Equilibrium Analysis of Combined Carbon Dioxide Reforming With Steam Reforming of Methane to Synthesis Gas. Int. J. Hydrogen Energy 2010, 35(23), 12821–12828.

26. Aasberg-Petersen, K. et al. Natural Gas to Synthesis Gas—Catalysts and Catalytic Processes. J. Natural Gas Sci. Eng. 2011, 3(2), 423–459.

27. Palo, D.R.; Dagle, R.A.; Holladay, J.D. Methanol Steam Reforming for Hydrogen Production. Chem. Rev. 2007, 107(10), 3992–4021.

28. Sá, S. et al. Catalysts for Methanol Steam Reforming—A Review. Appl. Catal. B Environmental 2010, 99(1–2), 43–57.

29. Singh, M.K. A Comparative Analysis of Alternative Fuel Infrastructure Requirements. SAE Technical Papers, 1989.

30. Badmaev, S.D.; Snytnikov, P.V. Hydrogen Production From Dimethyl Ether and Bioethanol for Fuel Cell Applications. Int. J. Hydrogen Energy 2008, 33(12), 3026–3030.

31. Bshish, A. et al. Steam-Reforming of Ethanol for Hydrogen Production. Chem. Papers 2011, 65(3), 251–266.

32. Haryanto, A. et al. Current Status of Hydrogen Production Techniques by Steam Reforming of Ethanol: A Review. Energy Fuels 2005, 19(5), 2098–2106.

33. Ni, M.; D.Y.C. Leung; M.K.H. Leung. A Review on Reforming Bio-Ethanol for Hydrogen Production. Int. J. Hydrogen Energy 2007, 32(15), 3238–3247.

34. Silveira, J.L. et al. The Benefits of Ethanol Use for Hydrogen Production in Urban Transportation. Renewable Sustainable Energy Rev. 2009, 13(9), 2525–2534.

35. Vaidya, P.D; Rodrigues, A.E. Glycerol Reforming for Hydrogen Production: A Review. Chem. Eng. Tech. 2009, 32(10), 1463–1469.

36. Achouri, I.E. et al. Diesel Steam Reforming: Comparison of Two Nickel Aluminate Catalysts Prepared by Wet-Impregnation and Co-Precipitation. Catal. Today 2013, 207, 13–20.

37. Morlanés, N. Reaction Mechanism of Naphtha Steam Reforming on Nickel-Based Catalysts, and FTIR Spectroscopy With CO Adsorption to Elucidate Real Active Sites. Int. J. Hydrogen Energy 2013, 38(9), 3588–3596.

38. U.S. Department of Energy. Benefits and Challenges. http://www.fueleconomy.gov/feg/fcv_benefits.shtml. 2013.

39. Peng, D.Y.; Robinson, D.B. Calculation of Three-Phase Solid-Liquid-Vapor Equilibrium Using an Equation of State. In Equations of State in Engineering and Research, K.C. Chao; R.L. Robinson, Eds.; Am. Chem. Soc.: Washington, DC, 1966; Vol. 182, pp. 185–195.

40. Kolios, G. et al. Heat-Integrated Reactor Concepts for Hydrogen Production by Methane Steam Reforming. Fuel Cells 2005, 5(1), 52–65.

41. Galletti, C. et al. CO-selective Methanation Over Ru-γ Al2O3 Catalysts in H2-Rich Gas for PEM FC Applications. Chem. Eng. Sci. 2010, 65(1), 590–596.

42. Bion, N. et al. Preferential Oxidation of Carbon Monoxide in the Presence of Hydrogen (PROX) Over Noble Metals and Transition Metal

Oxides: Advantages and Drawbacks. Topics Catal. **2008**, *51*(1–4), 76–88.

43. Park, E.D.; Lee, D.; Lee, H.C. Recent Progress in Selective CO Removal in a H2-Rich Stream. Catal. Today **2009**, *139*(4), 280–290.

44. Krishna, R. Adsorptive Separation of CO 2/CH 4/CO Gas Mixtures at High Pressures. Microporous Mesoporous Mater. **2012**, *156*, 217–223.

45. Sánchez, J.M.; Barreiro, M.M.; Maroño, M. Hydrogen Enrichment and Separation from Synthesis Gas by the Use of a Membrane Reactor. Biomass Bioener. **2011**, *35*(Suppl. 1), S132–S144.

46. Ehteshami, S.M.M.; Zhou, W.J.; Chan, S.H. Energy Analysis of an Electrolytic-Based Selective CO Oxidation System for On-Board Pure Hydrogen Production. Int. J. Hydrogen Energy **2013**, *38*(1), 188–196.

47. Francesconi, J.A. et al. Analysis of the Energy Efficiency of an Integrated Ethanol Processor for PEM Fuel Cell Systems. J. Power Sources **2007**, *167*(1), 151–161.

48. Ehteshami, S.M.M. et al. The Role of Electronic Properties of Pt and Pt Alloys for Enhanced Reformate Electro-Oxidation in Polymer Electrolyte Membrane Fuel Cells. Electrochimica Acta **2013**, *107*, 155–163.

49. Vogtländer, J.G.; Bijma, A.; Brezet, H.C. Communicating the Eco-Efficiency of Products and Services by Means of the Eco-Costs/Value Model. J. Cleaner Prod. **2002**, *10*(1), 57–67.

50. Vogtländer, J.G.; Brezet, H.C.; Hendriks, C.F. The Virtual Eco-Costs '99: A Single LCA-Based Indicator for Sustainability and the Eco-Costs: Value Ratio (EVR) Model for Economic Allocation: A New LCA-Based Calculation Model to determine the Sustainability of Products and Services. Int. J. Life Cycle Assessment **2001**, *6*(3), 157–166.

51. Lee, J.Y. et al. Life Cycle Cost Analysis to Examine the Economical Feasibility of Hydrogen as an Alternative Fuel. Int. J. Hydrogen Energy **2009**, *34*(10), 4243–4255.

15

The Impact of Energy Conservation Policies on the Projection of Future Energy Demand

AMIR HOSSEIN FAKEHI KHORASANI[1], SOMAYEH AHMADI[1]*, and MOHAMMAD ALI MORADI[2]

[1]*Department of Energy Engineering, Sharif University of Technology, Tehran, Iran*
[2]*Faculty of Mechanical Engineering, K. N. Toosi University of Technology, Tehran, Iran*

Abstract: In this article, the future trend of energy demand in Iran is analyzed using a long-range energy alternatives planning system (LEAP) model. First, the structure of Iran's energy consumption and its historical trend are evaluated. Then the key assumptions of a LEAP model are defined, which comprise different economic growth, population growth, and urban settlement perspectives in combination with several assumptions on energy-saving policy rules. These assumptions are categorized into three different scenarios: reference, high, and low. Results imply an ever-growing demand in all three scenarios, and the ratio of final energy demand in the year 2041 to that of 2010 in High, Reference, and Low growth scenarios are projected to be 2.24, 1.8, and 1.6, respectively. According to the reference scenario, if current trends of energy conservation continue, by the year 2041 the shares of natural gas, petroleum products, electricity, and renewable energy from Iran's energy demand basket are predicted to be, respectively, 53%, 34%, 10%, and 3%. Results show that Iran has a large potential for energy conservation, and, according to the reference scenario, the implementation of currently adopted rules may reduce its final energy demand by 21%. Additionally, renewable solar energy will play a more prominent role in the implementation of energy conservation policy acts in Iran.

Keywords: LEAP model, final energy demand, energy planning, scenarios analysis, energy economy

Nomenclature

Parameter

D	Final energy demand
A	Activity
as	Activity share
EI	Energy intensity
UI	Useful energy intensity
UE	Useful energy
FS	Fuel share
η	Efficiency
S	Stock
EC	Energy consumption
Sa	Sale
Su	Equipment survival curve
De	Degradation
FE	Fuel use per unit of vehicle
FD	Coefficient of vehicle aging
Mi	Annual distance traveled
MD	Vehicle aging coefficient on mileage
β	Household number–home number ratio
Ar	House area
DA	Dwelling area
HI	Household income
P	Price
FCS	Fuel cost share
FC	Fuel cost
TC	Total cost
V	Vehicle number
TF	Trend factor
VF	Virtual factor
ETrans	Transport energy demand
ECar	Car energy demand
EVan	Van energy demand
EMC	Motorcycle energy demand
EF	Freight energy demand
EPP	Public pasenger energy demand
EF	Fright energy demand
FCa	Full capacity
CF	Capacity factor
EID	Electricity demand

*Address correspondence to: Somayeh Ahmadi, Department of Energy Engineering, Sharif University of Technology, Azadi Ave, PO Box: 11365-11155, Tehran, Iran. Email: sepide.ahmadi2006@gmail.com

TEIP	Total electricity pump	
	Subtitle	
se	Sector	
b	Technology branch	
s	Scenario	
t	Time point	
Ag	Aggregate	
T	Type technology	
hn	Household number	
Ar	Area	
v	Life time of technology	
c	Supersede fuel	
$\delta, \alpha, \gamma, \lambda$	Elasticity	
elec	Electricity	
p	Person	
Hs	Household size	
y	Year	
i	Fuel type	
j	Subsector	

1. Introduction

An analysis of energy demand and its dynamic variations corresponding to socioeconomic developments is an essential part of any regional or countrywide energy planning studies.[1] For this purpose, several demand models were designed to investigate causal relations of growing demand for energy in various economic sectors. These models are based on statistical, econometrical, and engineering techniques, the latter being the most complex and sophisticated of all.[2]

Ranging from simple statistical models to complex combinatory ones, there are many methods and approaches to model energy demand. In statistical models, the value of energy demand is determined by simple extrapolations or multivariate statistical techniques such as discriminate analysis and taxonomy analysis, whereas in econometric methods these values are computed based on macro-economic theories. In econometrical and statistical models, the physical relations and engineering equations that are related to end-use energy appliances are usually neglected. On the other hand, by increasing the efficiencies of energy-intensive devices and appliances, the impact of technological changes and energy intensities has become more important in setting the framework for energy demand models. Thus, engineering or process analysis methods were devised. In the early process models, economic variables such as price and income are not considered explicitly—that is one of the critical flaws of these models.[3] So the combinatory group of end-use energy demand models has been developed in which technical and economic approaches are combined together, and mutual effects of economic and population growth as well as technical parameters on energy demand have been considered. Model for Demand of Energy for Europe (MEDEE), Model for Analysis of Energy Demand (MAED), Model for Analysis of Demand for Energy MADE II (MADE-II), and Long-range Energy Alternatives Planning System (LEAP) are the major tools for estimating energy demand with a bottom-up approach.[4–6] Among the mentioned methods, MEDEE and MAED are techno-economic models, and MADE-II consists of three parts: a model for income distribution (to project demand for useful energy in

household groups), a process-engineering model (the induction model), and the intermediate model for transportation sector.[7] LEAP is a comprehensive energy, economy, and environment tool for simulation of a whole energy system based on several defined scenarios. The scenarios are defined on the basis of the structure of consumption, conversion, and production of energy in a certain region or economy with significant diversity in population, economic growth, technology, price, and so forth. Due to a very flexible structure, LEAP enables its users to perform a detailed analysis of several social, economic, and technological scenarios on end-use energy consumption.

Earlier versions of LEAP were developed by Raskin in the 1980s.[6] The first study using LEAP on a global scale was conducted targeting the fossil-fuel-free future and implemented in 1992 by the World Bank to study Beijing's weather and air quality.[6] The software was used in 2001 by another research group with Fadel's supervision in Lebanon to investigate greenhouse gas (GHG) reduction scenarios in power generation with an emphasis on renewable energy.[8] The model was also recruited for a feasibility study of prediction scenarios of high and medium range for usage of biofuels in transportation and power sectors of Mexico.[9] Giatrakos then studied electricity of the island of Crete in the Middle East with sustainable energies and employing LEAP to analyze power generation expansion scenarios.[10] Shin studied effects of electricity generated by landfill on Korea's energy market with LEAP and also compared the potential to power produced by conventional ways. He finally inspected the effects of technology development on supply of energy demand.[11]

Studying the effects of pollutant emissions and their future trend are other features of this modeling software, which have been implemented in various demand sectors by Awami and Farahmandpour for Iran.[12] Similar studies have been done by the Environment Institute in Nepal[13] and by Bose on the transportation sector in Delhi.[14] A long-term forecast of national energy supply and demand in Thailand is another application of this modeling tool. In this study, present conditions of supply system and energy consumption are assessed—then the future trends are determined based on policy-making scenarios, and, ultimately, the quantity of energy demand and pollutant emission are calculated.[15] In addition, China's energy demand perspective in the coming 20 years has been simulated using the LEAP model by Shan. In this study, the modification of energy system structure and energy efficiencies have been investigated in different economic growth scenarios.[16] Also, a long-term energy-environmental model for developing countries' transportation sector has been developed by Sadri et al.[17] The objective of their study was to manage gasoline consumption for light-duty vehicles based on historical data for population and GDP. For this, they used the LEAP modeler tool; in the same research, Amirnekooei et al. developed a reference energy system to forecast energy consumption for Iran in a 25-year period and examined the effects of several demand and supply side management strategies on resource depletion and environmental emissions by the LEAP modeler.[18] Also, a review on sustainable energy scenarios in Iran has been done by Mohammadnejad et al.[19] In this research, an energy demand model is developed by LEAP in which some sectors and energy resources of Iran have been analyzed with a high level of aggregation.

The energy consumption pattern in Iran is undoubtedly unstable—an example of a populated, consumer-oriented oil manufacturing country. The prominent feature of this unstable pattern is low efficiency in final consumptions. In recent years, the Iranian government has started an energy conservation policy toward energy intensity decrement, especially in the demand side, by eliminating energy subsidies that leads to consumers substituting their energy-intensive appliances with advanced technologies. In this article, Iran's trend of energy demand has been studied using the LEAP modeler for two groups of scenarios: reference or business-as-usual scenario and a group of low, medium, and high developments with different levels of conservation policies. For this purpose, first, the historical trend and structure of Iran's energy demand are evaluated in five sectors: households, industry, agriculture, public services, and transportation. Then, after establishing the framework and modeling method, energy demand in various sectors is described. Finally, results of the proposed model are demonstrated and discussed.

2. Analysis of Iran's Energy Demand Behavior

Iran's energy consumption has had an ascending trend, increasing from 555.6 MBOE in 1994 to 1215 MBOE in 2010 (Fig. 1a).[19-23] According to the Iran's Center of Statistics, the country's total value-added during these years has risen up from 281,626 billion ($1 = 30000 rials) to 623,375 billion rials. This shows a growth in energy intensity from 1.79 to 2.13 BOE per million rials from 1995 to 2009 and a shrink in the last year—2010—to 1.95 BOE per million rials. The significant reason for the 3.2% drop in final energy consumption from 2009 to 2010 is due to the implementation of the "Targeted Subsidy Plan" act of 2010 in Iran. This subsidy reform act was passed by Iran's parliament and started by the government in January 2010. The aim of the plan was to gradually eliminate governmental subsidies on electricity and energy carriers in a five-year period. The act would make the government pay Iranian citizens a monthly stipend from the money that it would save from eliminating subsidies, along with the increase in energy prices. Before applying the subsidy reform plan, the Iranian government paid around 80% of total energy prices. The goal of the plan was to avoid domestic sales of gasoline, gas oil, fuel oil, kerosene, LPG, and other petroleum derivatives at less than 90% of Persian Gulf FOB prices.[25-27]

As it is depicted in Figure 1a, during the years 1994 to 2010, the household, transportation, and industrial sectors have had the greatest shares of country's total energy consumption. The increase in various sectors' energy consumption could be associated with public welfare improvement, reduction of energy efficiencies, and artificial regulated energy prices during the first years of the mentioned period. According to Figure 1b, natural gas, gas oil, and petroleum comprise the greatest portions of energy.

As previously stated, because of supply rationing of petroleum products and the administration of the targeted subsidy plan, final energy consumption was lowered by 3.2% in the first quarter of 2010 relative to the same period of the preceding year. According to Figure 2, the amounts of consumption in household, commercial, transportation, industry, agriculture,

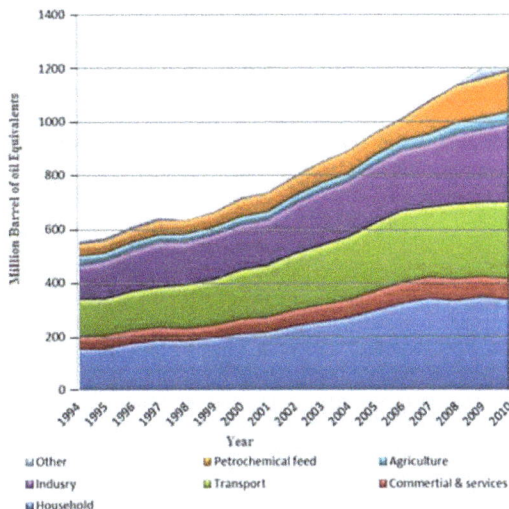

Fig. 1a. Energy demand trend between 1995 and 2010.

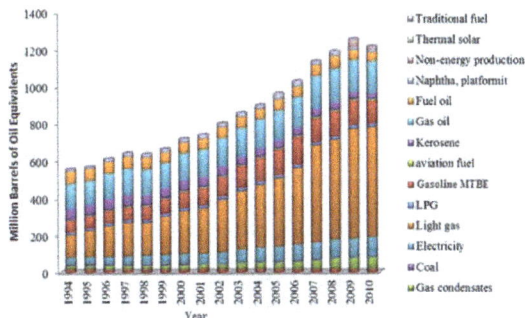

Fig. 1b. Share of energy carriers in Iran's energy demand perspective, 1995–2010.

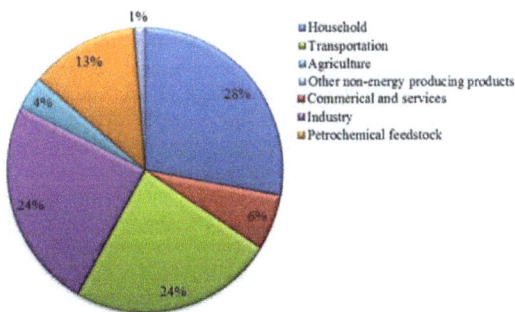

Fig. 2. Share of subsectors in energy consumption in Iran, year 2011.

petrochemical feedstock, and other non-energy producing products sectors, respectively, were 335.72 MBOE, 72.23 MBOE, 286.75 MBOE, 290.14 MBOE, 45.17 MBOE, 152.73 MBOE, and 15.31 MBOE.[23]

It is evident from Figure 2 that a major portion of energy is consumed in nonproductive sectors such as households that have no significant share in value-added of the country. It is worthwhile to note that although energy consumption of households decreased by 3% because of the government's targeted subsidy plan, energy consumption in other sectors, such as industries, transportation, agriculture, and power plants, increased.

Key energy, economic, and social variables of Iran are shown in Table 1. By following the country's energy consumption trend, one could deduce that the historical behavior of Iran's energy system in each period has been a function of the pursued energy policy and strategy in that period's preceding years. The adopted policies were based on fulfilling two main targets that are interdependent from each other: First, satisfying energy demand by increasing energy supply without changes in relative prices; second, keeping crude oil export capacity because of foreign currency requirements and development policies. The first target—to maximally supply energy demand with the least possible price—is subject to severe skepticism. This unsustainable policy is the main reason for consumers' prodigal and irresponsible behavior.[28]

3. Methodology

3.1 The Dimensions of the Model in Various Sectors

As was stated before, Iran's energy demand is divided into five sectors: household, industrial, agricultural, public services, and transportation. According to Figure 3, each of these sectors is then categorized into several subsectors.

The household sector is divided into urban and rural subsectors, which are subdivided into two categories: with and without natural gas service. The efficacious parameters in this sector are population growth, household size, energy price, climate variables, home appliance technologies, and gas piping development plans. Energy carriers used in the household sector are petroleum products (kerosene, gas oil, and liquid gas), natural gas, electricity, solar energy, and traditional fuels such as firewood and livestock waste. These could be superseded by one another in heating, cooking, and hot water subgroups. The activity parameter is measured by household size, and the modeling method is activity analysis. The consumption index is analyzed relative to square meters of living area, which has been implemented by a logistic model to evaluate the housing demand. Table A.1 shows the activity level of energy intensive appliances owned by urban and rural households in Iran.

The commercial sector refers to all activities that produce service instead of physical commodities. The subgroups structure is based on the available codes of International Standard Industrial Classification (ISIC), which indicates 10 main subgroups in the commercial sector. The energy consumption in this sector is regarded as useful energy demand, and the added value is designated to be the indicator of commercial activity. Energy carriers used in the commercial sector are petroleum products, natural gas, electricity, solar energy, and traditional.

In another classification, the industry was divided into two main categories of large industries (50 laborers or more) and medium/small industries (fewer than 50). The classification was conducted according to ISIC code, which includes 40 major

industries. Modeling was carried out by an activity analysis approach. The activity level is often measured by physical production, but at certain instances the fiscal added value is used. The essential energy carriers for industry are electricity, fuel oil, natural gas (fuel and feedstock), gas oil, coal, liquid gas (ceramic industries), kerosene, and petroleum. The share of each carrier in total consumption has been determined by the available data from the National Iranian Gas Company (NIGC) and the National Iranian Oil Products Distribution Company (NIOPDC).[29] National industrial production rate is an effective parameter in the modeling of this field, which is calculated by actual production divided by the production capacity in each industrial category. The relation of price and energy intensity and also that of economic growth to industrial production have been evaluated in the model. Moreover, in the categories that have not been identified, energy intensity is determined by internal and external energy auditing.

Modeling the transportation sector was accomplished by combining two methods of activity analysis and transport analysis. In the transport analysis, vehicles number, sales, and age are the most important parameters. By the activity analysis, activity level is measured in kilometer-per-capita and ton-kilometer. Main subgroups are light vehicles, motorcycles, freight vehicles, and public transportation. Cars, pickup trucks, SUVs, and vans are classified as light vehicles. Motorcycles are categorized into two-strokes and four-strokes, and taxis are considered semipublic transport stock. Freight transportation is divided into two main subgroups: internal and external. Both intercity and within-city freight transportation sectors are considered in the model. Additionally, the external freight transportation is classified into sea transportation and air transportation.

A major portion of the energy demand in agriculture is associated with horticulture, aviculture, and greenhouse production. The primary method to evaluate the demand is based on activity analysis with some adaptations from complex models. The major carriers are gas oil, natural gas, kerosene, liquid gas, and petroleum; alternative fuels such as solar, biomass, wind, and hydro energies will substitute in the future. Energy intensity is mostly estimated by technical and engineering calculations.

3.2 Modeling Methodology

In this study, energy demand in different sectors is calculated by three methods; activity analysis, stock analysis, and transport analysis. In activity-level analysis, energy consumption is calculated as the product as an activity level and an annual energy intensity. And in stock analysis, this parameter is considered by analyzing the current and projected future stock of energy-using devices and the annual energy intensity of each device. Also, energy consumption in transport analysis is calculated as the product of the number of vehicles, the annual average mileage, and the fuel economy of the vehicles.

In activity analysis, final energy and useful energy demand are calculated by energy intensity and activity level for each of technology, time point, and scenario according to Eq. 1–4. In this method, total activity level of technology is determined by multiplying all the activity levels of the technology chain from that point to demand level.

Table 1. Key energy, economy, and social variables of Iran in the year 2011.

	Rich gas	Gas conden-sate	Coal	Electricity	Natural gas	LPG	Gasoline	Aviation fuel	Kerosene	Gas oil	fuel oil	Naphtha	Thermal solar	Traditional fuel	CO_2 emissions	GDP
							MBOE								Million tons	Billion rial
Household			0.10	33.38	271.38	12.3			27.44	1.01			0.19	12.30	121.2	288021
Commercial and services				17.30	39.65	2.88	0.24		0.78	5.60	3.63					
Transportation				0.20	38.48	0.27	119.61	8.46		123.9	3.43				107.7	154121
Industry			1.94	37.40	237.47	0.72	0.30		0.15	14.22	22.8				94.7	74265
Agriculture				17.65	3.80		0.01		0.67	23.53	0.01				11.3	
Petrochemical feedstock	18.53	60.71			30.47							16.53				
Oil																48501

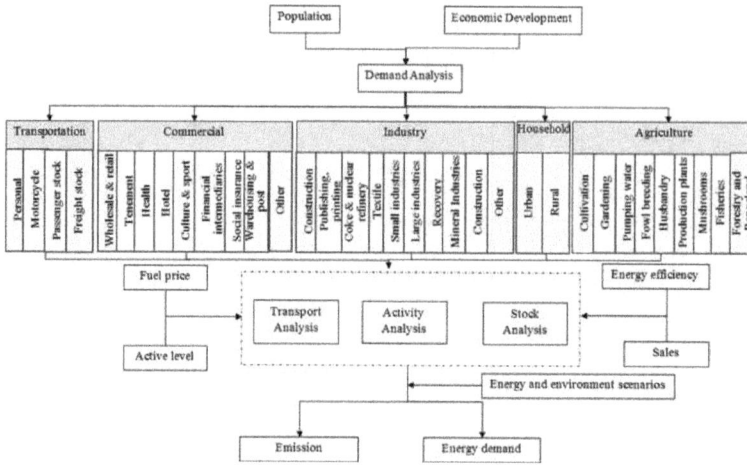

Fig. 3. Schematic structure of Iran's energy demand model.

$$D_{se} = \sum_b \sum_s \sum_t A_{b,s,t} \times EI_{b,s,t} \tag{1}$$

$$A_{b,s,t} = \prod_i A_{b,i,s,t} \tag{2}$$

$$EI_{se} = \sum_b \sum_s \sum_t UI_{Ag,s,t} \times AS_{b,s,t}/\eta_{b,s,t} \tag{3}$$

$$UE_{b,0} = \sum_b EI_{Ag,0} \times FS_{b,0} \times \eta_{b,0} \tag{4}$$

The stock analysis method is applied more in the household and commercial sectors, where activity analysis is more exercised. Energy demand in this method is estimated by Eq. 5–7.

$$EC_{se} = \sum_T \sum_y \sum_v S_{T,y,v} \times EI_{T,y,v} \tag{5}$$

$$S_{T,y,v} = \sum_v Sa_{T,y} \times Su_{T,y-v} \tag{6}$$

$$EI_{T,y,v} = EI_{T,y} \times De_{t,y-v} \tag{7}$$

The transport analysis method is applied to estimation of energy consumption in the transport sector according to Eq. 8–11. In this method, energy intensity is usually replaced by fuel economy of vehicles—that is, fuel consumption in 100 kilometers, or mileage per unit fuel consumed. Since the measured fuel economy is often associated with newly manufactured vehicles, its value changes due to their aging.

$$EC_{se} = \sum_T \sum_y \sum_v S_{T,y,v} \times Mi_{T,y,v} \times FE_{T,y,v} \tag{8}$$

$$S_{T,y,v} = \sum_v Sa_{T,y} \times Su_{T,y-v} \tag{9}$$

$$FE_{T,y,v} = FE_{T,y} \times FD_{t,y-v} \tag{10}$$

$$Mi_{T,y,v} = Mi_{T,y} \times MD_{t,y-v} \tag{11}$$

Formulation in the household sector:

$$EI_{Ar} = \frac{EI_{hn}}{Ar} \times \beta \tag{12}$$

$$EI_p = \frac{EI_{hn}}{Hs} \tag{13}$$

$$EF_{i,t} = \sum_{j=1}^{n} A_{i,j,t} \times EI_{i,j,t} \times Fs_{i,j,t} \tag{14}$$

$$EE_{i,t} = \sum_{j=1}^{n} A_{i,j,t} \times UF_{i,j,t}/\eta_{i,j,t} \tag{15}$$

$$\beta(t) = \frac{Hs}{1 + Exp\,(a + b.t)}$$

$$a = Ln\,\frac{Hs - \beta_0}{\beta_0} \tag{16}$$

$$b = \frac{1}{n} \cdot Ln\,\frac{\beta_0\,(Hs - \beta_1)}{\beta_1\,(Hs - \beta_0)}$$

$$E = a.HI^\alpha.DA^\delta.P_i^\lambda.P_c^\varepsilon$$

$$\lambda = \frac{\partial E/E}{\partial P_i/P_i} \tag{17}$$

Formulation in the commercial sector:

$$FCS_j = \frac{FC_{ij}}{TC_i} \tag{18}$$

$$TC_i = \sum_{j=1}^{10} FC_{ij} \qquad (19)$$

$$E_{ij} = FCS_j \times E_i \qquad (20)$$

$$EI_{ij} = \frac{E_{ij}}{v_j} \qquad (21)$$

$$v_j = c.GDP^{\tau}_{Commercial} \qquad (22)$$

$$GDP^{\tau}_{Commercial} = b.GDP^{\Phi} \qquad (23)$$

Formulation in the transport sector:

$$Ln(V_t) = Ln(600 - V_t) - 3/75$$
$$+ 0/26Ln(GDP_{pt}) + 0/033TF + 0/36VF \qquad (24)$$

$$ETrans_{j,t} = ECars_{j,t} + EVan_{j,t} + EMC_{j,t} + EPU_{j,t}$$
$$+ EPP_{j,t} + EF_{j,t} \qquad (25)$$

Formulation in the industry sector:

$$E_{i,t} = \sum_{j=1}^{n} FCa_{i,j,t} \times EI_{j,i,t} \times FS_{j,t} \times CF_{j,t} \qquad (26)$$

$$E_{j,t} = \sum_{i=}^{n} VD_{i,j,t} \times EI_{i,j,t} \times FS_{j,t} \qquad (27)$$

Formulation in the agriculture sector:

$$EI_{elec,i} = \frac{EID_i}{TEIP} \qquad (28)$$

$$EI_{Dies,i} = \frac{EI_{elec,i}}{\eta} \qquad (29)$$

$$Conv\_Cost = \frac{TC}{Conv\_No.} \qquad (30)$$

3.3 The Scope of Iran's Energy Demand Model

In the constructed model, the year 1999 was chosen as the base year, and the model's historical data encompass from 1999 through 2010. The year 2011 was selected as the first year of simulation, and 2041 was set as the end of the period. The simulation was completely performed in a LEAP environment, and all its features and abilities for energy system modeling were implemented.

4. Key Assumptions

4.1 Population

The key assumptions of the model consist of economic growth, population growth, and urban settlement forecasting. The Iran Center of Statistics (ICS) is the reference for population predictions in this study. Regarding the fact that ICS has envisioned Iran's population until the year 2026, and there is a need for the time series data of up to 15 years after the model, the data for after the year 2026 were extrapolated. In addition to ICS predictions, the UN Population Fund (UNFPA) has offered four scenarios for Iran's population, presented in Table 2. The average overall growth rate for Iran is expected to be between 0.51% and 1.71%.[27] In this study, we applied high, middle, and low scenarios for Iran's population forecast in high, reference, and low demand scenarios.

4.2 Economic Growth

Three scenarios were assigned to estimate the economic growth rate of the country. The first scenario is based on historical growth rate average,[26,28] and in the second scenario the growth rate is assessed by the World Bank.[26,29,30] The third scenario is the country's official target according to the department of planning and strategic supervision. Regarding the 5.5% average growth rate in recent decades, this trend was applied to the model as the baseline for the country's growth rate in the reference scenario. Further, apparently the quantity assessed by the World Bank is too low, and Iran's economic potential will be even higher if the imposed sanctions are revoked. On the other hand, in targeting for the country's growth rate (Table 3), values of the fifth row represent the necessary economic growth rate to reach the envisioned goals in the "20-year vision plan," which could be considered optimistic values for the growth rate of the country. Overall, the published value of the World Bank is taken as the pessimistic (lowest growth rate) scenario, the 5.5% scenario as the moderate scenario (reference growth rate), and the target growth rate by the 20-year perspective bill as the optimistic trend in the model.

Also, studying the urban development process shows an overwhelming rise in urban settlement from 31.2% in 1956, to 64.1% in 1996, and to a whopping 70.9% in 2006.[27,30] Extrapolation of historical data was used to assess the future of urban settlement. It predicts that the urban population percentage rises to 88.6% in 2041. To estimate the number of households (household size), regression of historical information was applied, which is presented in Eq. 12. The total number of households in 2041 was calculated by dividing the total population, according to ICS estimates, by household size, which resulted in a value of 48.728 million households.

The estimation results indicate that in 2041 the country's average household size will lower to approximately 2.28 persons per family. Further, by dividing the population by household size, the total number of households will be about 48.612 million that year.

The number of households in 2006 was 12.2% greater than the number of residential units. This means that available residential units will be increased by the factor of 2.8, and about 27 million new houses will be built.

5. Scenario Management

In energy planning studies, to take into account the uncertainty of system development, various scenarios must be defined. These scenarios are based on numerous logical assumptions about final

Table 2. Different scenarios on Iran's population growth.

Iran's population forecasts (Millions people)	2006	2011	2016	2021	2026	2031	2036	2041	Average growth
Statistics Center of Iran forecast	70.495	74.962	80.967	85.749	89.029	95.265	100.287	105.296	1.17
High scenario	70.495	74.962	81.571	87.14	91.958	96.229	100.499	104.905	1.1
Middle scenario	70.495	74.962	80.311	84.419	87.694	90.437	92.877	94.939	0.81
Low scenario	70.495	74.962	79.051	81.699	83.433	84.672	85.424	85.49	0.51
Constant fertility	70.495	74.962	80.638	84.85	88.179	90.944	93.403	95.492	0.82

Table 3. National growth rate perspective in various scenarios.

Scenarios	Economic growth forecast (%)	Method or organization estimator	Consideration
Middle	5.51*	Average growth of 1997–2007	Historical data (central bank)
Low	**Growth(2010;4.05%, 2015;3.63%, 2020;3.49%)	World Bank	World Bank
High	Growth(2011;5, 2012;7, 2013;7.9, 2014;9.1, 2015;10.5,2016;8)	Target of country***	Perspective writ

*This value was determined by averaging the achieved growth rate between the years 1997 and 2007.
**This value was determined by averaging the achieved growth rate between the years 1956 and 2007.
***Country's target growth rate to achieve the goals of the "20-year vision plan" according to the department of planning and strategic supervision.

(or useful) energy demand increase, prices of energy carriers, interest and inflation rates, technology development and penetration, energy conservation and optimization plans, cultivation of patterns, standards and novel technologies, and so forth. Since the results are plenty and important ones should be focused, a framework has been opted for a clear presentation and analysis of the model results. Thus, probable scenarios must be defined and clarified. To make a wholesome picture of the possible cases of the country's energy demand, and regarding the high uncertainty of growth rate compared to population growth, three scenarios were selected for growth rate. It should be pointed out that the reference scenario considers an average value-added growth rate (an average of the country's past 10-year growth rate). Moreover, for each case of growth rate, two scenarios were built in the model: base as usual and technology development.

5.1 Household sector

The probable scenarios for energy management in the household and commercial sectors are presented in Table 4. In the following sections, each of the mentioned scenarios will be analyzed and quantified to be implemented to the model. The generally accredited strategies for energy optimization in these two sectors are:

- Realization of energy carrier prices
- Full implementation of national building regulation, item 19
- Thorough consideration of performance-enhancing standards for equipment and facilities
- Speed-up of the current energy labeling of household devices
- Requiring buildings to meet mandatory energy efficiency acts
- Production of energy-efficient products and goods
- Regulation of the energy management bill

- Development of brand new technologies with minimal energy consumption (for example, new technologies based on implementation of personalized HVAC devices and similar new technologies to provide localized comfort that can reduce both space heating and cooling)
- Use of renewable energies with an emphasis on solar water heaters
- Adequate education and awareness

5.2 Transportation

The importance of transportation in the country's policy-making about energy is highlighted, mainly because of the huge share of petroleum products in the supply of this sector. Further, as the demand of this sector immensely influences the demand of crude oil, it is possible that national crude oil export is restricted because of the devouring demand of this sector, which has serious economic, social, and security-wise consequences for the country. The generally accepted policies to control transportation are classified as follows:

- Adopting pricing policies and modifying allocated fuel subsidies for this sector
- Promoting public transportation
- Development of alternative fuels and design of an appropriate fuel cart
- Requiring vehicles to meet mandatory fuel consumption and environmental standards
- Promotion of commute-reducing strategies, e-community, and traffic reduction
- Development of novel technologies, free trade, and competition among household auto-industry

Table 4. Probable scenarios for Iran's domestic energy management.

No.	Household subsector (final consumption)	Scenarios	Starting year in effect	Consideration
1	Space heating	Implementing mandatory national building regulation, item 19	2011	Be complete in a period of 8 years
2		The standard implementation of flue gas heater and water heaters	2011	In a 5-year process, 50% A label and 50% B label. Finally, 100% A label
3		Increase the efficiency of central heating systems	2011	From the current 60% to about 80% in outlook
4		Space heating using heat pumps	2016	To 3% until 2040
5	Space cooling	Implementing mandatory national building regulation, item 19	2011	Be complete in a period of 8 years
6		Solar absorption chiller systems – gas development	2012	To 3% until 2016 and 10% until 2040
7		Increase the water cooler efficiency and energy label from G to A	2012	40% reduction in electricity consumption
8		Promote the use of water coolers instead of air conditioners in wet areas	2011	
9	Heat water	Promotion of solar water heaters	2001	In 2001, the country has begun.
10		Improve efficiency of gas water heaters	2012	From the current 70% to about 90%
11	Cooking	Increase the share of electricity in cooking	2011	To 30% until 2040
12		Using solar cookers in remote villages	2012	3%
13		Improved cooking stove efficiency	2012–2042	To 40%
14	Lighting	Complete replacement of filament lamps	2011	
15		Complete replacement of fluorescent lamps	2012	
16		Promote LED technology	2014	
17	Refrigeration	Import licenses for refrigerators with label A	2012	
18		Improve refrigeration efficiency to A and B	2012	

Most, if not all, of these policies are in action or under study in Iran. To achieve desirable strategies and to move to a stable path of clean transportation, pricy fuel, cheap auto, and green technology, a number of probable scenarios that are discussed among the decision makers and analysts have been chosen, modeled, and studied. The target of these optimization measures is to reduce energy per passenger and unit of moved load. The energy management scenarios for this sector have been classified in Table 5.

5.3 Industry

At present, the energy intensity of Iran's industrial division is higher than the global standard. The Iran Fuel Consumption Optimization Company has performed extensive nationwide auditing projects to regulate mandatory criteria for operating and to-be-operating factories and set the energy consumption levels accordingly. It seems that without pricing regulations, these levels could not be met, and understandably no serious action has taken place to realize them. It is assumed in the model that with appropriate energy-management strategies, the current energy intensity could be reduced to the present global value. However, it seems that a significant potential is achievable by the mentioned pricing strategy. Likewise, it is assumed that new prices are the basis for administration of the newly regulated standards. The

rules of thumb to manage and optimize energy consumption in this sector are:

- Upgrading the efficiency and reducing the energy intensity of industrial processes
- Increasing carrier prices with a rational trend
- Upgrading the technology of highly consuming industrial equipment
- Recovering the losses in industrial units

The possible energy management scenarios for the industry that were used in the model are outlined in Table 6.

5.4 Agriculture

The major hydrocarbon carrier of use is gas oil, which is consumed in poultry farming, cultivation of crops and greenhouse plants, livestock, and fish farming. The share of poultry farms in the total consumption is high, e.g., about 40% of total gas oil consumption pertained to agriculture being consumed in poultry farms. Due to the low price of chicken relative to meat, it is customary to predict a high growth in chicken production. Well pumps and agricultural machinery are other important sinks of energy in this sector, for which no consumption criterion has been determined and regulated, and it is, of course, a running

Table 5. Probable energy management scenarios for Iran's transportation.

No.	Scenarios	Starting year in effect	Complete year
1	Mandatory implementation fuel basket and light transportation	2012	–
2	The development of public transport and fuel consumption management	2007	–
3	Forced to comply fuel consumption standards in new vehicle	2012	–
4	Development of hybrid vehicles	2012	–
5	Gradual reduction of import car tariff and allocate a portion of automobile demand to import up gradually to 35% in 2041	2012	–
6	Development of electric motorcycle	2012	To 50% until 2040

Table 6. Possible energy management scenarios for the industry.

No.	Scenarios	Starting year in effect	Complete year
1	Implementation of the fuel consumption standards in the industry	2011	2013
2	Adopt policies for converging consumption intensity to world standards	2012	2026
3	Development of CHP systems in large industrial energy	2012	–
4	Increasing the share of natural gas in the industrial sector	–	To 90% at the end of the fifth development plan
5	Increasing the industrial production rates	2012	2041

Table 7. Probable energy management scenarios in agriculture.

No.	Scenarios	Starting year in effect	Complete year
1	Decreasing the heating consumption (fossil fuel) to world standards until 2020 (the end of the Sixth Plan)	2011	2020
2	Replacement of filament lamps with energy-efficient bulbs	2011	2015
3	Increasing the efficiency of air conditioners (as much as 50%)	2011	2015
4	Heat pumps for space heating	2014	Gradually increase to 5% in 2041
5	The use of solar heating for hot water and space heating	2014	Gradually increase to 10% in 2041
6	Increase the share of solar energy up to 30% on outlook	From 2011	2041
7	Increasing the share of natural gas at least 30%	. . . . . .	2014
8	Heat pumps provide space heating share increased to 5%	From 2014	2041

project of IFCO. Among the mentioned unregulated consumers, agricultural machinery, consuming 40% more than global norm, has a major role on gas oil consumption. The generally accredited energy optimization plans for agriculture are:

- Energy audit of poultry farms
- Instant action on replacement of traditional heating systems to more intelligent heaters, installation of thermostats, and exhaust fan timers to control their on–off durations
- Insulation of poultry farm buildings and improving HVAC systems
- Implementation of intelligent building management systems (BMS) in poultry farms
- Application of renewable technologies in the greenhouse subsector
- Total efficiency enhancement of electric motors applied in agriculture

The more probable scenarios that have been included in the model are outlined in Table 7.

6. Results of the Model

In this section, the final total energy demand of Iran will be evaluated in high, reference, and low scenarios. Then the perspective of energy demands in subsectors will be discussed.

Iran's energy demand (from the results of the model) was approximated to be 1215 million BOE in 2010. It is expected that this parameter floats in different paths within various probable scenarios (high, reference, and low). It is clear from Figure 4 that the final demand of all energy carriers will be somewhat between 1289 MBOE and 2022 MBOE. The low end of the range refers to the scenario in which economic growth rate is low, and the high end represents outputs of the economic prosperity scenario. It is obvious that the transportation and domestic sectors have the

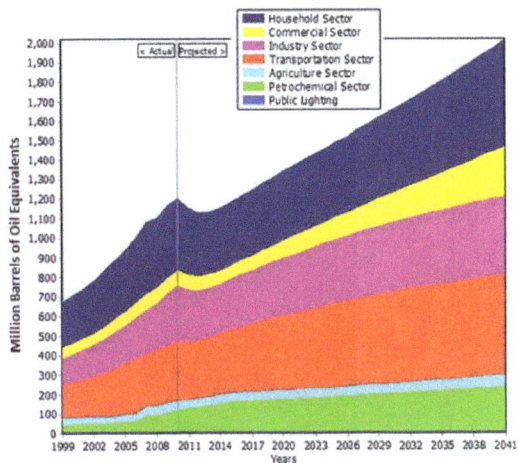

Fig. 4. Iran's final energy demand perspective by the referance scenario.

greatest shares in the final energy demand. This will be discussed further ahead.

6.1. Household Sector

Results of modeling (Fig. 5a) show that, despite significant improvements in recent and coming years due to the targeted subsidy plan, the demand for energy carriers in the household sector will continue to rise, and consumption, averaging a 1.33% annual growth, will increase from 367 MBOE in 2010 to about 545 MBOE in 2041. As is clear in Figure 5a, during the mentioned years (2010–2041), the shares of urban and rural domestic demand are, respectively, 84 and 16%. This is while the average population growth, according to the reference scenario, will approximate 1.14 %, and the average household growth will be 3.1%.

The household sector's energy demand perspective is depicted in Figure 5b divided by types of energy carriers.

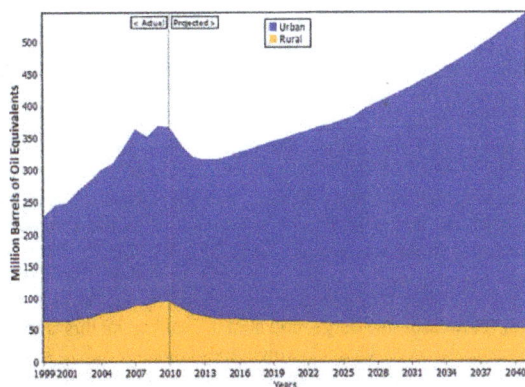

Fig. 5a. The domestic sector's final energy demand perspective.

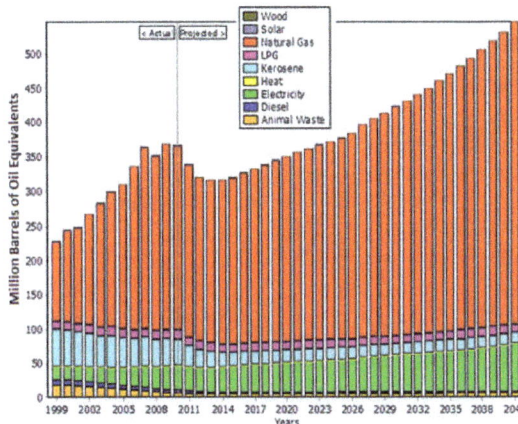

Fig. 5b. Domestic final energy demand perspective, under the reference scenario.

Natural gas, rising from 41.382 billion cubic meters (bcm and equal 254.91 MBOE) in 2010 into an estimated value of 47.947 bcm (295.35 MBOE) in 2041 (average annual growth of 1.61%), possesses the greatest share in supplying the demand. Kerosene, which because of the developing gas pipeline system has a decreasing trend, is the second most consumed fossil carrier in the domestic sector. Household demand for this carrier in 2009 was circa 6307 million liters (37.08 MBOE). If the demand for kerosene follows the expected trend, it will descend to 2661 million liters (15.65 MBOE) in 2041, which shows a cut to half of its previous value. The third fossil carrier in domestic use is LPG, rising to 2837 million liters per year by the first quarter of 2010. Results of the model indicate that there will be an increase in domestic LPG consumption, partially because of the reputation of LPG in regions with no supply of natural gas. It is also to some extent due to LPG low prices for domestic use after the enacted targeted subsidy plan. One of the most important carriers in domestic sector is electricity, whose consumption has been about 55629.6 million kWh (32.71 MBOE) per year in 2009. The demand of this carrier is expected to follow its rising trend in spite of significant increase in costs and enhancement of technologies, so that consumption is estimated to rise to 107250 million kWh (63.06 MBOE) in 2041. The major reason behind this rise despite the mentioned reductive factors is the drastic rise in the number of households and thereby the number of residential units (houses).

The household sector is the second greatest consumer, generating no added value per unit of energy as it is nonproductive. Figure 6 shows the demand of this sector to be about 369.4 MBOE in 2010, rising to 547.5 and 316 MBOE through the reference scenario with and without implementation of energy policies by 2041. As is illustrated in Figure 6, the 41.3% reduction of consumption in the coming years, which is due to the administration of policies, indicates a high potential in this sector. The significantly falling trend of household energy demand until 2026 in the reference scenario coupled with active policies is due to their strong effect, whereas the rising trend after 2026 up to 2041 indicates the prevalence of population growth and

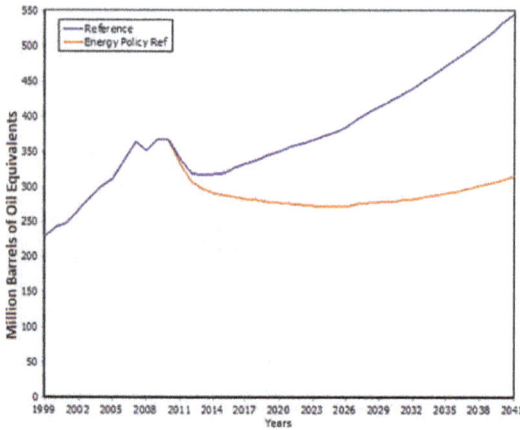

Fig. 6. Household energy demand perspective through two scenarios of reference with and reference without conservation policies.

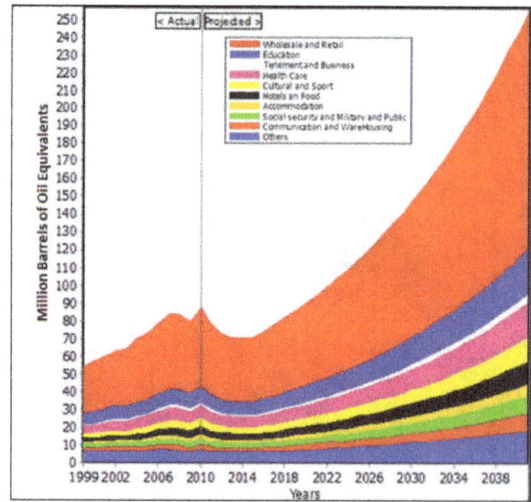

Fig. 7a. Commercial final energy demand perspective.

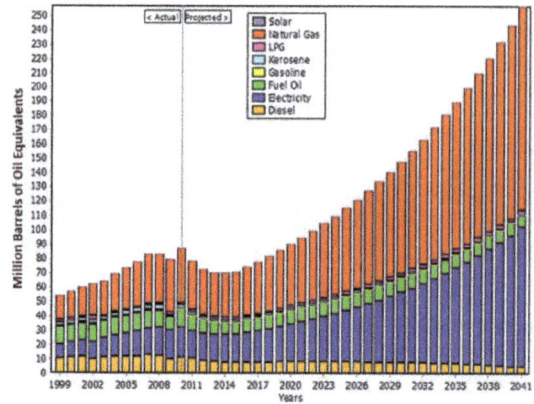

Fig. 7b. The commercial sector's perspective of final energy demand for various carriers.

saturation (completion) of some conservation policies. Among all hydrocarbon fuels, conservation policies have the most effect on natural gas consumption. It should be pointed out that realization of consumption cutbacks is due to full administration of the target subsidy plan, item 19 of national building regulations, and heating and water-heater equipment standards. Also the decline in kerosene consumption during the years indicates the spread of natural gas to households, in addition to the price increase. Furthermore, the model results portend a falling trend for electricity through time and implementation of conservation policies. This is while power demand is expected to grow because of the spread of electrical household devices, changing from fossil consumption to electricity in cooking devices, welfare upgrade using more HVAC devices, and so on. Solar energy is another energy carrier to be establishing its position in the coming years, mainly in heating and water heating and partly in cooking.

6.2 Commercial Sector

As it is depicted in Figure 7a (modeling result), commercial energy demand in the coming years will follow a rising trend, so that it will rise to 256.65 MBOE in 2041 from 77.23 MBOE in 2010. The greatest share of demand in subsectors during these years is possessed by wholesale and retail with an average of 51%. The second most consuming subsector is health care and education.

Results indicate that natural gas demand in the commercial sector was 5913 million cubic meters (36.42 MBOE) in 2010. Despite the decline of natural gas demand in the years of enacting the targeted subsidy plan, it will rapidly grow up to 22 bcm (135.52 MBOE) in the end of 2041 (Fig. 7b). This carrier comprises about 50% of the total demand in the commercial sector during the predicted period of model. Gas oil, second in the ranking, was demanded by approximately 1550 million liters by commercial sector in 2009. If the conditions of reference scenario prevail and natural gas supersedence continues, it is expected that the commercial sector's gas oil demand descend to 762 million liters by the end of the perspective period. Total

power consumption in the commercial sector was 32842 million kWh (19.31 MBOE) in 2009, and it is predicted to rise to 130697 million kWh (76.85 MBOE) in 2041. Compared to all other sectors, not only is the demand for electricity in this sector the fastest-growing one, but the policies of increasing price and energy optimization are predicted to be unable to stop the growing trend.

The commercial energy demand, like other sectors, has a rising trend, as shown in Figure 8. According to the reference scenario, indicate an increase from 88.1 MBOE in 2011 to 256.5 MBOE in 2041. On the other hand, administration of conservation policies will decrease the demand by 32.4% in 2041, which showcases the large potential of saving in this sector. Commercial power demand, which approximated 32842 GWh, is anticipated to grow at a staggering pace and rise to the range of 89000–325000 GWh, where the high end of the range is more than present national power consumption. Moreover, natural gas,

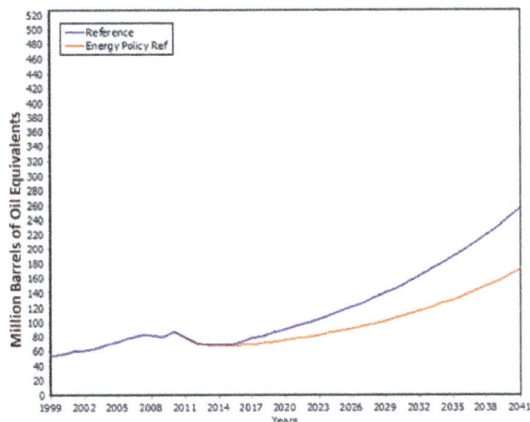

Fig. 8. Commercial energy demand perspective through two scenarios of reference with and reference without conservation policies.

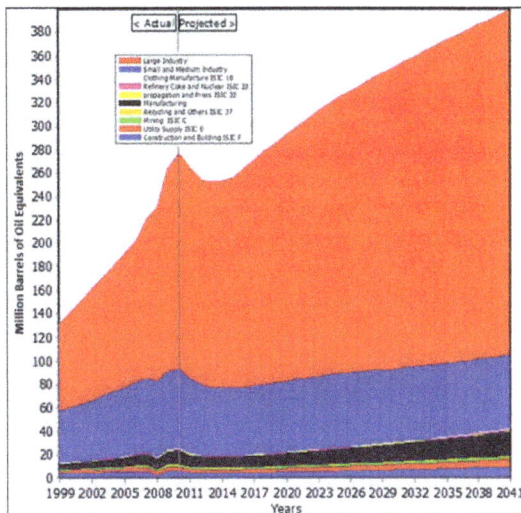

Fig. 9a. Industrial sector final energy demand perspective.

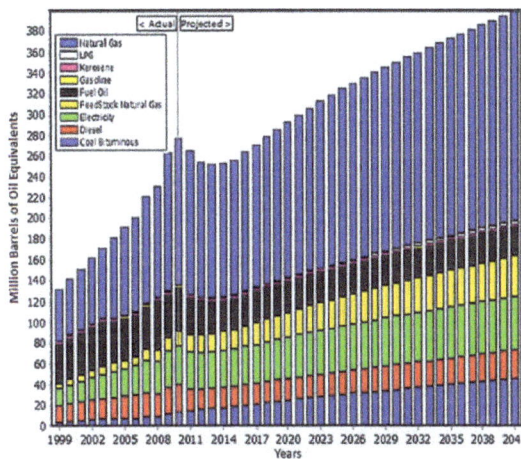

Fig. 9b. Final demand of various energy carriers in industry.

kerosene, LPG, and fuel oil, all following an ascending trend without active policies, are other important carriers of this sector. Otherwise, if policies are administered, their consumption will be reduced by an annual rate of 54, 36, 37 and 36%, respectively, which indicates yet another large potential for energy saving.

6.3 Industry

Results show that the primary energy carriers in Iran's industry are natural gas, gas oil, fuel oil, and electricity. With the increase in energy carriers' prices, energy intensity, which depends on energy demand flexibility in this sector, decreases. Regarding the enormous potential of energy saving in this sector and the continuing trend of technological development for improved performance, the demand is predicted to be affected by price variations in the short term, and then rise because of developments in technology such as energy recovery and so on. It is demonstrated in Figure 9a that demand follows a falling trend until 2014, then starts to rise from 2015 and reaches 299.4 MBOE in 2041. Heavy industries as the most consuming sector will possess a 72% share of this sector's total demand.

Natural gas consumption in Iran's industry was approximately 22530 mcm (138.78 MBOE) in 2010 in the industry sector results. This quantity will shrink at first, due to the shock of increased prices, and will increase gradually as industry develops. Figure 9b illustrates that natural gas consumption is expected to rise to 37 bcm (227.92 MBOE) by 2041. Since there is a large potential for energy saving in this sector, natural gas consumption is not to rise drastically. If, based on past policies, natural gas takes the place of oil products in this sector, its consumption will undoubtedly multiply. Studying the model results shows that the average natural gas demand in this sector will grow by a factor of 1.61% annually. Gas oil consumption in industry was about 3085 million liters (18.89 MBOE) in 2009. This value is expected to continue to increase despite short-term fluctuations and rise to ca. 4204 million liters (25.96 MBOE) annually by the end of the model timespan. Also, industry,

with 32% of the total power consumption share, ranks second in the country's electricity consumption. The results of the model indicate that total power demand boosts from 53.7 billion kWh (31.57 MBOE) per year in 2009 to ca. 85 billion kWh (49.98 MBOE) per year in 2041, mainly because of the country's industrial and technological development, and despite the rise of prices.

The ascending trend of demand for this sector throughout the studied years is depicted in Figure 10. As is clear in the figure, the 277.7 MBOE demand in 2011 will rise to 399.4 in 2041, according to the reference scenario. From another angle, since this sector is the third consumer in the country, effective energy policies could have a major impact on decreasing the energy demand. Activating the mentioned policies will reduce the demand by 7%

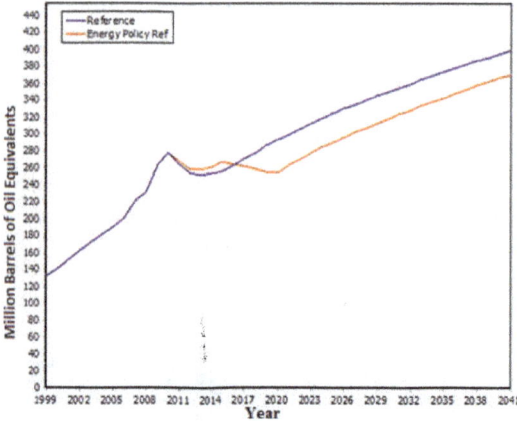

Fig. 10. Industrial energy demand perspective through two scenarios of reference with and reference without conservation policies.

in the end of the perspective period, viz. 2041. To be carrier-specific, the policies will reduce the demand for natural gas by 10%, electricity by 25%, and gas oil by 12%, while the demands for LPG and kerosene will grow by factors of 18% and 1.6%.

6.4 Transportation

It is clear in Figure 11a that, based to the result of the reference scenario, the average energy demand growth rate in transportation is 1.7% per year. In other words, the total energy demand in this sector will increase from 298.1 MBOE in 2009 to 512.5 MBOE in 2041. Also, compared to other key subsectors, light-duty vehicles (LDVs) comprise the greatest share of demand. The shares of technologies in this section are shown in Table A.2 in the appendix.

The transport sector results indicate gasoline, peaking at 73 million liters of daily (146 MBOE in year) consumption in 2006, is the most sensitive carrier in transportation. A slight

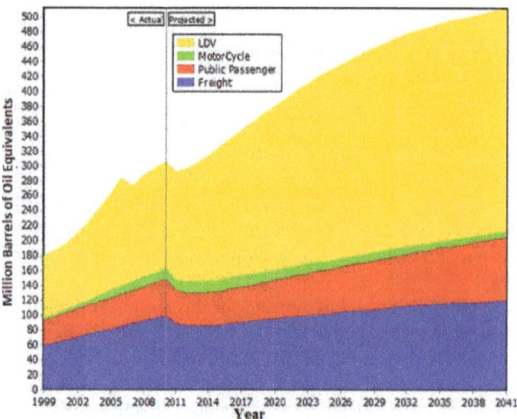

Fig. 11a. Transportation's final energy demand perspective.

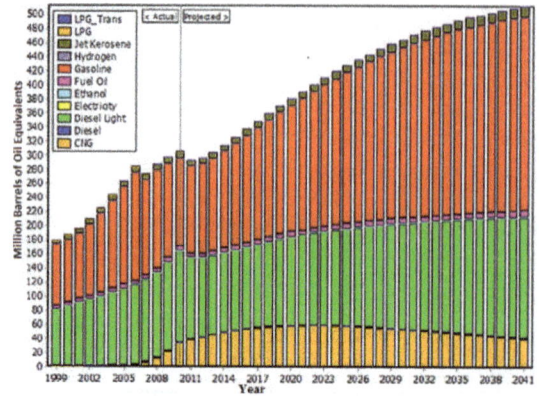

Fig. 11b. Final demand of various energy carriers in transportation.

rise of prices was the first step to change the severely ascending trend, followed by rationing policies, and a significant raise of the prices in December of 2010 (Fig. 11b). The consumption has now shrunk to 58 million liters per day (116.39 MBOE in year). Despite the high prices and technological advances in vehicles, because of LDVs' skyrocketing in the future, results portend yet a rise in petroleum demand. The demand will rise to 47.306 billion liters (260.08 MBOE) per year (130 million liters daily) in 2041. Also, more than 50% of total gas oil consumption in the country is allocated to transportation. Gas oil consumers in this sector are mainly heavy and medium freight, and passenger vehicles and marine vehicles. Diesel consumption in transportation was approximately 19318 million liters (119.3 MBOE) in 2009, which, despite its decline in the early years of the targeted subsidy plan, will increase again and reach a yearly value of 26510 million liters (163 MBOE) by 2041. It is evident in Figure 11b that CNG has had the greatest growth rate among all energy carriers in recent years. The relatively low price of CNG (3000 rials per cubic meter) compared to petroleum (5500 rials per liter) motivates consumers to turn to CNG fueled vehicles. (Every cubic meter of natural gas roughly equals a liter of petroleum.) This in turn causes an increase in CNG consumption, which will be a prevalent trend in the coming years. If the current trend prolongs, based on the reference scenario, CNG demand is expected to rise to 9 bcm in 2022, and then decrease gradually. The main reasons for demand reduction are the substitution of old vehicles by fuel-efficient cars, advances of technology, and performance enhancement.

It is shown in Figure 12 that the transportation energy demand follows an ascending trend, inasmuch as the value of consumption, according to the reference scenario, increases from 305.2 MBOE in 2011 to 512.5 MBOE by 2041. Further, the national dependency on petroleum, recent gas oil import, concurrent signals of petroleum sanctions, and people's daily affiliation with transportation necessitates strict plans to manage consumption. Results of the model mark a 21% reduction in total energy demand of transportation by 2041, down to 404.3 MBOE. The primary carriers in this sector are petroleum, gas oil, CNG, fuel oil, and, to a lower extent, LPG. Conservation policies in the reference scenario dramatically decrease petroleum and

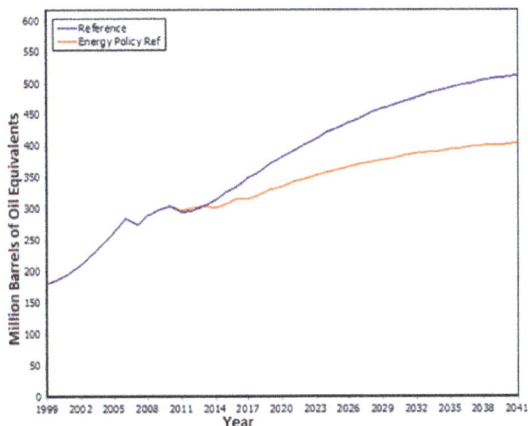

Fig. 12. Energy demand perspective of the transportation sector through two scenarios of reference with and reference without conservation policies.

CNG consumption, whereas they increase electricity and LPG consumption.

6.5 *Agriculture*

Results of the model indicate that the total energy demand of Iran's agriculture was 43.3 MBOE in 2009. With an annual growth of 0.65%, this value will rise to 57.44 MBOE by 2041 (Fig. 13a). Horticulture, water pumping, and aviculture are the main subsectors to demand energy carriers. According to Figure 13a, over 80% of energy consumption is attributable to these three subsectors.

The result of the commercial sector shows that the primary energy carriers in agriculture are gasoil, natural gas, and electricity. The shares of them are illustrated in Figure 13b. With an

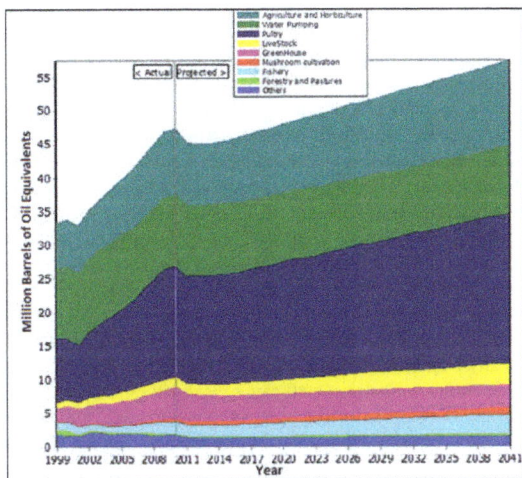

Fig. 13a. Final energy demand perspective of agriculture.

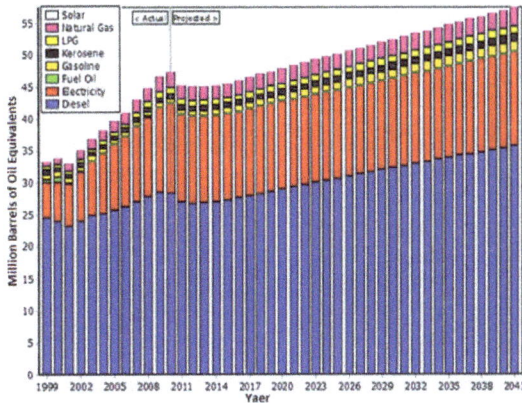

Fig. 13b. Perspective of final demand of various carriers in agriculture.

average annual growth of 0.7%, diesel consumption in agriculture is expected to rise from about 4491 million liters (24 MBOE) in 2009 to 5489 million liters (30.18 MBOE) in 2041. Presently, about 13% of total power consumption pertains to agriculture. Major electricity consumers are water pump-age electro-pumps and poultry farms. Consumption in 2009 was about 21411 GWh (about 1412.6 MBOE). A total of 23757 GWh (13.96 MBOE) for agriculture is predicted under the reference scenario. On the other hand, natural gas has not established itself as a major carrier in agriculture, mainly because of the long distance of farmlands from the gas pipeline network. Nevertheless, as the pipeline has spread throughout recent years, natural gas consumption, especially in poultry farms, has elevated. An amount of 400 million cubic meters (2.46 MBOE) was reported to have been consumed in 2010, while it is anticipated to spike at 453 million cubic meters (2.79 MBOE) by 2041.

The energy demand variations in the agricultural sector (Fig. 14) indicate a 20.8% growth by 2041 compared to that of 2011. However, implementation of conservation policies will reduce the anticipated 57.4 MBOE in 2041 (according to the reference scenario) to 43.9 MBOE. Main carriers in agriculture are LPG, natural gas, and electricity, whose consumption are predicted to reduce by 56%, 20%, and 15%, respectively, if the policies are administered within the reference scenario. This is while throughout the years of study, solar energy were utilized as a major carrier to supply the energy demand and helped to conserve more energy demand.

7. Conclusion

Iran's energy demand trend in recent and coming years is depicted in Figure 15, under the reference, high, and low economic growth scenarios. Results imply an ever-growing total energy demand in all three scenarios—the ratio of demand in 2041 to that of 2010 in high, reference, and low growth rate scenarios are 2.24, 1.8, and 1.6, respectively. Furthermore, in this timespan, the average share of household, transportation, and industrial sectors are 26, 27, and 21% of the total demand.

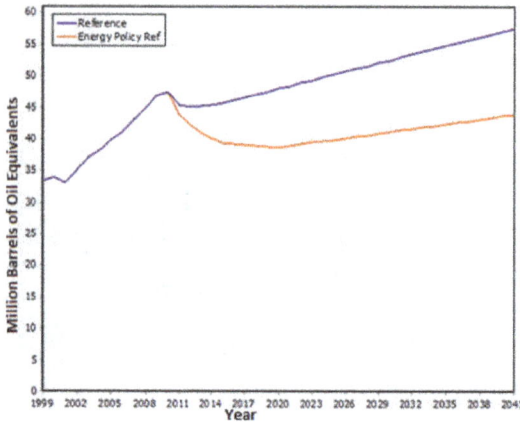

Fig. 14. Energy demand perspective of agriculture through two scenarios of reference with and reference without conservation policies.

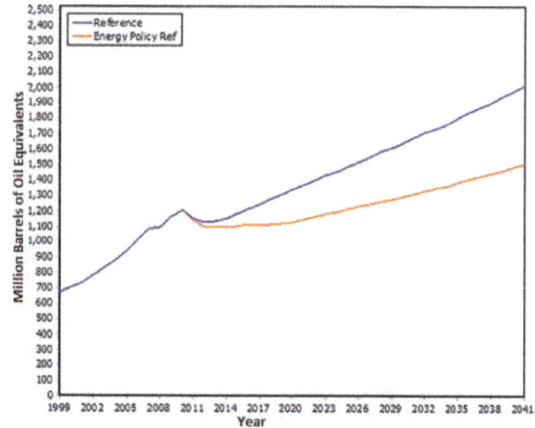

Fig. 16. Iran's energy demand perspective, through two scenarios of reference with and reference without conservation policies.

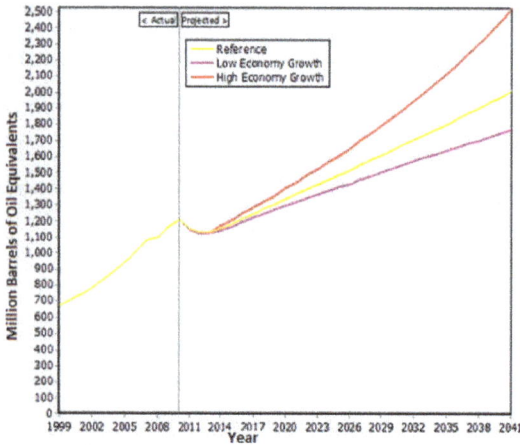

Fig. 15. Iran's final energy demand perspective under different growth rate scenarios.

Thus, considering the high share of nonproductive sectors, which leads to no added value by unit consumption of energy, this trend will cause worries. Additionally, the high share of transportation that can partially be due to multiplication of transport vehicles (stocks) is environmentally troubling. Also, according to the reference scenario, between 2012 and 2041 natural gas will supply 53% of total energy demand, while oil products and electricity will each supply 34% and 10% (Fig. 15). The ongoing increase in demand can put Iran's position as a leading oil and gas exporter at risk. The decrement of consumer devices, lack of energy management, low efficiency equipment, relative upgrade in social welfare, low prices of energy carriers, and inappropriate consumption attitudes are the most important factors that contribute to the rising trend of consumption in

the coming years. Therefore, regulation of price and non-price policies such as approbating standards and criteria of energy consumption, implementation of novel high-efficiency technologies in consumption sectors, education, and cultural motivation for energy conservation, substitution of fossil and low-efficiency fuels by other energy carriers, improving the performance of current industrial equipment, leveling prices of energy carriers, and governmental support for consumption reduction plans and energy service companies are the underlying recommendations to mitigate the country's seemingly irreducible energy consumption.

Iran's energy demand trend from 1999 to 2041 is depicted in Figure 16 with two scenarios, reference with and reference without conservation policies. It is clear in the figure that the large potential of energy saving in various sectors have caused the policies to reduce the energy demand by 21% each year, relative to the scenario without active policies. Implementing effective energy conservation and management strategies will cause the household, commercial, industrial, transportation, and agricultural sectors to consume, respectively, 40, 28, 7, 16, and 23% less energy than they would consume without those strategies, on a yearly basis. Therefore, with regard to the policies, transportation, industrial, and household sectors, respectively, rank as the first, second, and third largest energy consumers of Iran in coming year. However, in the reference scenario the household sector is the largest energy consumer in Iran. Therefore, energy policy reduces energy demand in the nonproductive sector. Also, the results, gained from running the two mentioned scenarios, show the share of each energy carrier in the country's demand basket. It stands to reason that the policies are most effective in reducing demand for natural gas and, to a lesser extent, kerosene, petroleum, and CNG. On average, implementation of the policies causes the share of electricity to decrease by 13%. Moreover, implementing the policies will cause clean energies such as solar energy to develop as prime carriers. Nevertheless, the policies will increase the share of LPG in the demand cart by 2% each year.

Appendix

Table A.1. Activity level of some equipment in the household sector.

Equipment in household sector	Household		Average of power	Average utilization hours
	Urban %	Rural %	Power	hour
Radio	45.78	29.54	30	4
TV	97.44	91.75	120	6
Refrigerator	63.37	83.67	150	12
Freezer	29.22	12.11	200	10
Refrigerator/freezer	34.93	16.26	300	12
Vacuum	84.69	49.69	1200	0.5
Washing machine	73.37	34.82	2000	0.5
Computer	34.29	8.92	200	5
Water cooler	59.37	24.26	500	4
Dishwashing machine	2.39	0.14	1500	0.5
Flatirons	73.4	53.1	1500	0.2
Cooler	10.71	10.67	2000	4
Electric water-heaters	3.8	6.3	2000	1
Tea maker	9.4	4.3	500	1
Electric slow cooker	1.6	1.1	1000	0.5
Electric rice cooker	6.3	2.3	800	0.5
Fan	26.6	2	20	6
Electric samovar	9.4	8.5	1000	0.5
Black and white television	1.63	3.26	50	6
Sewing	62.96	44.1	100	0.1
Treadmills	3.1	0	1000	0.25
Hair dryer	40.2	30.3	1200	0.2
Printer	1.6	1	80	0.1
Meat	14.6	5.6	500	0.1
Strainer	9.4	5	300	0.1
Mixer	6.3	2.3	100	0.1
Electric heaters	14.3	4.8	2000	2
Plates	2.4	3.2	1500	1.5
Sandwich maker	12.2	2	800	0.2
Microwave	10.9	3.1	900	0.2
Red electric	10.8	2	1000	0.2

Table A.2. Share of technologies in the transportation sector.

	2011	2014	2017	2020	2023	2026	2029	2032	2035	2038	2041
Freight	27.6	25.8	26.5	25.3	24.5	22.4	21.4	20.8	20.8	21.6	21.6
Public passenger	17.2	12.9	14.1	12.9	13.0	14.2	14.0	14.2	15.9	16.4	16.7
Motorcycle	5.2	6.5	3.5	3.9	2.0	2.3	0.4	1.9	0.6	0.6	0.6
LDV	50.0	54.8	55.9	57.9	60.5	61.2	64.1	63.1	62.7	61.4	61.2
Sum	100	100	100	100	100	100	100	100	100	100	100

References

1. Fakehi, A. H.; Shafie, S. E.; Soobohi, Y.; Ghofrani, M. M. Useful energy demand modeling in different economic sectors. Sixth National Conference on Energy (Iran). 2007, *13*, 25–38.
2. Swan, L. G.; Ugursal, V. I. Modeling of end use energy consumption in the residential sector: A review of modeling techniques. *Renew. Sustain. Energy Rev.* 2009, *13*, 1819–1835.
3. Institute of International Energy Studies (IIES). Energy demand and the share of natural gas in the country, a review of energy demand modelling, natural gas projects in twenty years. 1997, *15*, 26–90.
4. United Nation ESCAP: Environment and Natural Resources Development Division: Scrotal Energy Demand Analysis and Long-term Forecast: Methodological Manual. MEDEE-S.No: ST/ESCAP/1521,1995.
5. International Atomic Energy Agency. IAEA: Computer Tools for Comparative Assessment of Electricity Generation Options and Strategies. Vienna, Austria.1995.
6. *User guide for long-range energy alternative planning system.* Boston, MA, 2011.
7. Saboohi, Y. Model for Analysis of Demand for Energy—MADE-II. Institute für Kernenergetik und Energiesysteme (IKE), University of Stuttgart, Technical Report, IKE 8-19, 1989: 0173–6892.
8. El-Fadel, M.; Zeinati, M.; Ghaddar, N.; Mezher, T. Mitigating energy-related GHG emissions through renewable energy. *Energy Policy* 2001, *29*, 1031–1043.
9. Islas, J.; Manzini, F.; Masera, O. A prospective study of bioenergy use in Mexico. *Energy* 2007, *32*, 2306–2320.
10. Giatrakos G. P.; Tsoutsos T. D.; Zografakis N. Sustainable power planning for island of Crete. *Energy Policy* 2009, *37*, 1222–1238.
11. Shi, H. C.; Park, J. W.; Kim, H. S.; Shin, E. S. Environmental and economic assessment of landfill gas electricity generation in Korea using LEAP model. *Energy Policy* 2005, *33*, 1261–1270.
12. Awami, A.; Farahmandpour, B. Analysis of environmental emissions and greenhouse gases in Islamic Republic of Iran. Iran: International Institute for Energy Studies (IIES); 2008, 13, 57–63.
13. Institute for Global Environmental Strategies. Urban transportation and environment in Kathmandu valley, Nepal. Integrating global carbon concerns into local air pollution management. Japan: Institute for Global Environmental Strategies; 2006.
14. Bose, R. K. Energy demand and environmental implications in urban transport—Case of Delhi. *India Atmospheric Environ.* 1995, *30*, 403–412.
15. Rabia, S.; Sheikh, S. A. Monitoring urban transport air pollution and energy demand in Rawalpindi and Islamabad using leap model. *Energy* 2010, *35*, 2323–2332.
16. Shan, B. G.; Xu, M. J.; Zhu, F. Z.; Zhang, C. L. China's energy scenario analysis in 2030. 2nd international conference on advances in energy engineering. *Energ. Procedia* 2012, 1292–1298.
17. Sadri, A.; Ardehali, M. M.; Amirnekooei, K. General procedure for long-term energy-environmental planning for transportation sector of developing countries with limited data based on LEAP (long-range energy alternative planning) and Energy PLAN. *Energy* 2014, *77*, 831–843.
18. Amirnekooei, K.; Ardehali, M. M.; Sadri, A. Integrated resource planning for Iran: Development of reference energy system, forecast, and long-term energy-environment plan. *Energy* 2012, *46*, 374–385.
19. Mohammadnejad, M.; Ghazvini, M.; Mahlia, T. M. I.; Andriyana, A. A review on energy scenario and sustainable energy in Iran. *Renew. Sustain. Energ. Rev.* 2011, *15*, 4652–4658.
20. Institute of International Energy Studies (IIES). Hydrocarbon balance 2005: 2007.
21. Institute of International Energy Studies (IIES). Hydrocarbon balance 2007: 2008.
22. Institute of International Energy Studies (IIES). Hydrocarbon balance 2008: 2009.
23. Institute of International Energy Studies (IIES). Hydrocarbon balance 2009: 2010.
24. Eghtessade-e- Energy. Iranian association for energy economics. 2012, 150, 65–66.
25. Statistical Center of Iran website, http://www.amar.org.ir.
26. Statistical Center of Iran. Iran's national account 1969–2007. 2009
27. Statistical Center of Iran. Population and Housing Census 1956–2011, 2012.
28. Rostamihozori, N. Development of Energy and Emission Control Strategies for Iran. Der Universität Fridericianazu Karlsruhe (TH); 2006, 15–27.
29. Ghalandari, A. Patterns of natural gas consumption in household sector. Fourth National Conference on Energy, Tehran, 2004, 18–27.
30. The provincial population projections to the year 1405: Urban and rural, Statistical Center of Iran, 2009.

16

Comparative Analysis of Hybrid GAPSO Optimization Technique With GA and PSO Methods for Cost Optimization of an Off-Grid Hybrid Energy System

DEEPALI SHARMA[1]*, PRERNA GAUR[2], and A. P. MITTAL[2]

[1] Guru Tegh Bahadur Institute of Technology, New Delhi, India
[2] Netaji Subhas Institute of Technology, New Delhi, India

Abstract: In this study, a new methodology, hybrid GAPSO (HGAPSO), has been developed to design and achieve cost optimization of an off-grid hybrid energy system (HES). Since standard particle swarm optimization (PSO) algorithm suffers from premature convergence due to low diversity, and genetic algorithm (GA) suffers from a low convergence speed, in this study modification strategies have been used in GAs and PSO algorithms to achieve the properties of higher capacity of global convergence and the faster efficiency of searching. This improved algorithm HGAPSO described and implemented in a MATLAB environment has been compared with GAs and PSO algorithms in finding the optimum minimum annual cost of a real off-grid energy system (a group of villages in India). The optimization process resulted in HES, utilizing photovoltaic (PV) arrays, batteries, a diesel generator, and other renewable sources, which, in turn, may prove to be a feasible and sustainable power supply alternative for a remote unelectrified rural area. The superiority of HGAPSO algorithm over GAs and PSO algorithms for the problem at hand is shown in terms of convergence generations and computation time.

Keywords: Optimal solution, GA, PSO, HGAPSO, hybrid energy system

1. Introduction

The computational drawbacks of existing numerical methods in the context of optimization problems have compelled researchers to repose their trust in computational intelligence-based techniques. Genetic algorithms (GAs) and particle swarm optimization (PSO) algorithms provide a robust and efficient approach for solving complex real-world problems. In this study, a procedure based on hybridization of GAs and PSO algorithms has been exploited to find the optimum cost of HES.

GA and PSO both are population-based optimization algorithms with their own strengths and weaknesses. For example, while the GA method is more suitable for solving multiple objectives problems and is quite robust, its convergence speed is slow. The reason behind this is that the GA method requires evolutionary operators such as selection, crossover, and mutation for generation of solutions, thereby leading to a large number of

*Address correspondence to: Deepali Sharma, M. Tech, Guru Tegh Bahadur Institute of Technology, ECE, G-8 Area, Rajouri Garden, New Delhi 110064, India. Email: deepalisharma2@gmail.com

function evaluations. The strength of the PSO algorithm, on the other hand, is its fast convergence. This is because it simply uses mathematical operators instead of evolutionary operators for generation of solutions. This also leads to easier coding of the PSO algorithm in comparison to the GA method. The main drawback of the PSO algorithm is its premature convergence as it lacks diversity. Thus, a better algorithm would be the one that incorporates the strengths of these two algorithms and overcomes the weaknesses of both—in other words, an algorithm that has fast convergence as well as high diversity. A hybrid algorithm, coupling GA and PSO algorithm, was thereby proposed because combining the two search techniques seems to be a feasible approach. Since both have a very similar working methodology of starting with an initial population and then finding and ignoring non-dominant solutions, this approach seems to fall in place.

The problem of cost optimization of HES, containing a varied mix of resources, has been addressed by many authors/researchers. In most cases, conventional optimization techniques have been used, resulting in lower calculation efficiencies. In Borowy and Salameh,[1] least squares method has been used to find the optimum size of a PV array for a hybrid wind/PV system. In Wies et al.,[2] the MATLAB/Simulink model has been used for economic analysis of a hybrid PV/diesel electric power system. In Bagul et al.,[3] the optimum size of PV array and battery storage has been obtained for a standalone wind/PV hybrid system using a three-event probability density function. In Barsoum and Vacent,[4] HOMER software has been used for balancing cost, operation, and maintenance of a hydrogen hybrid energy system. In Liu and Islam,[5] the effect of site and size factor on the reliability of the wind/diesel/battery hybrid energy system has been examined. A deterministic algorithm has been used in Belfkira et al.[6] for modeling and optimization of a standalone wind/PV hybrid system. Khan and Iqbal[7] have analyzed a small wind-hydrogen standalone hybrid energy system. Dondas et al.[8] solve a unit commitment problem that defines the organization of starting/extinction of every power plant to satisfy constraints at a lower cost for a hybrid system. Saif et al.[9] have used linear programming for multi-objective capacity planning of a PV-wind-diesel-battery hybrid power system. Other than these conventional techniques, intelligent optimization techniques have also been used by many authors for cost optimization of HES. Although speed is seen to have improved to a certain extent in these papers, the possibility of trapping in local minimum is always there.[10]

Various authors have carried out hybridization of evolutionary algorithms.[10–14] PSO/GA-hybrid algorithm has been proposed in Mohammadi and Jazaeri,[10] where PSO is used to generate the initial population of GA. In Esmin et al.[11] GA-PSO coupling has been done with a mutation operator. However, both these algorithms are only partially coupled and thus do not make full use of hybridization. With the same objective, Jeong et al.[12] were tempted to investigate configurations for GAs and PSO algorithms fully coupled in a hybrid algorithm in order to achieve improvements in diversity and convergence simultaneously.

In Jeong et al.,[12] Kao and Zahara,[13] and Premalatha and Natarajan,[14] analysis and comparison of hybrid GAPSO (HGAPSO) with GA and PSO for several standard functions have been carried out, and it has been established that HGAPSO gives better results in terms of average error and average time for simulation. Also, the capability of searching optima seems to have improved in terms of diversity and convergence in Jeong et al.,[12] Kao and Zahara,[13] and Premalatha and Natarajan,[14] which has been exploited in this study to optimize cost of a standalone hybrid energy system.

In this study, the computational intelligence-based technique HGAPSO has been applied using MATLAB to optimize an off-grid HES. The results of the technique have been displayed by means of real site conditions at a cluster of nine villages named Narendra Nagar in northern India, which has been chosen as exemplary remote area for an off-grid energy system.[15] The focus has been on finding the optimum cost of the HES suitable enough to be employed for this particular site. The line diagram of HES is as shown in Fig. 1.[15] HGAPSO has been used in this study to decide the optimum mix of all these resources in such a manner that it results in the lowest cost of the HES.

The article has been divided into four subsections. In the next section, the methodology has been explained, including the overview of the HGAPSO algorithm. The subsequent section describes the optimization problem, and in the last section, results obtained by the application of the HGAPSO algorithm for optimization of a hybrid energy system are discussed, and its comparison with the results obtained from GAs and PSO algorithms are presented.

2. Methodology

As elucidated above, the optimization process within this study has been implemented using the HGAPSO algorithm, and its

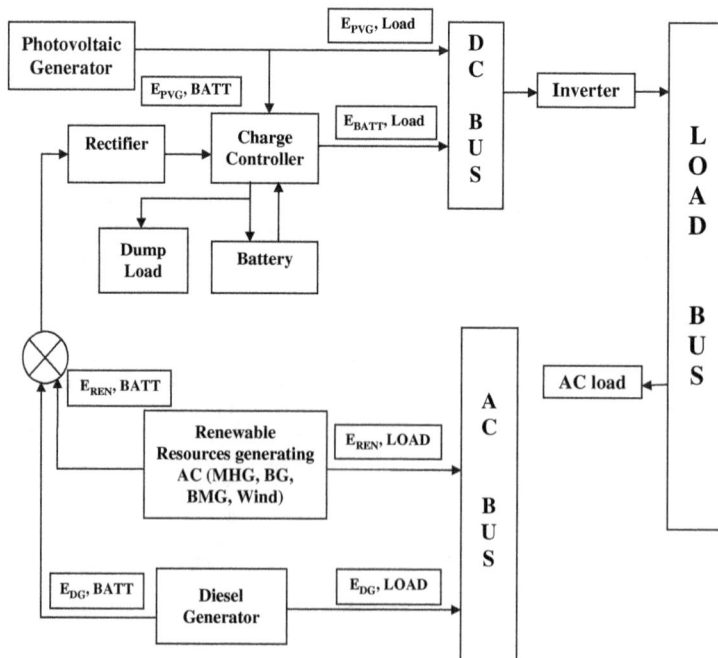

Fig. 1. Energy system for off-grid remote area.[15]

results have been compared with GA and PSO methods. The following sections provide a brief overview of these optimization techniques.

2.1 Genetic Algorithm (GA)

The GA belongs to a family of computational models inspired by Darwin's reproduction and survival of the fittest theory. The GA uses the basic reproduction operators such as crossover and mutation to produce the genetic composition of a population. Some exchange and reordering of chromosomes, producing offspring that contain a combination of information from each parent, is often referred to as *crossover* because of the way strands of chromosomes cross over during the exchange. Diversity in the population is achieved by mutation. A typical genetic algorithm procedure takes the following steps: A population of candidate solutions (for the optimization task to be solved) is initialized. New solutions are created by applying genetic operators (mutation and/or crossover). The fitness (how good the solutions are) of the resulting solutions is evaluated, and suitable selection strategy is then applied to determine which solutions will be maintained into the next generation. The procedure is then iterated. After several iterations or generations, the algorithm converges to the best individual that represents the optimal solution to the problem at hand. Genetic algorithms have been successfully applied to numerous problems from different domains, including optimization, automatic programming, machine learning, operations research, bioinformatics, and social systems. In some cases, GA tends to take a fair bit of time to converge. Hence, to achieve improved efficiencies, GA is sometimes used in conjunction with other optimization approaches.

2.2 Particle Swarm Optimization (PSO)

PSO is a population-based stochastic optimization technique developed by Eberhart and Kennedy in 1995, which is inspired by social behavior of organisms such as fish schooling and bird flocking. The PSO algorithm conducts a search using a population of particles that corresponds to individuals in a genetic algorithm. A population of particles is initially randomly generated. Instead of applying evolutionary operators to manipulate the individuals, each individual in PSO flies in the search space with a velocity that is dynamically adjusted according to its own flying experience and its neighbors' flying experience. Based on these values, local best values are obtained called *pbest*, out of which a global best value is obtained, called *gbest*. Each particle tries to modify its position using the following information: the distance between the current position and pbest and the distance between the current position and gbest. This modification can be represented by the concept of velocity. Velocity and position of each particle can be modified by the following equations (1) and (2):

$$V_{id}(t + 1) = \omega \times V_{id}(t) + C_1 \times R_1 \times (pbest_{id}(t) - X_{id}(t))$$

$$+ C_2 \times R_2 \times (gbest_{id}(t) - X_{id}(t)) \qquad (1)$$

$$X_{id}(t + 1) = X_{id}(t) + V_{id}(t + 1) \qquad (2)$$

where ω is the inertia factor influencing the local and global abilities of the algorithm and controls the influence of previous velocity on the new velocity, $V_{i,d}$ is the velocity of the particle i in the d_{th} dimension, and C1 and C2 are weights affecting the cognitive and social factors, respectively. R1 and R2 are two uniform random functions in the range [0, 1] and are used to maintain the diversity of the population. Each particle has to change its position X_{id} toward the position of the two guides, pbest and gbest, which must be selected from the updated set of non-dominated solutions. From equations (1) and (2) it may be deduced that a particle's decision regarding its movement depends on its own experience, which is the memory of its past position and the experience of its most successful particle in the swarm.

Compared with the GA method, the PSO algorithm has some attractive characteristics. It has memory, so knowledge of good solutions is retained by all the particles; whereas in GA, previous knowledge of the problem is discarded once the population changes. But as stated before, it suffers from the problem of premature convergence. The main reason for this problem is the fast rate of information flow between particles, resulting in the creation of similar particles with a loss in diversity that increases the possibility of being trapped in local optima. Therefore, some strategies to improve the performance of this algorithm for a given problem must be considered.

2.3 Hybrid GAPSO (HGAPSO)

The HGAPSO algorithm combines the advantages of swarm intelligence of the PSO algorithm and the natural selection mechanism of the genetic algorithm in order to increase the number of highly evaluated agents at each iteration step. As discussed before, this hybridization results in high diversity and low computational cost.

In literature, there are three different hybrid approaches proposed[13]:

1. PSO-GA (Type 1): The *gbest* particle does not change its position over some designated time steps; the crossover operation is performed on *gbest* particles with chromosome of GA.
2. PSO-GA (Type 2): The stagnated *pbest* particles change their positions by mutation operator of GA.
3. PSO-GA (Type 3): In this model, the total numbers of iterations are equally shared by GAs and PSO algorithms. In this model, both GAs and PSO algorithms are run in parallel.

Based on experiments performed in Kao and Zahara,[13] the Type 3 approach seems to give the best results for the majority of the functions and is thus applied in this work.

The algorithm of the HGAPSO optimization technique as used in this study is given in Fig. 2.[12]

First, multiple solutions are generated randomly as initial population and objective function values are evaluated for each solution. After the evaluation is done, the population is divided into two subpopulations. One of these subpopulations is updated by the GA operation, while the other is updated by the PSO operation. New solutions created by each operation are combined in the next generation, and non-dominated solutions in the combined population are archived. The archive data is shared between the GA and PSO, i.e., non-dominated solutions created by the PSO can be used as parents in GA, while non-dominated solutions created by GA can be used as global guides in PSO.

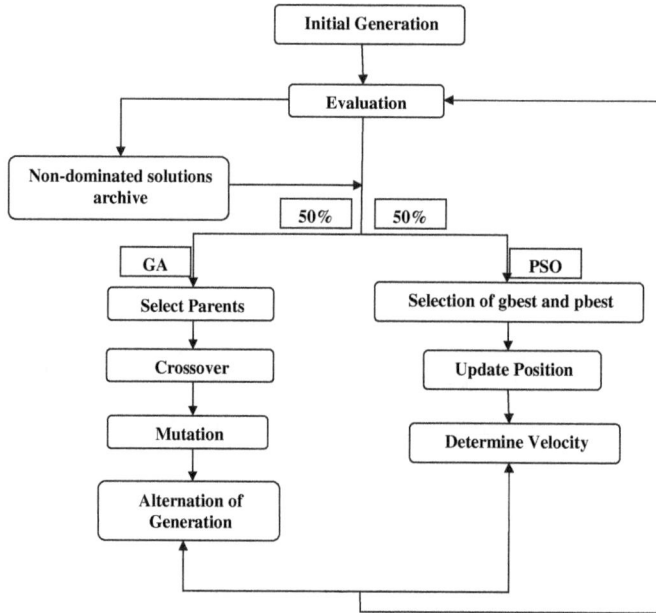

Fig. 2. Flowchart of HGAPSO algorithm.[12]

3. Problem Description

A majority of the population in India resides in villages, and a large number of villages are unelectrified. Their electrification by grid extension is not economically viable as the cost involved is quite high in comparison to low power consumption in such areas.[16–17] Moreover, the decreasing production costs of renewable energy technologies, expanding research in electric storage devices, and increasing environmental concerns have made off-grid hybrid energy systems a potential solution for rural electrification. Such systems are more dependable and have reduced life-cycle cost. Hybrid energy systems are combinations of two or more energy conversion devices (e.g., electricity generators or storage devices) or two or more fuels for the same device, which, when integrated, overcome limitations that may be inherent in either. One of the main advantages of HES is that it reduces storage requirements when compared to individual RES. Meeting the electricity demand using renewable sources in a reliable manner is a challenging task because of their noncontinuous characteristic. The complexity of nonlinear integral planning in HES on account of the random nature of renewable sources and loads requires that it be solved by intelligent optimization techniques.[18–21] Optimization of HES via computational intelligence search techniques can easily overcome the drawbacks of conventional optimization techniques, such as the requirement of large running-time and large running-space.

3.1 Resource and Load Data

In this study, the problem of off-grid HES optimization using the HGAPSO algorithm has been taken up. As a case study

for HES optimization, the Narendra Nagar block of the district Tehri Garhwal, which is in the Uttarakhand State of northern India, has been considered. It consists of nine villages without access to the national power line.[15] The electrification of this hilly and remote area by grid extension seems economically unviable. The solar radiation in this area is, on average, 5.08 kWh/m^2 and goes as high as 6.06–6.95 kWh/m^2 in the months from April to June. Annual total solar energy potential of 1854.18 kWh/m^2/yr is available to be converted to electricity by a photovoltaic (PV) system.[24] Average wind speed in this area is found to be around 5.2 m/s. The total micro-hydro generation (MHG) potential of this area is found to be around 15 KW.[25] Total biogas (BG) availability for electricity generation is estimated to be 133.2323 m^3/day, and total biomass generation (BMG) potential for electricity generation is estimated to be 147.757 ton/year.[15] This resource availability data is also listed in Table 1.

Using the above data, the following equations have been used to calculate the electricity generation potential of each resource, and the results thus obtained have been tabulated in Table 2.

Total electricity by MHG in KWh in an hour is given by

$$E_{MHG} = P_{MHG} \times \eta_{MHG} \tag{3}$$

$$P_{MHG} = 9.81 \times Q \times \rho \times h \tag{4}$$

where P_{MHG} is electrical power generated by the micro-hydro unit, η_{MHG} is the efficiency of micro-hydro generation that is taken to be 60%,[15] Q is discharge in m^3/s, ρ is density of water (1000 kg/m^3), and h is head in meters.

Table 1. Renewable resource availability data.

Resource	PV	Wind	MHG	BG	BMG
Availability	1854.18 KWh/m²/year	5.2 m/s average speed	15 KW	133.2323 m³/day	147.757 ton/year

Table 2. Calculated energy production potential of all selected technologies.

Technology	MHG	PVG	WT	BG	BMG
Energy produced per unit per day (KWh/day)	183.878	213.972	15	266.4646	303.6103

Table 3. Economic parameters of all selected technologies.[15,22]

Technology	MHG	PVG	WT	BG	BMG	DG	Battery
Cost of Energy (Rs./KWh)	1.45	15.68	33.56	3.98	4.78	11	3.26

The hourly delivered electricity energy output of a biomass gasifier energy system is given by

$$E_{BMG} = P_{BMG} \times \eta_{DFEG} \tag{5}$$

$$P_{BMG} = \frac{\text{Total fuel wood available (ton/yr)} \times 1000 \times CV_{BM} \times \eta_{CBM}}{365 \times 860 \times (\text{Operating hours/day})} \text{ KW} \tag{6}$$

where CV_{BM} is calorific value of biomass (fuel wood) that is 4015 kcal and η_{CBM} is biomass conversion efficiency that is taken to be 25% (typical value). Operating hours per day are taken to be 12 hours,[15] and η_{DFEG}, which is efficiency of the dual-fuel engine generator, is taken to be 90%.

Hourly energy output of biogas generator is given by

$$E_{BGG} = P_{BGG} \times \eta_{DFEG} \tag{7}$$

$$P_{BGG} = \frac{\text{Total biogas generated (m}^3/\text{day)} \times CV_{BG} \times \eta_{CBG}}{860 \times (\text{Operating hours/day})} \text{ KW} \tag{8}$$

where CV_{BG} is calorific value of biogas, which is 4700 kcal, and η_{CBG} is biogas conversion efficiency, which is taken to be 30% (typical value). Operating hours per day are taken to be 10 hours,[15] and η_{DFEG}, which is efficiency of the dual-fuel engine generator, is taken to be 90%.

Hourly energy output of a PV generator can be calculated by following equation:

$$E_{PVG} = G \times A \times \eta_{PVG} \tag{9}$$

where G is hourly irradiance in KWh/m², A is surface area of PV modules in m², and η_{PVG} is efficiency of a PV generator taken to be around 12%.

Hourly energy output of a wind turbine (WT) generator can be calculated as

$$E_{WT} = P_{WT} \times \eta_{AD} \tag{10}$$

$$P_{WT} = \frac{1}{2} \times \rho \times A \times v^3 \times C_p \tag{11}$$

where ρ is the air density (kg/m³), A is the swept area of the rotor (m²), V is the wind speed (m/s), C_p is the efficiency of the wind turbine, and η_{AD} is the efficiency of the AC/DC converter (assumed to be 95% in this study).

The energy produced by various technologies is based on their maximum available potential per year in this area.

Table 3[15] provides cost of energy in Rs./KWh for available resources, viz. MHG, BMG, BG, PV, WT, and DG (diesel generator), which have been used in this work.

The load profile is as shown in Fig. 3. The highlights of the load profile, as per the study conducted in Gupta et al.,[15] are as follows. For this area, there is nearly no demand for electricity in the night, a morning peak with up to 69 kW from 8:00 a.m., a peak of 111 kW at 1:00 p.m. and a long-lasting evening peak from 6:00 to 11:00 p.m., going up to 114 kW at 9:00 p.m. The three biggest consumer devices in this energy system are lighting, fans, and televisions. In the winter period, fans are not used to the same extent as in the summer season; thus, the load profile is taken from the summer period in order to enable the energy system to deal with peak loads in the summertime. Total average daily load required is given to be 1271.61 KWh. The maximum hourly load demand is 113.824 kWh at 1800–2100 h, and the minimum is 2 kWh at 2300–2400h and 0000–0400h on a typical summer day. In Gupta et al.,[15] wind energy resources have not been considered. In this study, for this particular area wind energy resources have also been considered, although its contribution in the optimized result comes out to be negligible only because of low wind speed in this area.

The line diagram of a hybrid energy system for an off-grid remote area used in this study is already shown in Fig. 1.

As shown in Fig. 1, the HES consist of a PV subsystem, other renewable resources such as MHG, BG, BMG, and wind, and a diesel generator. The PV subsystem generates DC voltage, part of which is exchanged with a battery bank and part is used to serve AC load through an inverter. Other renewable sources generate AC, so they can directly serve the load. In this case for battery power exchange, AC is first rectified and sent to DC bus. For intermittent increase and decrease in energy, dump load and diesel generator are used, respectively. The HGAPSO algorithm has been used in this article to decide the optimum mix of all these resources in such a manner that it results in the lowest cost of HES.

The following section provides in detail the information about the assumptions of the model, objective function, and constraints.

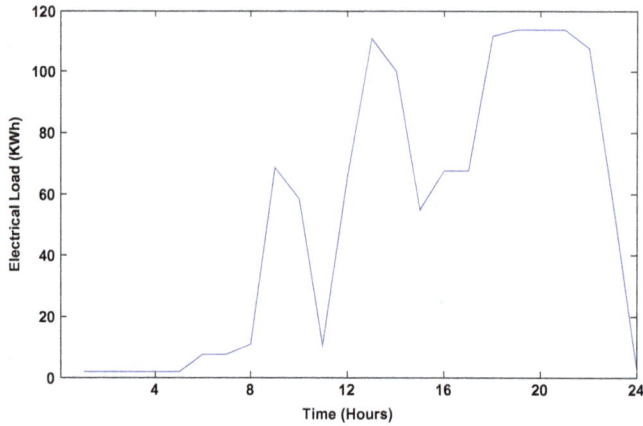

Fig. 3. Load profile of the area.

3.2 Assumptions

1. Prices of all resources over the lifespan of the project, i.e., 20 years, are assumed to be constant. Also, it is assumed that the output of the PV array and the wind turbine linearly depends only on meteorological data and not on the age of the facility. Also, constant efficiencies for the diesel generator and for the battery charging or discharging are assumed, which are 40%, 90%, and 95%, respectively. The capacity of the battery is assumed to be constant over its lifetime.

2. Weather and load data are assumed to be constant within each one-hour time step. Load profile is also assumed to be constant during this one-hour period.

3. The available load data is representative of the whole economic lifetime.

4. The total PV electricity generation is estimated by applying full-load hours while only the global horizontal solar radiation has an influence on the hourly PV power output.

5. Technology for dump load and its cost are not considered in this model.

6. The cost function of the system includes investment cost, operation, and maintenance cost for a total lifespan of 20 years.

3.3 Optimization Approach Used

This system consists of a number of renewable energy sources; thus there are various load dispatch strategies that could be used. The one used in this work is as follows. First, the load demand is subtracted from the total energy generated from these renewable resources to calculate the net load for each slot of time, which here is an hour. If the net load is zero or positive, then the excess energy is supplied to charge the batteries, until the battery state of charge (SOC) reaches 80%. If battery SOC reaches 80%, the excess energy may be given to a dump load. If the net load is negative for a particular hour, first the battery energy is used to meet the excess load demand, until the battery SOC is above 20%. Battery is chosen as the first alternative to meet load demand

as the unit cost of the battery is less than that of a diesel generator. When the battery SOC goes below 20%, the diesel generator is turned on to meet the net load demand and to charge the battery. The advantage of using a diesel generator for meeting net load demand and to charge the battery is that this enables the diesel generator to operate at high output power, thus improving its operating efficiency and reducing the cost involved in its operation.

The diesel generator is made to run for a minimum run time, which in this article is taken to be 20 minutes (typical value), to reduce engine wear and minimize the maintenance cost. If the net load is greater than the diesel generator rated capacity, the battery will contribute to meet the net load demand. These conditions are programmed in the developed algorithm so that it checks the net load and battery SOC at each time step, i.e., one hour, and decides the energy flow accordingly to calculate the minimum cost of the system.

The implemented HGAPSO algorithm concerns the distribution of load demand over the available units in order to minimize the operational cost. Thus, the unit cost of various resources, as given in Table 1, is very important as the objective function is formed using this data such that the overall system operating cost comes out to be minimum. The common strategy used generally is that the cheapest generating unit is used first, and as load increases, the more expensive units are brought in. To implement it in the simulation algorithm, the various inputs as given in Table 1 and various constraints as already stated above are to be implemented, which makes it quite a complex and computationally expensive simulation program if done by conventional computational approaches. In this study, it is done using the HGAPSO algorithm and as validated by the results, the HGAPSO algorithm does this more efficiently compared to other popular techniques.

The objective function for cost to be minimized is obtained using Table 1, Table 2, and Table 3.

This cost function is to be minimized subject to the constraint of meeting load demand. The total of energy provided by each resource must be greater than or equal to the load demand. Some

Table 4. GA and PSO parameters used in this study.

GA parameters	PSO parameters
Size of population: 100	$C1 = 1.5$
Fitness scaling: proportional	$C2 = 4\text{-}C1$
Probability of crossover: 0.7	
Crossover function: Heuristic	
Mutation function: Gaussian	
Elitism: 2 chromosomes	

other simulation control parameters taken are battery charge and discharge limits to be 80% and 20%, respectively, diesel generator minimum and maximum loading to be 80% and 100%, respectively, and diesel generator starting point and stopping point as a percentage of battery capacity to be 20% and 80%, respectively. In this work, the installation cost for all items of hybrid power plant is considered to be per unit for each item.

After many tests using various parameter values, the GA and PSO parameters used in this study are as given in Table 4.

4. Results and Discussion

The results obtained for optimization of installed capacity and the total cost of electricity for various algorithms are as tabulated in Table 5. It gives a summary of the optimized renewable energy contribution of the various energy resources used in order to cover the load demand by all the three methods of optimization, i.e., HGAPSO, GA, and PSO. As is evident from this table, the minimum optimized cost of HES as well as the minimum cost of per unit electricity generated by HES is obtained from HGAPSO algorithm, thereby establishing the superiority of the HGAPSO algorithm over the GA and the PSO algorithm for the problem at hand.

Also, it can be seen from the results, in all the three cases, the wind energy contribution is coming out to be zero, which is justified as wind energy is not viable for the given wind velocities (around 5.2 m/s) for the region under consideration. Also, it can be observed from the obtained results that the average least annual system unit cost is coming out to be Rs. 3.98 as computed by the HGAPSO algorithm, which is 10.56% less than that obtained from the PSO algorithm, and that obtained from the GA method is 7.41% less than that obtained from the PSO algorithm (taking cost obtained from the PSO algorithm as the reference to calculate cost savings as it is the highest in the studied three cases). Of the total primary energy requirement of the villages, the renewable generators produced 81.86%, while the diesel generator produced 18.14% of the energy, which can easily be calculated from Table 5. (The total contribution of DG is 20 KW of the total 82 KW load required, which results in 81.86% of renewable energy resources and 18.14% of DG.)

Moreover, comparison of different optimization strategies used, i.e., HGAPSO algorithm, GA, and PSO algorithm, as shown in Table 6 shows the superiority of the HGAPSO algorithm in terms of computation time as well as average number of convergence generations, as expected and proposed earlier. This is more clearly evident in Fig. 4, which shows the convergence curves of the HGAPSO algorithm, PSO algorithm, and GA. It can be clearly seen that HGAPSO converges to the optimum fitness value at around 30 iterations and is less than those for the GA and PSO algorithm. Thus, HGAPSO is superior to PSO and GA in terms of termination criterion required, for the problem under consideration.

5. Conclusion

An optimum mix of resources to achieve cost optimization of HES is carried out in this research. From the results obtained, as tabulated in Table 5, it is concluded that the HGAPSO algorithm

Table 5. Optimization results using HGAPSO, GA, and PSO for various installed capacities.

Technology / AI technique	Installed capacity (KW)						Battery (KWh)	Total cost of electricity (Rs./KWh)	Annual optimal cost of HES (Rs.)	% Cost savings
	MHG	BMG	BG	PVG	Wind	DG				
HGAPSO	19.26	51	16	1	0	20	106	**3.98**	**2,214,343.5**	**10.56%**
GA	19.50	54	16	2	0	21	98	4.12	2,268,190.4	7.41%
PSO	19.30	54	16	2	0	21	97	4.45	2,395,115.3	0.00%

Note. As shown by boldfaced numbers, least system cost is given by the HGAPSO algorithm.

Table 6. Comparison of different optimization strategies used.

Optimization technique	Running time (sec)	Convergence generations			Population size
		Minimum	Maximum	Average	
HGAPSO	**26.76**	6	53	**30.55**	80
GA	30.87	5	57	40.74	100
PSO	29.75	10	50	45.40	80

Note. As shown by boldfaced numbers, the HGAPSO algorithm is superior in terms of computation time and average number of convergence generation as compared to the other two algorithms.

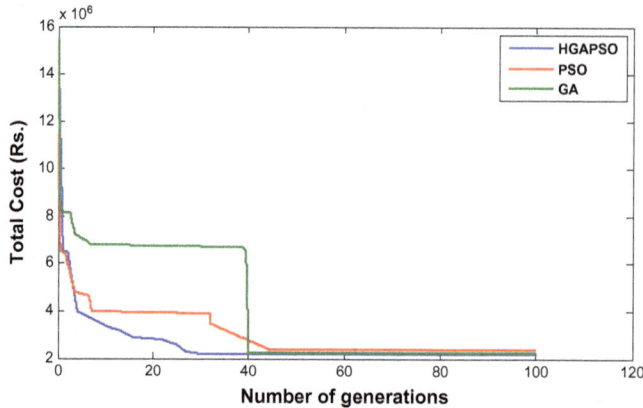

Fig. 4. Convergence of optimization of algorithms.

is giving better results than the GA and the PSO algorithm in terms of reduced cost (10.56% less than PSO) and hence validates the efficacy of the HGAPSO algorithm over the GA and the PSO algorithm for achieving cost optimization of HES. The main reason for choosing the method of HGAPSO, which comes under computational intelligence-based optimization techniques, is the chance to overcome the computational limitations of the existing numerical techniques, such as high computation time and inability to converge at optimum point in case of complex constraints, which is shown to be achieved by the results obtained as tabulated in Table 6 and Fig. 4. It is tested on a case study as presented in Gupta et al.[15] It is Narendra Nagar in northern India, consisting of nine villages, and has a peak load demand of 113.8 kW. The average wind speed is 5.2 m/s, and 1800 h/a of PV full-load hours can be achieved. The output of the model's optimization process is a diesel-PV hybrid energy system including a battery, BMG, BG, and MHG. Hybrid energy systems that include renewable energy help to reduce the cost of electricity much more efficiently. Thus, the algorithm presented is capable of designing a low-cost electrification system for the villages by using renewable resources and keeping the output constant with the help of a diesel generator. Also, it can be extended to include other resources as well, such as fuel cells, and to explore the comparative economic feasibility of a HES with various different types of energy storage systems.

Appendix: Nomenclature

Acronyms

GA: Genetic algorithm
PSO: Particle swarm optimization
HGAPSO: Hybrid GAPSO
HES: Hybrid energy system
PV: Photovoltaic
DG: Diesel generator
BG: Biogas
BMG: Biomass generator
MHG: Micro-hydro generator

Symbols

BATT: Battery
$E_{PVG, Load}$:   Energy given by PV source to load
$E_{BATT, Load}$: Energy given by battery to load
$E_{DG, Load}$:   Energy given by DG to load
$E_{REN, Load}$:   Energy given by renewable source to load
$E_{PVG, BATT}$:   Energy given by PV source to charge battery
$E_{DG, BATT}$:   Energy given by DG to charge battery
$E_{REN, BATT}$:   Energy given by renewable energy source to charge battery

References

1. Borowy, B. S.; Salameh, Z. M. Optimal Photovoltaic Array Size for a Hybrid Wind/PV System. *IEEE Trans. Energ. Convers.* **1994**, *9*, 482–488.

2. Wies, R.W.; Johnson, R. A.; Agrawal, A. N.; Chubb, T. J. Simulink Model for Economic Analysis and Environmental Impacts of a PV With Diesel-Battery System for Remote Villages. *IEEE Trans. Power Sys.* **2005**, *20*, 692–700.

3. Bagul, D.; Salameh, Z.M.; Borowy, B. Sizing of a Stand-Alone Hybrid Wind-Photovoltaic System Using a Three Event Probability Density Approximation. *Solar Energ.* **1996**, *56*, 323–335.

4. Barsoum, N.N.; Vacent, P. Balancing Cost, Operation and Performance in Integrated Hydrogen Hybrid Energy System, Proceedings of the First Asia International Conference on Modelling and Simulation (AMS '07), 2007.

5. Liu, X.; Islam, S. Wind–Diesel–Battery Hybrid Generation System Reliability Analysis on Site and Size Factors, 4th International Conference on Electrical and Computer Engineering ICECE, 2006; 19–21.

6. Belfkira, R.; Nichita, C.; Barakat, G. Modeling and Optimization of Wind/PV System for Stand-Alone Site, Proceedings of the 2008 International Conference on Electrical Machines, 2008.

7. Khan, M.J.; Iqbal, M.T. Analysis of a Small Wind Hydrogen Stand-Alone Hybrid Energy System. *Appl. Energ.* **2009**, *86*, 2429–2442.

8. Dondas, M.; Alkhalil, F.; Degobert, P.; Colas, F.; Robyns, B. Supervision of a Hybrid System Based on Photovoltaic Arrays and Supercapacitors, IEEE Conference POWERENG2009, Lisbon, Portugal, March 18–20, 2009; 342–347.

9. Saif, A.; Gad Elrab, K.; Zeineldin, H.H.; Kennedy, S.; Kirtley, J.L. Multi-objective Capacity Planning of a PV-Wind-Diesel-Battery

Hybrid Power System, IEEE International Energy Conference, 2010; 217–222.

10. Mohammadi, A.; Jazaeri, M. A Hybrid Particle Swarm Optimization-Genetic Algorithm for Optimal Location of SVC Devices in Power System Planning, In *Proceedings of the 42nd International Universities Power Engineering Conference*, 2007; 1175–1181.

11. Esmin, A. A. A.; Lambert-Torres, G.; Alvarenga, G. B. Hybrid Evolutionary Algorithm Based on PSO and GA Mutation, In Proceedings of the 6th International Conference on Hybrid Intelligent Systems, 2006.

12. Jeong, S.; Hasegawa, S.; Shimoyama, K.; Obayashi, S. Development and Investigation of Efficient GA/PSO-Hybrid Algorithm Applicable to Real-World Design Optimization, IEEE Congress on Evolutionary Computation, 2009.

13. Kao, Y.-T.; Zahara, E. A Hybrid Genetic Algorithm and Particle Swarm Optimization for Multimodal Functions. In *Applied Soft Computing 8*; Elsevier, 2008, pp 849–857. doi:10.1016/j.asoc.2007.07.002.

14. Premalatha, K.; Natarajan, A.M. Hybrid PSO and GA for Global Maximization. *Int. J. Open Problems Comp. Math.* **2009**, *2*, 597–608.

15. Gupta, A.; Saini, R.; Sharma, M. Steady-state Modelling of Hybrid Energy System for Off Grid Electrification of Cluster of Villages. *Renew. Energ.* **2010**, *35*, 520–535.

16. Nayar, C.V. Recent Developments in Decentralized Minigrid Diesel Power Systems in Australia. *Appl. Energ.* **1995**, *52*, 229–242.

17. Rajoriya A.; Fernandez E. Sustainable Energy Generation Using Hybrid Energy System for Remote Hilly Rural Area in India. *Int. J. Sustain. Eng.* **2010**, *3*, 219–227.

18. Hameed, A.-M. H.; Elhagri, M. T.; Shaltout, A. A. Optimum Sizing of Hybrid WT/PV Systems via Open-Space Particle Swarm Optimization, Second Iranian Conference on Renewable Energy and Distributed Generation, 2012.

19. Zhanga, B.; Yangb, Y.; Ganb, L. Dynamic Control of Wind/Photovoltaic Hybrid Power Systems Based on an Advanced Particle Swarm Optimization, IEEE International Conference on Industrial Technology—ICIT, 2008.

20. Bansal, A. K.; Gupta, R. A.; Kumar, R. Optimization of Hybrid PV/wind Energy System using Meta Particle Swarm Optimization (MPSO), IEEE Indian International Conference on Power Electronics, 2011, New Delhi, India.

21. Zhao, Y. S.; Zhan, J.; Zhang Y.; Wang, D. P.; Zou, B. G. The Optimal Capacity Configuration of an Independent Wind/PV Hybrid Power Supply System Based on Improved PSO Algorithm, 8th International Conference on Advances in Power System Control, Operation and Management, Nov 8–11, 2009; 1–7.

22. Huneke, F.; Henkel, J.; Benavides, J. A.; Erdmann, G. G. Optimisation of Hybrid Off-Grid Energy Systems by Linear Programming. Energy, Sustainability and Society, 2012.

23. Magyar, G.; Johnsson, M.; Nevalainen, O. An Adaptive Hybrid Genetic Algorithm for the Three-Matching Problem. *IEEE Trans. Evol. Comp.* **2000**, *4*, 135–146.

24. Mani, A.; Rangarajan, S. *Solar Radiation over India*. 2nd Ed. Allied Publishers: New Delhi, India; 1982; pp 302–303.

25. Main Project Report–2005. Unelectrified Villages—Surveys, Potential Sources and Electricity Demand, vol. II–4/12. Roorkee, India: A.H.E.C, I.I.T; 2005.

17

European End-User's Level of Energy Consumption and Attitude Toward Smart Homes: A Case Study of Residential Sectors in Austria and Italy

TAMER KHATIB[1]*, ANDREA MONACCHI[1], WILFRIED ELMENREICH[1], DOMINIK EGARTER[1], SALVATORE D'ALESSANDRO[2], and ANDREA M. TONELLO[2]

[1]*Institute of Networked & Embedded Systems/Lakeside Labs, Alpen-Adria-Universität Klagenfurt, Klagenfurt, Austria*
[2]*WiTiKee s.r.l., Udine, Italy*

Abstract: This article presents a quantitative assessment of the level of energy consumption of inhabitants located in Carinthia and Friuli-Venezia Giulia. In addition, an analysis for the current structural barriers for smart powered homes and smart energy management systems is conducted. A questionnaire consisting of 43 questions is used to address the aforementioned issues. In particular, a sample size of 385 respondents with a confidence of 95% and marginal error of 5% is found to be representative of the adopted area. Based on the results, we modeled the average energy consumption of a typical 110 m² area household with 16.8 kWh/day, a 2.6 kW peak, and a load factor of 27%. Furthermore, an average of 46% of the respondents expressed the willingness to exploit tariff systems for operating their electrical appliances, and about two thirds of the respondents declared that they care about the energy efficiency at their households. However, low renewable energy utilization is observed due to some existing structural barriers. Therefore, an analysis and a discussion are carried out to investigate these barriers. Finally, some recommendations are provided according to the obtained results.

Keywords: Energy consumption level, smart energy management system, feed-in tariff

1. Introduction

Carbon dioxide emission problems encouraged researchers and governments to investigate smart homes, which utilize sustainable energy sources at a high level of energy efficiency together with a smart energy management system (EMS). Smart homes are renewable energy–powered homes that exploit computer-based technologies to control a home's electrical appliances. Such systems can range from simple remote controlling of lighting and other simple loads to complex micro-controller-based networks with different levels of automation and intelligence. Smart homes are promoted for reasons of energy security and efficiency.[1,2,3]

*Address correspondence to: Tamer Khatib, Institute of Networked & Embedded Systems/Lakeside Labs, Alpen-Adria-Universität Klagenfurt, Khevenhullerstabe 35, Klagenfurt 9020, Austria. Email: tamer_khat@hotmail.com

The share of renewable energies in 2007 in Austria was about 25%.[1] However, three quarters of the demanded energy is by conventional energy sources. Due to the lack of oil resources, Austria brings about 70% of the needed oil by imports. Hence, reduction of fossil fuel imports is an important goal in Austria. EU member states are supposed to save 1% of their final energy use per year through energy efficiency improvements by 2016.[2] Moreover, overall energy efficiency should be improved by 20% until 2020 as required by the Europe 2020 strategy. The Europe 2020 renewable energy strategy aims to be the start of the transition to renewable and sustainable energy. In general, the aims of this strategy are to limit the contribution of conventional energy, to improve the greenhouse gas performance of biofuel production processes, to encourage a greater market penetration of low carbon technologies, and to improve the reporting of greenhouse gas emissions by obliging member states. In 2009 in Austria, residential heating was considered the biggest energy end-use category, causing three quarters of private households' energy use and 18% of total final energy use.[2] Bittermann[4] depicts that for a residential load in Austria, 40% of the total energy consumption is utilized for thermal systems, including water heating (16%), space heating (14%), and cooking (10%). In addition, cooling systems, including refrigerators and freezers, account for 12% of the total demanded energy. Meanwhile, lighting, consumer electronics, other large household appliances, small appliances, and standby modes account for 8%, 7%, 9%, 4%, 4% of the total demand, respectively. Households in rural areas such as Carinthia consume 21% more energy than urban regions such as Vienna. Furthermore, the costs of energy for water heating and space heating in rural areas are 106% and 172%, respectively, of the average amount in urban cities, such as Vienna.[4] These statements are further supported[5] where space heating, cooling, and water heating is considered the second largest share of the energy consumed in Austria in 2008.

In Italy, the total electrical energy consumption in 2009 was about 320 TWh, while the total energy production was about 275 TWh.[6] Furthermore, it is reported that Italy uses mainly oil (16%), gas (50.7%), and coal (16.8%) for electricity production; all of these resources are imported.[7] Moreover, Italy is currently buying electricity from other countries to supply 2% of its total energy demand.[8] Aste[9] depicts that the Italian residential sector, as a part of the European residential sector, is responsible for 40% of the final energy consumption. Therefore, the building sector has a very high potential in terms of reducing consumption. Italy, as a member of the European Union, committed to have at least 20% of the overall gross final energy consumption coming from renewable sources by 2020.[9]

However, before enforcing energy efficiency and smart home acts, the gap between energy users represented by the public, energy producers, and the government and energy efficiency needs to be studied. To this end, it is important to let the citizens be aware of the detailed energy consumption in their households.[1] Smart homes and energy efficiency technologies are challenging due to many gaps and wrong practices. In this context, the INTERREG-IV Italy-Austria program through the European regional development fund (ERDF) and national public resources have co-funded the MONERGY research project (ICT solutions for energy saving in Smart Homes). The project focuses on measuring the attitude of the public toward energy-efficient technologies, on increasing public awareness regarding energy-efficient technologies, and on developing innovative solutions that will have an impact on the reduction of energy consumption in Friuli-Venezia Giulia (IT) and Carinthia (AT) households.

The main objectives of this article are, first, to estimate the level of energy consumption in the homes of the studied areas by collecting information about the available electrical appliances and, second, to measure the attitude of the end user toward a smart home concept and the adoption of smart EMSs. Finally, we analyzed the reasons that may cause a gap for the adoption of EMSs by the end-users and for the inefficient management of the energy resource.

2. The Questionnaire

Carinthia is the southernmost Austrian state. Located within the Eastern Alps, the largest metropolitan area consists mostly of the Klagenfurt basin. The Carnic Alps and the Karawanks make up the border to the Italian region of Friuli-Venezia Giulia (Friuli-VG) and Slovenia. The population of Carinthia is about 558,300 with a density of 59 inhabitants per km². The other region observed, Friuli-VG, is one of the 20 regions of Italy. It has an area of 7,858 km² and about 1.2 million inhabitants.

In this research, a questionnaire consisting of 43 questions was developed. The questions were formulated to study the characteristics of households, type of devices/appliances, and end-user behavior. The targeted sample consists of people who are older than 18 years and who live in the considered regions. The survey was offered in two languages—Italian and German—in order to address the native language of the majority in both regions. A sample of the survey can be found online at http://tinyurl.com/questionnaire-monergy.[11] In order to have a random sample of the participants and to ensure a wide and balanced distribution of the respondents, the survey was simultaneously announced via mailing lists that include addresses for employees, students, companies' members, and other inhabitants across the two regions.

The research sample plays a major role to ensure the validity of any study. The research sample size is influenced by a number of factors, including the purpose of the study, the total population size, the desired statistical precision and the accepted sampling error, and the limitations of the study. The volume of a representative sample (S) can be given by[12]

$$S = \frac{x^2 NP(1-P)}{d^2(N-1) + x^2 P(1-P)} \qquad (1)$$

where N is the size of the population; P is the ratio of the population and it is estimated to be 0.5 to give the maximum sample size. The parameter d is extracted from a chi-square distribution with 1 degree of freedom at the desired confidence level (3.841).

Krejcie and Morgan[12] provide a supporting table based on Equation 1 to be used for obtaining the size of the representative sample. The table includes the population size and the corresponding sample size that should be selected from it. Following that, the representative sample size must be about 385 with a confidence level of 95% and a marginal error of 5% as the total number of inhabitants in the case study is about 1.7 million. By analyzing the responses obtained, we found that more than 70% of these responses were from the cities of Klagenfurt and Udine, which together have a population of 200,000 inhabitants. In this research, we addressed households with two to three inhabitants. Therefore, considering these two important aspects, the representative sample size will have a higher confidence level and a lower marginal error.

The aim of this article is to measure the level of energy consumption using the end-user feedback. In addition to that, the article aims to analyze the behavioral and structural barriers that contribute to the gap between the energy efficiency and the public. Therefore, the following research questions are formulated:

Q1: What is Carinthia and Friuli-VG citizens' level of energy consumption?

Q2: What is the Carinthia and Friuli-VG citizens' attitude toward the smart home concept and the energy management systems?

We approach the aforementioned questions by investigating different aspects as illustrated in Table 1. In order to answer the first research question, we asked for information about the availability of some electrical appliances, some related behavioral practices, and information about the electricity bill. Furthermore, to measure the attitude toward the adoption and usage of

Table 1. Aspects used to answer the stated research questions.

Research question	Variables
Q1	1. Laundry times per month
	2. Use of dishwasher, drier, washing machine
	3. Presence of electric hob
	4. Presence of electric oven
	5. Number of consumer electronic devices
	6. Habit to leave devices in standby
	7. Use of electrical heaters
	8. Number of air conditioning units
	9. Use of air conditioner during summer
	10. Number of electric boilers
	11. Average monthly energy bill
Q2	1. Knowledge of home automation (HA)
	2. Ownership of HA system
	3. Willingness to purchase
	4. Usefulness of energy awareness
	5. Switched to energy-saving light bulbs
	6. Habit to leave lights on with no one
	7. Replaced devices in the last 4 years
	8. Devices used in lower-price periods
	9. Devices used in lower-demand periods
	10. Absence of tariffs promoting energy shifting
	11. Presence of RE system (PV, wind, geothermal, or SWH)

Table 2. Survey demographic data.

Demographic variable	Carinthia	Friuli-VG
Location	59.0%	41.0%
Gender		
Female	52.0%	45.0%
Male	48.0%	55.0%
Age		
18–35	37.1%	59.7%
36–45	29.6%	17.3%
46–65	31.2%	20.1%
>65	2.1%	2.9%
Education		
School	25.3%	37.4%
Bachelor's	4.8%	7.2%
Master's	35.5%	38.1%
PhD	29.0%	14.4%
Other	5.4%	2.9%

smart home energy management systems, some questions about awareness, positive actions, existing policies, and beliefs were formulated.

3. Questionnaire Results

In this research, 397 responses were received. However, these data were screened in order to remove any anomalies or inconsistencies. For example, the number of floors with respect to the household size was cross-checked. After that, 340 valid responses were obtained, with a completion rate of 85.64%. This sample size is acceptable as a representative sample size compared to the theoretical representative simple size (385). As for the demographic balance of the sample, 193 responses were received from Carinthia, while 139 responses were from Friuli-VG (1.34:1). Furthermore, 167 responses were sent by females from both regions, while 166 responses were sent by males (1:1.04). Table 2 shows a detailed demographic analysis of the conducted survey.

In this research, we applied a duplicating questions policy by rephrasing certain questions in order to check some critical aspects. Moreover, anonymity was ensured, and it was also ensured that the questionnaire was not filed more than once.

A preliminary analysis of the proposed survey was previously conducted in Monacchi et al.[11] This analysis includes principal component analysis giving a good overview of the dataset by the biplot produced as shown in Figure 1.

From the figure, Carinithians' use of electric hobs, heaters, and boilers accounts for a greater share of their energy profile

being accounted for by these devices, but this is less apparent in Friuli-VG, as a greater proportion of residents use gas-powered devices. On the other hand, people from Friuli tend to have more air conditioners. In Carinthia, certain households use a night meter to manage the main electrical boiler using a cheaper tariff. Given a confidence level of 95%, the Spearman's ρ for the number of residents and the average monthly electricity bill is 0.408 for Friuli and 0.308 for Carinthia.[11]

3.1 Q1: What Is the Carinthia and Friuli-VG Citizen Consumption Level of Energy?

In order to address this question, the respondent was asked about 11 issues as illustrated in Table 3. However, considering the fact that heating devices are the most energy-consuming devices,[13,14] we focused on this issue by asking the respondent to provide information on the heating systems that are used in houses. Table 3 shows the results obtained regarding the type of energy sources used in space and water heating. It is worth mentioning that some of the respondents utilize two types or even more of energy sources for space and water heating.

From Table 3, it is clear that few people in both zones use electricity for space heating. However, in Carinthia, 41.4% of the respondents use electricity for water heating. In addition, the respondents were asked about the adoption of air conditioning units at homes as well as the frequency of using these units. According to the data, only 2.16% of the Carinthian respondents have air conditioning units at home with an average of one device per home and a usage frequency of two times per day. In Friuli-VG, 45.19% of the respondents have air conditioning units with two devices per home and a usage frequency of two times per day.

In order to estimate the average consumption of a home, the respondents were asked to list their existing appliances, so that the consumption of these appliances can be estimated. Table 4 shows the results obtained from the survey. The average operating hours and average rated power records are according to Basu et al. and Tewolde et al.[13,14]

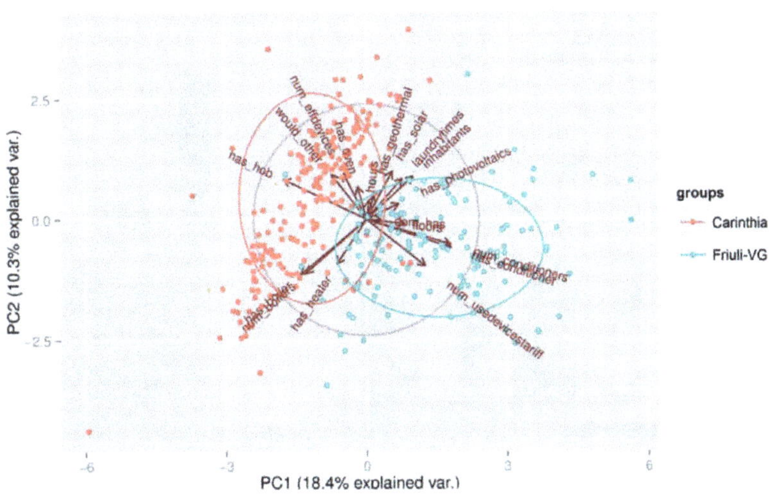

Fig. 1. Biplot of the principal component analysis applied to the conducted survey.[11]

Table 3. Energy sources used for space and water heating.

Energy source	Carinthia	Friuli-VG
	Space heating	
Electricity	10.2%	6.5%
Gas	39.8%	63.3%
Gasoline	21.5%	8.6%
Wood/pellet	18.8%	18.7%
Solar power	2.7%	2.9%
Geothermal	7.0%	0.0%
	Water heating	
Electricity	41.4%	12.2%
Gas	24.7%	82.0%
Gasoline	22.0%	6.5%
Wood/pellet	13.0%	3.5%
Solar power	15.6%	13.0%
Geothermal	7.5%	0.0%

Table 4. House electric energy audit.

	Carinthia	Friuli-VG	Typical rating power [W]	Typical operating hours per month
Refrigerator	98.9%	99.3%	150	300
Electrical oven	100.0%	88.0%	2,400	10
Electrical hob	98.4%	5.2%	950	60
Freezer	40.9%	27.3%	400	300
Microwave oven	60.8%	61.2%	1,500	11
Hood	69.9%	82.7%	450	60
Dishwasher	85.0%	68.4%	1350	25
Washing machine	92.5%	87.1%	450	17
Dryer	28.0%	5.8%	3000	17
Iron	74.7%	77.0%	800	5
TV	85.5%	89.2%	190	180
Computer	96.2%	97.8%	300	75
Cordless phone	31.7%	66.9%	70	10
Air conditioner	2.2%	45.2%	1,500	200
Printer	73.1%	79.9%	50	5
DVD player	69.4%	69.8%	400	60
HiFi stereo	63.4%	54.0%	1000	40
Home theater	7.5%	13.0%	1000	180
Game console	34.4%	28.1%	200	60

On average, the electricity price in Austria is 0.155 €/kWh. Italy has a multi-tariff system that starts with a tariff of 0.129 €/kWh for a consumption up to 900 kWh per month. The tariff increases up to 0.314 €/kWh with consumption above 4440 kWh per month. In order to estimate the monthly bill, the data in Table 4 can be used to calculate the average monthly consumption. Then the monthly bill can be estimated considering the aforementioned prices. In our further analysis, only appliances with a presence in more than 50% of the households are considered.

In order to generalize a daily load demand, a prediction of human behavior is required. Due to the extreme difficulty in predicting such a variable, some intuitive roles can be applied for the usage time of the listed appliances as given in Table 4. For example, the refrigerator and the freezer are assumed to be operated periodically considering the typical operation time. Oven, hob, hood, and microwave oven are assumed to be operated at breakfast and dinner times only. Appliances for laundry, dishwashing, and similar activities are assumed to be operated in the evening. Finally, some other devices such as televisions, computers, and air conditioners are assumed to be operated at different times such as late morning, in the afternoon, and mainly in the evening. Statistical models presented in the literature[14,15] have been utilized in generating an estimated load demand. Figure 2 shows an estimated daily load demand for both zones. The load

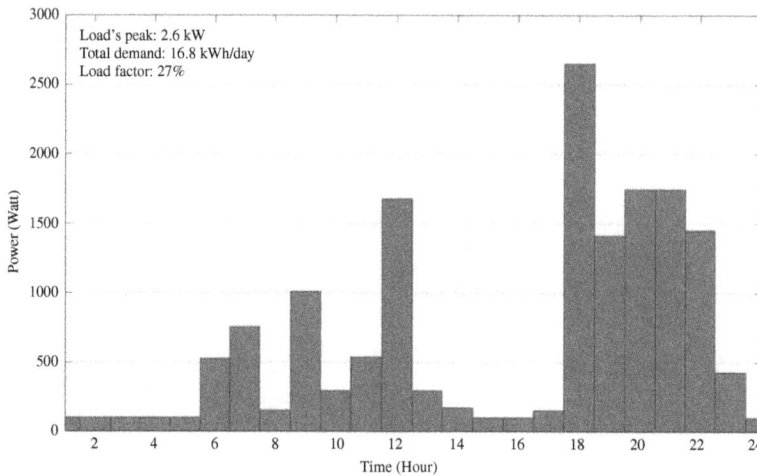

Fig. 2. Predicted daily load demand for Carinthia and Friuli-VG households.

demand has a daily consumption of 16.8 kWh, peak power of 2.6 kW, and load factor of 27%. According to the survey, the average area of the households is 110 sqm with average inhabitants of 2.6 persons. The register-based census 2011 from Statistik Austria yields an average size of 98.7 sqm for Carinthia with an average of 43.2 sqm per person. Thus, in our survey we have a slight bias toward larger households with more people in them. As a result of our survey, the average energy monthly consumption accounts for 4.6 kWh/(sqm month) or 197 kWh/(inhabitant month). For comparison, Statistik Austria reports on 2011 a value of 2050 kWh/inhabitant annually, yielding 170 kWh/(inhabitant month).[3,15]

According to the aforementioned electricity prices, the expected amount of the electricity monthly bill is €8.12 and €65 in Carinthia and Friuli-VG, respectively. This result matches closely the average amounts of the electricity bill declared by the respondents, which are €71.4 and €76.4 in Carinthia and Friuli-VG, respectively. In other words, the collected information by the respondents can be used to generalize a daily load demand as both amounts (expected and declared) are close to each other. However, it is of high importance to derive a more accurate load demand model based on actual measurements.

To this end, a part of the MONERGY research activity is dedicated to a measurement campaign for eight selected households located in Carinthia (AT) and Friuli-VG (IT). These households are monitored continuously for a one-year period. In particular, four households are currently monitored across the province of Udine (Italy) and four more households in the area of Klagenfurt (Austria). The dataset will be used in future research work to develop a model for energy demand in these regions. Online monitoring can be seen using the following link (http://tinyurl.com/monergy-campaign). An example for one of the monitored houses is illustrated in Figure 3. From the figure, the average consumption of this house is about 14.8 kWh day, indicating that the proposed load demand can be used as a typical load demand for the adopted case study.

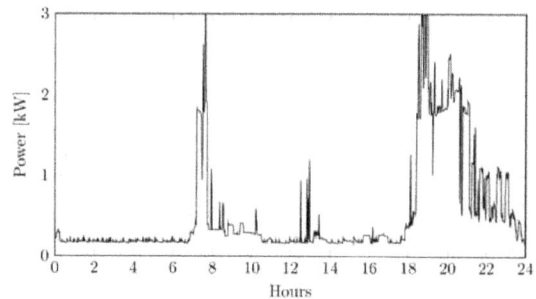

Fig. 3. Load demand example for one of the monitored houses in MONERGY project.

3.2 Q2: What Is Carinthia and Friuli-VG Citizens' Attitude Toward the Smart Home Concept and the Energy Management Systems?

In this part of the survey, we measured the attitude of citizens toward the deployment and use of energy management systems. According to the results, inhabitants from Carinthia expressed the willingness to exploit this kind of tariff for operating the washing machine (48%), the electrical boiler (23%), and the dryer (20%). 67.20% of users from Carinthia declared to have replaced an electrical device with a more energy-efficient one during the last 4 years to reduce the consumption at home. Among those, energy-efficient light bulbs (51%), washing machines (32%), televisions (19.89%), electrical hobs (15%), refrigerators (13.44%), and dryers (9.14%) account for most of replacements. Households of Friuli usually can exploit multiple pricing conditions. Users declared they considered the current cost of electricity when using their washing machine (62.59%), lights (24.46%), iron (22.3%), electric oven (21.58%), dryer (10.79%), conditioner (10.07%), and dishwasher (9.35%). Similarly to Carinthia, lights (38.85%), washing machines (17.99%), and televisions (9.35%) are the most replaced devices with more efficient ones.

In general, low renewable energy utilization was observed—in particular, a very low penetration of geothermal heating. The number of photovoltaic systems in Friuli is higher than Carinthia, with a percentage of penetration equal to 7.91% and 2.69%, respectively. The situation is opposite when looking at the solar-thermal heating, accounting for 16.67% in Carinthia and 13.67% in Friuli deployment.

4. Discussion and Recommendations

Based on the results of both research questions, we can conclude that the citizens who live in Carinthia and Friuli-VG households typically have comparably high energy consumption for space heating at high energy prices. Consequently, some citizens are motivated to deal with multi-tariff systems or to consider energy-saving devices. However, only a minor part of them are actually willing to deal with smart homes' energy management systems or smart-powered homes.

In general, smart home and energy efficiency gaps are attributed to low public awareness and market and policy failures. Therefore, understanding the public attitude, for example, is extremely important to propose suitable technical decisions and governmental policies. In Hirst and Brown[16] and Attari et al.,[17] the authors concluded that there are mainly two reasons for the gap between the public and smart homes and energy efficiency: behavioral and structural barriers. The behavior and the practice of the public and private organizations is the reason for the structural barrier, while the individual energy end-user is less responsible for that. An example of this kind of barrier is fuel price uncertainty. Fuel prices are typically fluctuating, which gives a very fuzzy image about future fuel prices. This situation actually prevents consumers from investing in new energy technologies such as smart homes. In addition to the potential price subsidization and fluctuation, high costs of energy-efficient technologies are considered one of the most important causes of low energy efficiency. Moreover, the current policy of applying high discount rates to make tradeoffs between the initial investment and savings also prevents the customer from any investment in energy-efficient technologies. On the other hand, approved government policies are one of the structural reasons for the energy efficiency gap. The applied government policies usually encourage energy consumption rather than energy efficiency as the profit of selling the electricity, for example, is a function of government income. The lack of the technical standards behind technology development is considered a structural barrier that prevents the consumer from any investment in EMSs.[16] Furthermore, there are many factors that restrict the deployment of energy-efficient technologies such as infrastructure, geography, and human resources. Regarding the behavioral barriers, they can be defined as negative characterization of the end-user decision-making relating to energy consumption. There are many reasons for this negative decision-making characterization of the end user, such as end-user attitude toward smart homes and energy efficiency. Better awareness of smart homes and energy efficiency leads to a positive attitude toward these technologies and, consequently, could greatly affect their energy-related consumption and purchase behaviors. However, the risk of smart home investments is considered one of the behavioral barriers. In fact, the fluctuation of fuel prices and current high discount rates for conventional energy systems' operating costs have made smart home investments risky for many of the end users. Furthermore, the lack of the nontechnical information about these systems caused some negative attitudes with the consumers. Nontechnical information on systems feasibility and reliability may greatly encourage consumers to change their energy consumption behavior. In addition to the lack of information, misplaced incentives is considered one the behavioral barriers. The most classical example for misplaced incentives for PV investment is the landlord–tenant relationship. In fact, decisions about the energy features of a building are often made by people who will not be responsible for the energy bills. For example, landlords often buy the air conditioning equipment and major appliances, while the tenant pays the electricity bill. As a result, the landlord is not generally rewarded for investing in energy efficiency. Conversely, when the landlord pays the utility bills, the tenants are typically not motivated to use energy wisely. As a result, tenants have no incentive to install efficient measures benefiting the landlord, and the landlord has little incentive to invest in measures that benefit the tenant. Additionally, the lack of life-cycle thinking on costs and savings has imposed barriers for energy conservation.

4.1 Smart-Powered Home Feasibility and Current Structural Barriers

In general, the adopted case study does not have high solar energy potential, but it is still acceptable for smart-powered homes by photovoltaic (PV) system. In Klagenfurt, for example, the average daily solar energy received by a horizontal surface with an area of 1 m^2 is 3.43 kWh. This means that a domestic PV system with a 5 kWp PV array may produce 16 kWh in average per day. However, according to the predicted load demand illustrated in Figure 1, about 9 kWh are consumed during the daytime only. This is to say that the reaming energy generated by PV should be utilized in the nighttime to mitigate the peak power as well as the energy consumption. To store the excess energy produced by the proposed PV system, a battery unit with a capacity of 180Ah/48V is required. The battery must be able to inject currents of up to 30 Amp in order to meet the required power peak at some points. For such a system is able to power about 90% of the load demand with an availability rate of 85%.[18] The cost of this system is mainly given by the cost of the PV array, batteries, and other power conditioning and mounting stuff. Considering the current prices of PV systems, such a system may cost €10,000 + 20% sales tax. In this case, the customer will save up to 90% of the consumed energy, which means €91.8 per month or €1,101.6 per year. Consequently, the expected payback period of such an investment is about 11 years.

There are many avoidable causes for this long payback period. The sales tax—to start with—increases the payback period two years. In addition, the storage unit capital and replacement price also contributes negatively to the payback period. It is assumed that the customer can act as a *prosumer* (producer and consumer), whereas he is only able to utilize the energy generated by the system in the daytime and to sell the excess energy directly to the grid without the need to store it. This option reduces the

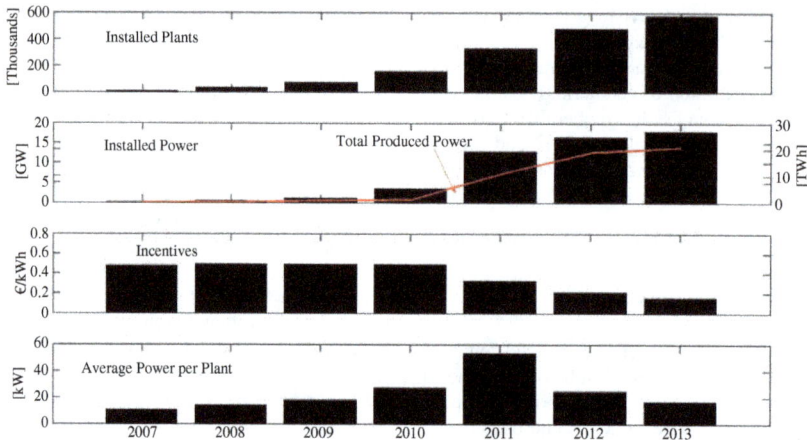

Fig. 4. Photovoltaic system investment status in Italy in the period between 2007 and 2013.

payback time by up to 1.5 years. However, an acceptable payback period should not exceed 5 years for such an investment in order to encourage the people to go with it. Feed-in tariff policies may address this point by applying good retail prices for the energy generated from PV for a specific period. For example, the payback period will be reduced by one year if the utilities decided to buy the energy units generated by PV systems in the first two years at 140% of the energy price they sell to the customer.

Italy, to start with, is the country in Europe with the second highest energy production from PV plants.[19] So far, about 18 GW of nominal power of PV plants have been installed. Figure 4 shows the PV investment status in Italy. The first two plots from the top of the figure, respectively, depict the evolution of the number of total installed PV plants and the correspondent nominal power. As it can be seen, these two curves increase exponentially between 2008 and 2012. Such behavior arises from a series of energy incentive programs that have been promoted by the Italian government. In particular, five different decrees have been issued in the years 2007–2012. These five decrees can be simplified in the concept of the incentive mechanism, which is based on a reward that lasts 20 years for each produced kWh. In particular, the first decree,[20] which was issued in 2005, defined a reward per kWh that was produced and consumed in loco. The second decree,[21] which was issued in 2007 and which applied to all plants that started working by the end of 2010, extended the reward to the overall produced energy and defined different tariffs for different size of plants and kind of installation (on the ground, partially integrated, or integrated on the roof). With the third decree,[22] which was issued in 2010 and which applied to all plants that started working between January and May of 2011, some more categories of plants and corresponding rewards were defined: plants installed on buildings or other plants, integrated plants with innovative characteristics, concentrated solar plants. Both the fourth decree,[2] which was issued in 2011 and which applied to all plants that started working after May 2011, and the fifth decree,[24] which was issued in 2012, defined new rewarding tariffs. The fifth decree applied to plants that started working by the July 6, 2013, and in particular when the cumulative budget of yearly incentives reached €6.7 billion. Although a deep analysis

of the decrees is out of the scope of the present work, we can simplify the concept by saying that the incentive mechanism is based on a reward that lasts 20 years for each produced kWh. As an example, in the third plot of Figure 4, the evolution of the incentives for a PV plant of nominal power comprised between 20 and 200 kW is illustrated. It is clear that there is a correspondence between the increase in the number of installed PV plants and the incentives. In particular, it is concluded that in 2013, when the incentives have reached the minimum value also the correspondent new installations of PV plants will be minimum. Finally, it is worth mentioning that the huge increase in PV plants observed in 2011 is mainly due to the installation of PV plants that, on average, show nominal power of about 50 kW (see the last sub-plot of Figure 4). This means that the largest part of plants was not installed by residential users that in general have a power peak of about 3 kW. Since the government has decreased the amount of the subsidy paid to the energy generated by PV system in order to control the commercial use of the systems, residential systems have been greatly and negatively affected.

As for Austria, according to Mayr et al.,[25] in 2001 the Austrian photovoltaic market experienced a good boost due to the approval of the green electricity bill again and feed-in tariff act. However, these rules somehow collapsed again in 2004. As a result, grid-connected plants with a total capacity of 175,493 kWp and stand-alone systems with a total capacity of approximately 220 kWp were installed. Hence, in 2012 the total amount of installed PV capacity in Austria increased to 175,712 kWp with a production up to 337.5 GWh. The average system price of a grid-connected 1 kWp photovoltaic plant in Austria decreased from 3,579 €/kWp in 2010 to 2,698 €/kWp in 2012, i.e., a reduction of 24.6 %. In 2002, the feed-in tariff act was approved in Austria. The type of approved feed in tariff (FIT) policy at that time can be described as a constant, periodically updated, administratively defined tariff for a certain duration and system size,[26] though it is claimed[27] that the aim of a FIT policy is to enable the customer to have a rapid and substantial growth in a PV investment. Anyway, in addition to the approved FIT policy investment, co-funding (ICF) has been implemented to better support the PV investment in Austria. Such a funding scheme provides investors initial financial support for the construction and installation of a PV system. There are three concerns about the Austrian subsidy policy. First, the approved policy does not consider the future prices of PV technology. Second, the approved policy is not competitive.[27] Third, these subsidies are currently approved via a Web-based first-come, first-served application procedure. The eligibility, subsidization, or site of installation is not considered. Moreover, a FIT-subsidized PV system usually feeds all generated electricity into the grid, because the FIT tariff is higher than the end-user electricity price.[28] PV system initial funds provide an incentive to owners to self-consume the highest amount of produced energy and sell the excess amount to the electricity retailer, as market prices for selling electricity are usually much lower than for buying it. Finally, similar to Italy the ministry of finance in Austria has recently voted to introduce a grid fee for mid-size PV systems (>15 kWp) installed for self-consumption (undercover commercial systems). A levy of 1.5 euro cent per kWh[29] will be charged to any PV system installed from March 1, 2014 that generates more than 5,000 kWh of solar power per year. However, this act also greatly affects those who have installed large residential systems (probably in remote house clusters).

5. Conclusions

In this research, a quantitative assessment was done of the level of consumption as well as the attitude of EU inhabitants who are located in Carinthia and Friuli-VG toward the use of smart energy management systems. The representative sample size was found to be 385 respondents with a confidence of 95% and marginal error of 5%. A questionnaire with 43 questions was used to address two main issues, namely, the level of energy consumption and the attitude toward the use of smart energy management systems. Based on the results, we were able to derive a model for a typical 110 sqm area household in terms of energy consumption. In this model, the daily consumption is 16.8 kWh, with a 2.6 kW peak and a load factor of 27%. Furthermore, it was found that an average of 46% of respondents expressed the willingness to exploit tariff systems for operating their electrical appliances. Moreover, about 67% of respondents declared that they care about the energy efficiency at their households in order to reduce the monthly bill. However, low renewable energy utilization was observed due to some existing structural barriers. There are some structural barriers for the adoption of smart energy management systems and renewable energy sources in the zone of the study—for instance, the FIT amount decrease due to the increase in PV system installation without any consideration to the desired payback period, the outnumber of commercial PV system compared to residential PV systems, the constant FIT scheme policy, applied taxes, the lack of consideration of the fluctuation in PV technology prices and aging factor, and, finally, lack of consideration of the optimal design and placement of these systems. As a final conclusion, a uniform, constant subsidy is inefficient, while a simple discriminative first-price reverse auction comes with several advantages. First, this type of an "auction" ensures market prices and prevents inefficiencies due to inaccurate market price. Second, the competition in PV system production caused by this auction leads to lower levels of public subsidy. Third, having a PV

production auction has transparent and objective advantages for potential private investors. In particular, this research work falls in the scope of a specific research project called MONERGY. At the stage, we aimed to collect energy consumption for selected houses in Austria and Italy and measure the attitude of the household toward smart homes and energy management systems. Thus, future research work must start with extracting appliance usage models that can be analyzed for the purpose of appliance usage prediction and management. On the other hand, wider study of current structure and behaviors barriers that affect the attitude of the household negatively must be conducted. These two issues will lead to worthwhile technical information on smart homes and energy management systems as well as nontechnical issues regarding clean energy policy such feed-in tariff, incentive, and current governmental initiatives.

Funding

This work was supported by the European Regional Development Fund (ERDF), the Carinthian Economic Promotion Fund (KWF) under grant KWF 20214|23743|35470, and the Italian Project Interreg IV Italia-Austria MONERGY.

References

1. Stocker, A.; Großmann, A.; Madlener, R.; Wolter, M. Sustainable Energy Development in Austria Until 2020: Insights From Applying the Integrated Model "e3.at." *Energ. Policy* **2011**, *39*, 6082–6099.
2. Holzmann, A.; Adensam, H.; Kratena, K.; Schmid, E. Decomposing Final Energy Use for Heating in the Residential Sector in Austria. *Energ. Policy* **2013**, *62*, 607–616.
3. Khatib, T.; Elemenreich W. Novel Simplified Hourly Energy Flow Models For Photovoltaic Power Systems. *Energ. Conv. Manag.* **2014**, *79*, 441–448.
4. Bittermann, W.; Gollner, M. *Modelling of Power Consumption in Private Households in Austria According to Type and Usage Statistics Austria*, 30304.2009.003-2009.697, Vienna, 2011.
5. Kranzl, L.; Kalt, G.; Müller, A.; Hummel, M.; Egger, C.; Öhlinger, C.; Dell, G. Renewable Energy in the Heating Sector in Austria With Particular Reference to the Region of Upper Austria. *Energ. Policy* **2013**, *59*, 17–31.
6. Bergamasco, L.; Asinari, P. Scalable Methodology for the Photovoltaic Solar Energy Potential Assessment Based on Available Roof Surface Area: Application to Piedmont Region (Italy). *Solar Energ.* **2011**, *85*, 1041–1055.
7. Esposto, S. The Possible Role of Nuclear Energy in Italy. *Energ. Policy.* **2008**, *36*, 1584–1588.
8. Aste, N.; Adhikari, R.; Compostella, J.; Del Pero, C. Energy and Environmental Impact of Domestic Heating in Italy: Evaluation of National NOx Emissions. *Energ. Policy* **2013**, *53*, 353–360.
9. Monforti, F.; Huld, T.; Bódis, K.; Vitali, L.; D'Isidoro, M.; Lacal-Arántegui, R. Assessing Complementarity of Wind and Solar Resources for Energy Production in Italy. A Monte Carlo Approach. *Renew. Energ.* **2014**, *63*, 576–586.
10. Dall'O', G.; Sarto, L.; Sanna, N.; Martucci, A. Comparison Between Predicted and Actual Energy Performance for Summer Cooling in High-Performance Residential Buildings in the Lombardy Region (Italy). *Energ. Build.* **2012**, *54*, 234–242.
11. Monacchi, A.; Elmenreich, W.; D'Alessandro, S.; Tonello, A.M. *Strategies for Domestic Energy Conservation in Carinthia and Friuli-Venezia Giulia*, Proceedings of IEEE Industrial Electronics Society 39th Annual Conference, 4791, 4796. 2013.

12. Krejcie, R.; Morgan D. Determining Sample Size for Research Activities. *Edu. Psych. Measur.* **1970**, *30*, 607–610.

13. Basu, K.; Hawarah, L.; Arghira, N.; Joumaa, H.; Ploix, S. A Prediction System for Home Appliance Usage. *Energ. Build.* **2013**, *67*, 668–679.

14. Tewolde, M.; Longtin, J.; Das, S.; Sharma, S. Determining Appliance Energy Usage With a High-Resolution Metering System for Residential Natural Gas Meters. *Appl. Energ.* **2013**, *108*, 363–372.

15. McLoughlin, F.; Duffy, A.; Conlon, M. Evaluation of Time Series Techniques to Characterise Domestic Electricity Demand. *Energy* **2013**, *50*, 120–130.

16. Hirst, E.; Brown, M. Closing the Efficiency Gap: Barriers to the Efficient Use of Energy. *Res. Conserv. Recycl.* **1990**, *3*, 267–281.

17. Attari, S.; DeKay, M.; Davidson, C.; Bruine de Bruin, W. Public Perceptions of Energy Consumption and Savings. *Proc. Natl. Acad. Sci.* **2010**, *107*, 16054–16059.

18. Khatib, T.; Mohamed, A; Sopian, K. A Review of Photovoltaic Systems Size Optimization Techniques. *Renew. Sustain. Energ. Rev.* **2013**, *22*, 454–465.

19. Gestore Servizi Energetici (GSE). *Rapporto Statistico 2012, Impianti a Fonti Rinnovabili Settore Elettrico*, 2012.

20. Ministry of Economic Development, Decree 28 July 2005, Gazzetta Ufficiale della Repubblica Italiana, 5 Aug. 2005.

21. Ministry of Economic Development, Decree 19 February 2007, Gazzetta Ufficiale della Repubblica Italiana, 23 Feb. 2007.

22. Ministry of Economic Development, Decree 6 August 2010, Gazzetta Ufficiale della Repubblica Italiana, 24 Aug. 2010.

23. Ministry of Economic Development, Decree 5 May 2011, Gazzetta Ufficiale della Repubblica Italiana, 12 May 2011.

24. Ministry of Economic Development, Decree 5 July 2012, Gazzetta Ufficiale della Repubblica Italiana, 10 Jul. 2012.

25. Mayr, D.; Schmidt, J.; Schmid, E. The Potentials of a Reverse Auction in Allocating Subsidies for Cost-Effective Roof-Top Photovoltaic System Deployment. *Energ. Policy* **2014**, *69*, 555–565.

26. Lesser, L.; Su, X.; Design of an Economically Efficient Feed-In Tariff Structure for Renewable Energy Development. *Energ. Policy* **2008**, *36*, 981–990.

27. Richardson, I.; Thomson, M.; Infield, D.; Clifford, D. Domestic Electricity Use: A High-Resolution Energy Demand Model, *Energ. Build.* **2010**, *42*, 1878–1887.

28. KLIEN, 2012. Leitfaden Photovoltaik-Anlagen 2012. Eine Förderaktion des Klima- und Energiefonds der Österreichischen Bundesregierung. http://www.umweltfoerderung.at/uploads/pv2012leitfaden.pdf.

29. PV magazine. 13 March, 2014. http://www.pvmagazine.com/news/details/beitrag/austria-follows-german-lead-on-solar-self-consumption-fee_100014506/#axzz3EyQELp68.

18

Heat Transfer through Glazing Systems with Inter-Pane Shading Devices: A Review

ANURANJAN SHARDA[1]* and SUDHIR KUMAR[2]

[1]*Department of Mechanical Engineering, Rayat and Bahra Institute of Engineering and Bio-Technology, Punjab, India
[2]Department of Mechanical Engineering, National Institute of Technology, Kurukshetra, Haryana, India

Abstract: In order to evaluate thermal performance of a glazing system, the determination of fundamental properties of glazing system is necessary. To analyze the involvement of these properties, the heat transfer phenomena and the major factors affecting that transfer needs to be elaborately understood. The main aim of this article is to provide insight into different aspects of estimating heat transfer through various glazing systems, with special reference to ones with shading devices, and to critically discuss and identify areas of work that have been neglected or that need further investigation. The authors conclude that there is a need of exploring the potential of double-glazed windows with inter-pane blinds in the Indian climate and also for its justification through experimentation. The absence of quantification of percentage contribution of the control parameters in influencing the measured response is also highlighted as a gap in the literature review.

Keywords: Thermal transmittance, double-glazed windows, guarded heater plate (GHP), inter-pane blinds

1. Introduction

The solar energy passing through a clear glass glazing warms up the internal surfaces by absorption, and these surfaces then become heat radiators of low frequency and start re-emitting trapped heat, causing the temperatures to rise. Various types of energy interactions taking place across the window are depicted in Figure 1.

Total instantaneous energy flowing through a window, Q can be determined by the equation

$$Q = UA_{pf}(t_{out} - t_{in}) + (SHGC) A_{pf} G_t \qquad (1)$$

where U-value is the overall coefficient of heat transfer, t_{in} is the interior air temperature, t_{out} is the exterior air temperature, A_{pf} is total projected area of fenestration, SHGC is solar heat gain coefficient, and G_t is the incident total irradiance. Heat transfer through any glazing system is completely represented by its thermal properties (first term) or solar optical properties (second term). U-value or the thermal transmittance is determined by

*Address correspondence to: Anuranjan Sharda, Department of Mechanical Engineering, Rayat and Bahra Institute of Engineering and Bio-Technology, S.A.S. Nagar, Punjab, India. Email: anuranjansharda@gmail.com

$$U = \frac{U_{center} A_{center} + U_{edge} A_{edge} + U_{frame} A_{frame}}{A_{pf}} \qquad (2)$$

where, A_{center} is area of center section, A_{edge} is area of edge section, and A_{frame} is area of frame section.

The SHGC has two components: first, the direct solar transmittance of the glazing system, $T_{dir-sol}$ and, second, the inward-flowing portion, N_i. Here α_s is the solar absorptance of a single pane of glass. SHGC is hence expressed as

$$SHGC = (T_{dir-sol} + N_i\alpha_s) = (T_{dir-sol} + q_i) \qquad (3)$$

The secondary heat transfer toward the inside, i.e., q_i, arises from absorption of solar radiation by the fenestration system. The solar radiation that is absorbed by the fenestration system is transferred to the atmospheric environment by convection and infrared radiation.

The inclusion of a shade to any glazing system (referred to as complex glazing system) adds to uncertainty in the occurrence of radiative and convective heat transfer between inner glass layer and the shade. With these extensions, the equation for the solar heat gain coefficient of a system with L layers now becomes

$$SHGC(\theta,\phi) = T_{1,L}^{fH}(\theta,\phi) + \sum_{k=1}^{L} N_K A_{k:(1,L)}^{f}(\theta,\phi) \qquad (4)$$

where

- θ = incident angle relative to normal layer
- Φ = azimuthal angle (in plane of layer, about normal)

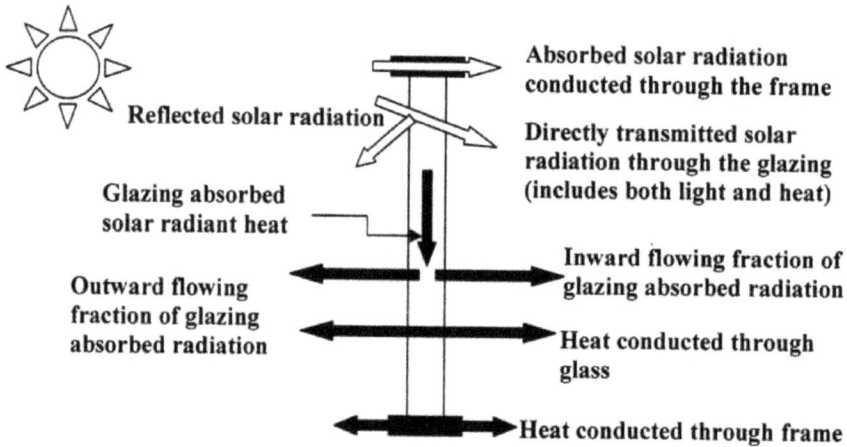

Fig. 1. Basics of heat transfer through windows.

- $T^{fH}_{1,L}(\theta,\Phi)$ = directional-hemispherical front transmittance of system
- $Af_{k:(1,L)}(\theta,\Phi)$ = directional absorptance of kth layer in system
- N_k = inward-flowing fraction of absorbed energy for k^{th} layer
- L = last layer of the system

Hence it is concluded that the heat transfer through any glazing system is completely represented by its U-value and SHGC.[1–2]

2. Literature Review

In this study, four main areas have been identified relating to research in the field of heat transfer through glazing system: (a) heat transfer through glazing systems, (b) heat transfer through glazing systems with shading devices, (c) modeling and measurement of solar/optical properties of shading devices, (d) heat transfer through double-glazed windows with between-the-glass shading devices, and (e) heat transfer through glazing systems in Indian conditions.

2.1 Heat Transfer through Glazing Systems

2.1.1 Experimental Studies on Heat Transfer through Glazing Systems

To validate various theoretical models and to standardize their results, various window-testing methods were introduced, the main ones being calorimeter method, guarded hot-box method, guarded heater plate methods, etc.[3]

Elmahdy and Bowen[4] used a calorimeter to compare the thermal characteristics of various types of glazing units, including the ones with coatings. The authors compared the experimental results with those obtained from the VISION program and concluded that accurate simulation and modeling of glazing units can be used to evaluate the thermal performance of different glazing units without going in for expensive test apparatus. Alvarez et al.[5] used a solar simulator test lamp and a similar calorimeter apparatus to evaluate the thermal performance of glazing. It was concluded from the results that there was a reduction

in the radiant heat inflow using Reflectasol glass-6 mm (RG-6) when compared to a clear glass-3 mm. This was due to higher optical reflectance and lesser thermal transmittance of the RG-6 glass. However, as RG-6 transmits lesser visible energy, the requirement of energy for interior lighting increased.

By making use of the guarded hot-box method, Christensen et al.[6] conducted experimental studies to investigate the surface temperature profiles of the glass panes as well as of the air present inside the cavity of a double-glazed window. A new thermal performance index was defined that remains independent of overall temperature difference for constant convection conditions and decreases by recessing of windows into the wall, or by usage of blinds or drapes. It was also observed that conductance decreases with thickness.

Elsherbiny et al.[7] conducted experiments on guarded heater plate apparatus to determine the heat transfer by natural convection across vertical as well as inclined layers. The results depicted the complexity of the effects of aspect ratio and Rayleigh number on the heat transfer. The authors presented the correlations to determine the convective heat transfer, and results were found to be within 3% (for vertical layers) and 5% (for inclined layers) of the corresponding measured values. Wright and Sullivan[8] also used a guarded heater plate apparatus to determine the U-values of glazing systems consisting of two glass panes separated by air and up to two layers of fluorinated ethylene propylene (FEP) films as intermediate glazing. The results were compared with the modified VISION software values, and 3% variation for most of the cases and 8% variation for all the cases was reported. It was also shown that the heat flux to the bottom of the warm side glazing was higher than the heat flux to the top of the same glazing. The authors[9] also developed an experimental procedure for measuring the thermal resistance of edge seals by the use of a similar guarded heater plate apparatus and highlighted the importance of edge seals in cold climates where condensation and frost readily occurs.

A similar guarded heater plate apparatus was used by Baker et al.[10] to experimentally investigate the effect of pane spacing of double-glazed and air-filled windows on the U-values and compared the results with those predicted by the modified

VISION software. The authors concluded that the optimum pane spacing was reduced with (i) the addition of low-e coating, (ii) change from AAMA (American Architectural Manufacturers Association) to ASHRAE conditions, (iii) reduction of glazing panes from the system, and (iv) with a change in the fill gas.

Varshney et al.[11] presented a rapid, inexpensive, and easy-to-use thermal-box method to determine the U-value in the field by measuring interior/exterior air temperatures and interior/exterior window surface temperatures. The results were found to be within 8% of the rated U-values of the concerned windows. The authors also concluded that the U-values varied with exterior air temperatures.

2.1.2 Analytical and Theoretical Studies on Heat Transfer through Glazing Systems

Rubin[12] developed a general energy-balance procedure for calculating the net energy flux through the glazed area of the window having any number of layers including thin film coating or enclosed gas. The author also verified the results by comparing them with experimental measurements of heat flow using a calibrated hot-box.

Different software such as VISION, FRAME, WINDOW, etc. has been developed and used for studying heat transfers associated with windows. Using VISION, Wright and Sullivan[13] reviewed the heat transfer phenomenon involving convection in a vertical slot with detailed overview of conditions of laminar and turbulent flow, hydrodynamic stability, and local heat transfer rates. It was concluded that for conventional double-glazed windows when air or argon was used, the motion of the fill gas remains laminar and free of secondary cells under most conditions. Baker et al.[14] used the VISION and FRAME programs to model different windows and observed that for accurate analysis of heat flow, the model should be detailed in areas of high heat-flow but in areas of low heat-flux, complexity can be ignored. It was concluded that the outside film coefficient has very little impact on the double-glazed windows compared to single-glazed systems. Finlayson et al.[15] revised the calculations procedures by Rubin and presented an improved WINDOW 4.0 computer program for calculating the thermal and optical properties necessary for heat transfer analyses of fenestration products. This program has further evolved into a WINDOW 6.3 version[16] developed at the Lawrence Berkeley National Laboratory (LBNL).

Erhorn et al.[17] reviewed the test methods available for determining the U-value of a glazing system and observed that for hot plate methods, the temperature difference and, to a lesser extent, the temperature level influenced the measured U-value. It was also reported that for hot-box method the U-values have to be presented along with the actual heat transfer coefficient. The authors explained the precautions to be taken to avoid deviations from the true results, which could be caused by following faulty procedure or by the effect of ambient or test conditions.

Various numerical models were developed and experimentally validated. Wright and Sullivan[18] formulated a 2D numerical model to simulate natural convection in a tall, vertical, rectangular cavity. This model was validated using experimental results and data from other simulations. It was reported that for Ra $< 10^4$ the average Nusselt number (Nu) calculated was within 5% agreement with measured values. Even the local Nusselt-numbers were predicted generally to within 5% and at times

to within 2% when compared with experimental results. The authors[19] also formulated a 2D numerical-control-volume model that theoretically determined the values of local and average heat transfer rates through insulated glazing units (IGU) under the conditions of a guarded heater plate test. The calculated values of heat flux closely resembled the measured ones using guarded heater plate apparatus. The effect of various factors such as combination of different low e coatings, fill gases and edge seal designs on heat patterns, and the temperature profiles for a number of glazing systems was studied through simulations.

Studies related to solar heat gain coefficient (SHGC) of center glass as well as frame were also carried out by the researchers. Wright[20] examined several methods to determine the solar gain through windows and recommended the use of spectral calculations to accurately determine the amount of solar radiation directly transmitted and absorbed in the glazing system. The validity of applying SHGC of center glass to the full area of the window was reiterated by the author, and it was suggested to neglect SHGC of frames for sufficiently higher thermal resistance of frame materials. Wright and McGowan[21] preferred to determine the solar heat gain of a window frame through calculations rather than by practical measurements. The authors developed a simple expression to estimate the SHGC of a window frame and compared the results of this simplified model with those of the detailed 2D numerical model. It was concluded that simpler calculations introduced a small error into the calculation of solar gain for the entire window, which could be safely neglected while comparing various window frame designs.

For a double-pane window smaller pane-spacing, the heat transfer is dominated by conduction through air. However, as the spacing increases there is a small decrease in heat transfer due to the reduction in heat conduction through air in proportion to the thermal conductivity of air. As the pane spacing further increases, there is an increase in convection as outer circulation flow grows stronger, which significantly increases the heat transfer. Thus there exists an optimum pane-spacing for a specific double-glazed window. Aydin[22] reported that the heat loss through a double-pane window was minimized using an optimum value of the thickness of the air layer between the panes. Though the author in this study assumed that the temperature of each plane surface remains constant, realistic temperature conditions were suggested for future study.

2.2 Heat Transfer through Glazing Systems with Shading Devices

For complex fenestrations, SHGC can be calculated using equation (4) by calculating the solar-optical transmittances and absorptances by the matrix method of Klems[23] and inward-flowing fractions by Klems and Kelley.[24] To overcome the non-availability of data used for calculating SHGC, efforts were made to either model or to determine experimentally the solar optical properties of common shading devices.

2.2.1 Experimental Studies of Heat Transfer through Glazing Systems with Shading Devices

Collins and Harrison[25] experimentally determined the inward-flowing fraction of absorbed solar energy in interior venetian blinds, by simulating the absorbed radiations through electrically heated blinds. The effect of blind slat angle, interior/exterior temperature difference, absorbed irradiance, and exterior air film coefficient on the inward-flowing fraction of a calibration transfer standard was also investigated. It was concluded that inward-flowing fraction of an open blind is decreased with an increase in interior/exterior temperature difference and is marginally affected by the absorbed irradiance. In a similar study, Collins[26] investigated the effect of blind geometry and calorimeter tilt on the inward-flowing fraction. It was reported that inward-flowing fraction decreased with increasing tilt, and this decrease was more significant for tilt angles greater than 60°. The heat transfer from the interior glass increased with the increase in nominal distance from 15–25 mm. Collins and Harrison[27] also experimentally determined the effect of calorimeter tilt on the SHGC and U-values of a single-glazed window with an interior shade and concluded that SHGC was marginally decreased, whereas the U-value increased with calorimeter tilt from vertical to 45°.

Collins et al.[28] empirically validated the average convective heat transfer coefficients obtained from the earlier developed steady state, 2D, and conjugate finite element model. The authors also reported the close agreement of the measured and predicted blind slat temperatures.

Various experimental studies on SHGC using shading devices along with glazing system were also carried out by the researchers. Collins and Harrison[29] experimentally investigated the amount of solar heat gain through a double-glazed window with two types of interior blinds—one of high reflectivity and the other of high absorptivity. The authors concluded that SHGC was reduced significantly for highly reflective blinds and, to a lesser extent, for highly absorptive blinds. However, no significant effect of the presence of shade was reported on the U-value of the considered system. Loutzenhiser et al.[30] empirically validated solar gain models for glazing systems with exterior and interior venetian blind assemblies constructed in building-energy-simulation programs, Energy-Plus and HELIOS. It was concluded that results predicted were as per expectations, and the deviations were in accordance with program inputs and the experimental uncertainties. Simmler and Binder[31] experimentally determined the total solar energy transmittance (TSET) values for a number of shading–glazing combinations and reported satisfactory agreement of measured and numerically calculated values. The developed numerical model was stated to be suitable for the calculation of solar gain of multiple glazing and louver-type shading devices with non-specular surfaces, whereas for complex constructions, testing of real-scale components with a more detailed model was recommended. Kotey et al.[32] also experimentally determined the SHGC, solar transmittance, and interior attenuation coefficient of a double-glazed window attached with various interior-shading devices. Results were reported to agree with those from ASHRAE window attachment (ASHWAT) models. The effect of various shading devices and their location on the solar heat gains was also investigated by the authors.

2.2.2 Analytical and Theoretical Studies of Heat Transfer through Glazing Systems with Shading Devices

Rosenfeld et al.[33] reviewed the progress made in development and validation of various models to predict the properties of complex glazing systems. The authors used the measured optical properties of individual components at normal incidence as input data for these models to accurately predict the thermal and solar transmittance in the range compared with already-published experimental results. The authors concluded that these models were suitable for integration into building energy simulation tools, whereas an improved model was required for scattering components.

Software and programs for analyzing glazing systems with shading devices were also developed for evaluating the thermal performance of windows. Pfrommer et al.[34] developed a program called GLSIM-BLIND to investigate, compare, and optimize blind arrangements during the design phase of a building. The influence of slat curvature was observed to decrease with increased radius of curvature and for most practical cases; this influence was small. The calculation model used by GLSIM-BLIND was verified by comparing the results with those calculated by ray-tracing software RADIANCE, but no validation was done against the results with full-scale field measurements. Datta[35] investigated the thermal performance of buildings using TRYNSYS software for four different cities in Italy and incorporated an external shading model (consisting of external fixed horizontal louvers with different slat lengths and tilts in it). The author concluded that external fixed horizontal louvers of proper design for south windows were effective in reducing not only cooling loads of a building in summer but also the overall annual primary energy loads of a building. In addition, the author observed that optimum shading system depends on the location and weather considerations in which it was to be used. Collins and Wright[36] suggested new equations for calculating SHGC, U-value, and other window performance indices for window systems with a diathermanous layer and highlighted the errors inherent in applying existing standard calculation methods including those for WINDOW and FRAMEplus software. The authors developed a more stable procedure for determining the radiative heat between the layers called the R^{∞} method to evaluate center-glass performance indices of windows with diathermanous layers.

CFD studies involving finite element as well as finite volume models were conducted for better insight into the effect of various parameters on the thermal performance of glazing systems with shading devices. Ye et al.[37] developed a 2D numerical model for free convection from the window surface adjacent to a venetian blind, for cases where blinds with low emissivity surfaces were used. The authors concluded that the effect of blade-to-plate spacing on the flow pattern and on convective heat transfer coefficients was very strong; however, it almost disappeared when the blind was located far from the surface. The effect of blade-angle orientation was also reported to be stronger for smaller blade-to-plate spacing. Using a steady state, laminar, 2D, and conjugate finite element model of a vertical isothermal surface with heated, horizontal, and rotatable blinds, Collins et al.[38] examined the sensitivity of radiative and convective heat transfer from the surface with respect to blind to glass spacing, blind slat angle, irradiation level, glass temperature, and emissivity of blind and glass. Analysis made was limited to the effect of selected parameters on heat transfer rate and not of their interaction. The authors thus recommended the use of a 3-level factorial design to evaluate the interactions effects. Collins[39] used a steady, laminar, 2D, and conjugate conduction/convection finite element model to examine the convective heat transfer from an irradiated venetian blind adjacent to the indoor surface of a window. The convective heat transfer coefficients for the same were determined. The authors concluded that the presented methodology successfully determined the solar and thermal performance of glazing system with internal shading. Collins and Harrison[40] developed a steady-state, laminar 2D, and conjugate finite element model of an isothermal indoor glazing surface with adjacent heated, horizontal, and rotatable louvers and presented useful correlations to predict heat transfer through it. The authors concluded that results from this model correlated well with calorimetric data. Shahid and Naylor[41] developed a 2D finite volume model, validated it with published experimental and numerical results, and used it to study the thermal performance of venetian blind positioned adjacent to single- or double-glazed window for coupled convection and radiation heat transfer. The authors concluded that the presence of blinds reduced the overall heat transfer rate through the windows by reducing the thermal radiations from the indoor glazing.

Fang[42] presented the equations for calculating the U-factor of the window with the long-wave high-reflectivity venetian blind with different type of glazing systems. These equations were experimentally proved to be within 64.4% and 66.8% of the experimental data. The author reported that the U-factor increased by 28% for the single-glazing system and 20% for the double-glazing system when the reflectivity was decreased from 0.93 to 0.1. It was also concluded that the heat loss increased by 25% for the single-glazing system and 10% for the double-glazing system when the slat angle was increased from 5° to 30°.

Wright and Kotey[43] presented the methods for extending the already-existing solar optical models for systems of specular glazing layers. Though different techniques such as ray tracing, net radiation analysis, matrix reduction, and iterative numerical processing existed, the authors preferred a recursion algorithm devised by Edwards[44] in a modified form due to it being simple, easily programmable, and computationally fast and efficient.

2.3 Modeling and Measurement of Solar/Optical Properties of Shading Devices

2.3.1 Experimental Studies to Determine Solar/Optical Properties of Shading Devices

Papamichael and Winkelmann[45] developed a mathematical model based on matrices to calculate the bidirectional solar-optical properties of multilayer fenestration systems by the use of similar properties of each individual layer. The authors concluded that the model was advantageous in terms of time and cost. In addition, such a model also provided information on the net absorptance of each layer rather than the total absorptance of the complete fenestration system. Keeping in view the above procedure, Klems and Warner[46] measured the bidirectional properties of each layer of a complex fenestration system using a scanning radiometer. These properties along with the inward-flowing fractions for each layer were used to calculate the SHGC for any shading layer and glazing pane combination. As the data necessary to use this method was not readily available for all shading systems, the authors recommended the use of quick responding integrating sphere for measuring the complete system rather than individual layers. It was also suggested to use a simpler and

less accurate method for most design purposes. Klems[47] then presented a simplified procedure for estimating performance of fenestration systems containing shading from minimal data.

However, later on, Milburn and Hollands[48] established the capability of broad area irradiation (BAI) in producing accurate measurements after making certain corrections relating to sample reflectance. The authors discussed the methods and correction schemes to minimize ESR (external sample reflectance) errors and to eliminate ISR (internal sample reflectance) errors. The transmittance measurements on a simple sample were also carried out, and experimental results were compared with a benchmark transmittance measurement. Good agreement between these results was reported. Kotey et al.[49] presented simplified models to calculate the effective solar optical properties of a drapery for both incident beam and diffuse radiation, which were useful in comparing performances of different draperies. The authors reported the disagreement of results from similar apparatus BAI-IS (broad area irradiation with integrating sphere) with previous research and emphasized that the solar optical properties of a drapery do not always decrease by a constant factor with respect to folding and/or incidence angle as proclaimed earlier. It was concluded that these simplified solar optical models produced results that can serve as useful input to building load and annual energy calculation tools. *Collins and Jiang[50] validated the already-developed theoretical model that predicted solar performance of louvered shading layers.* The authors reported almost perfect agreement of the predicted total solar directional hemispherical transmittances from the *Yahoda and Wright[51]* model with those measured experimentally with BAI-IS. The authors also explained the deviations from the experimental results at the peak transmission point, on the basis of assumptions made in the model.

Kotey et al.[52–53] used a double-beam, direct-ratio recording and rapid-scanning spectrophotometer with an integrating sphere to obtain the off-normal solar optical properties of roller blind as well as insect screen samples. The authors reported no variation of roller blind reflectance at incidence angles within 60° of normal. However, its transmittance varied with incidence angle and was represented by a cosine power function. The authors also presented a set of models for calculating off-normal properties of any roller blind or any insect screen. The reliability of the measurement technique was ascertained by comparing the test results to already-developed analytical and ray-tracing models, including those of software Energy Plus.

2.3.2 Analytical and Theoretical Studies to Determine Solar/Optical Properties of Shading Devices

Rubin[54] presented a set of calculation procedures for determining the solar optical properties of even a complex window consisting of partially transparent layers (e.g., multilayer coating, antireflection coating, solar control film). The author also presented the results of sample calculations over the range of incident angles for conventional and advanced energy conservation window design. In a similar study, Rubin et al.[55] presented the procedures for calculating different properties of specular glazing with an aim of reducing the number of measurements. The authors also emphasized the need of more realistic models to determine detailed spectral and directional properties of glazing systems.

Tzempelikos et al.[56] reviewed and compared different experimental procedures and analytical and numerical modeling techniques to accurately determine the bidirectional transmittance distribution functions of complex glazing systems and presented the limitations in existing models and standards. The authors recommended the use of bidirectional transmittance distribution functions to model the optical properties of venetian blinds followed by the use of any of the available methods (e.g., ray tracing) to evaluate illuminance distributions in rooms. The authors concluded that though roller shades were modeled as perfect diffusers, the direct component do have an effect and the direct-to-diffuse transmittance might not have been accurately modeled. Hence a need of future research on this topic was identified.

Many studies on investigation of properties of blinds were carried out. Yahoda and Wright[51, 57] performed detailed modeling of the effective long-wave radiative and solar/optical properties of a louvered blind layer that could be placed anywhere in the window system. This model was based on fundamental radiant exchange analysis and treated solar beam and diffuse radiation separately. Finally, a moderately successful, center-glass model of thermal transmittance was proposed by combining the long-wave radiation model with some simple convection correlations. Kotey et al.[58] also presented a simple method to determine the effective solar optical properties of a venetian blind. The authors concluded that the results of the flat and curved slat models readily agreed with experimental data for commercially available venetian blinds, except when the profile and slat angles were aligned. Here, the flat slat model predicted very large blind transmissions. These models provided useful input to multilayer glazing/shading models used for rating or for building energy simulation.

Kotey and Wright[59] calculated the hourly transmitted, reflected, and absorbed fluxes for both summer and winter conditions for a window with light- and dark-colored venetian blinds using simplified models and computational procedures. The authors observed that these procedures produced results that were useful in providing input to building load and annual energy calculation tools and also in comparing various complex fenestration systems. The effect of window configuration and slat angle on building energy requirements at different times of year was also investigated.

Laouadi and Parekh[60] later on presented models capable of computing the optical characteristics of complex fenestration systems, including ones with both clear as well as specular glazing layers. In these models, in addition to the optical properties of the individual glazing layers of window, the haze and gloss properties of the glazing layers were also considered. A general optical model was also developed for screen-like glazing layers such as insect/shading screens, roller blinds, drapery sheets, honeycomb transparent insulation, and fiberglass translucent glazing.

Laouadi[61] presented and validated a general methodology to compute the thermal performance of complex fenestration systems using the available measurement and CFD simulation results for the U-factor of double-glazed windows with between-pane and internal blinds. The effect of assumed porous layers on the convective film coefficients of adjacent gas spaces was accounted for by a newly developed thermal penetration length model, which was particularly applied to various types of shading devices. The authors recommended further validation so as to include other complex layers, specifically the ones that generate turbulent flow in cavities between shades and glazing.

2.4 Heat Transfer through Double-Glazed Windows with Inter-Pane Shading Devices

2.4.1 Experimental Studies on Heat Transfer through Double-Glazed Windows with Inter-Pane Shading Devices

A guarded heater plate apparatus has been used extensively to investigate the thermal performance of glazing systems with inter-pane shading devices. Garnet et al.[62] used this apparatus to obtain the center-glass overall heat transfer rates in a double-glazed window with inter-pane blinds. The author studied the effect of blinds, slat angle, and slat tip-to-glass spacing and reported enhanced performance for such types of windows. It was also concluded that condensation and frost resistance was improved because of reduced local heat transfer effects at the top and bottom of the cavity due to the weakened primary flows. Huang et al.[63] also used a guarded heater plate apparatus to determine the amount of heat transferred across the center glass region, for a double-glazed window with venetian blinds in between the panes. The effect of various parameters on U-values was studied, and the measured U-values were validated through comparison with earlier measurements and also with the theoretically validated models in open literature.

In addition to studies involving double-glazed windows with inter-pane shading devices, various studies on similar configurations have also been conducted. Onur et al.[64] experimentally investigated an air window collector having vertical black blinds made of cloth under actual outdoor conditions. Various control parameters were measured/recorded using a data acquisition system, and the effect of these parameters on the thermal performance was determined experimentally. Naylor and Lai[65] measured the local and average convective heat transfer coefficients in a vertical window-like enclosure with an internal venetian blind. The authors concluded that blinds affect the local as well as overall convective heat transfer rates in the enclosure. The authors also investigated the effect of glazing spacing on thermal performance and recommended the avoidance of narrow spacing of double-glazed windows with internal blinds.

Breitenbach et al.[66] measured the spectral bidirectional transmittance functions for a range of incidence angles for double-glazed windows with venetian blinds at different positions. Authors also developed a model to calculate the properties of the double-glazed window using the optical properties at normal incidence of the component elements. The authors concluded that this model was capable of being integrated into building energy simulation programs.

2.4.2 Analytical and Theoretical Studies on Heat Transfer through Double-Glazed Windows with Inter-Pane Shading Devices

Rheault and Bilgen[67] developed a theoretical radiation model to calculate the heat transfer through double-glazed windows, which contains pivoting louvers hermetically sealed between two glass panes. The results were compared with those of an ordinary double-glazed unit, and it was concluded that by using this system, the predicted auxiliary load for heating and cooling was reduced by up to 36% and 47% for winter and summer, respectively. Zhang et al.[68] further examined the insulation effects of closed blind slats contained in between two window panes, by assuming them to be a permeable screen. The authors conducted a numerical study to determine the effect of this permeable screen on the temperature field, the flow field, and the overall heat transfer rate. It has also been shown how critical conductance can be used to calculate the air gap that can be tolerated between two consecutive strips in the screen.

Bulow-Hube et al.[69] determined the total solar energy transmittance (g-values) of various external, internal, and inter-pane shading devices by measuring in a real climate using a double hot-box arrangement. These measured results were compared with those of ParaSol software, and it was reported that g-values of inter-pane fabrics as predicted by Parasol were higher than those of outdoor measurements. The authors explained the reasons for these deviations, but to obtain true values it was recommended to multiply the g-values from the software by 1.1 for internal fabrics and by 1.2 for internal blinds.

Wright et al.[70] developed a simplified model to investigate the thermal resistance of a glazing system with venetian blinds as an inter-pane shading device. The model predicted accurate results including the effect of pane spacing, slat angle, gas fill, and a low-e coating on the panes. The authors also developed a reduced slat length (RSL) model to evaluate the heat transfer through the complex fenestration system, which was validated by comparing its results with those of CFD studies performed by Tasnim.[71] It was concluded that correlations developed for cavity film coefficients by the RSL model accurately predicted the thermal performance of double-glazed windows with inter-pane shades; thus intensive and time-consuming CFD simulations may be safely avoided. Almeida et al.[72] conducted an experimental study to examine natural convective heat transfer through a double-glazed window with between-panes louvered blinds. The authors concluded that for laminar and steady flow, the lateral convective heat transfer was dominated by conduction, and the RSL model closely predicted the experimental results. However, for highly unsteady and turbulent flow, as expected the RSL model gave poor predictions of the convective heat flux for all slat angles. Collins et al.,[73] too, have applied the RSL model to a between-panes blind that is heated by absorbed solar radiation, i.e., in this model a third temperature was prescribed to the sunlit blinds so that simulation of heat transfer can be performed even for cases where the shade is hotter than the glass. The authors conducted a numerical study of convective heat transfer in the cavity and developed a correlation to predict the same. The authors concluded that this model does not predict the U-factor better than 0.70 of value given by Huang et al.,[63] but it performed well in providing insight into the flow structures occurring in the system.

The effect of various parameters on the thermal performance of double-glazed windows with inter-pane shading devices has also been studied. Yahoda and Wright[57] determined the U-factors for different pane spacing for windows with cavity blinds through a heat transfer analysis using effective long-wave radiative properties of venetian blinds and crude calculations involving convective heat transfer inside the cavity. The results thus obtained agreed to within 10% with those corresponding to the other experimental work. Thus it was concluded that detailed complicated models were of little value while dealing with the interaction between a cavity blind and the convective flow in the same cavity. Collins et al.[74] developed and validated a 2D finite-volume model of a glazing cavity with rotatable baffles located at the midway of two panes. The authors examined the effects of wall spacing, baffle angle, and temperature and the wall-to-wall temperature difference. It was observed that the local Nusselt number reached a steady-periodic state at the ends of the cavity over a very short distance and thus supported the 1D center-glass analysis conducted by various researchers. The authors also concluded that for all practical conditions convective heat transfer can be considered to be independent of Rayleigh number.

2.4.3 Computational Fluid Dynamics (CFD) Studies on Heat Transfer through Double-Glazed Windows with Inter-Pane Shading Devices

Studies involving completely coupled heat transfer models for double-glazed windows with inter-pane shading devices have been carried out. Naylor and Collins[75] developed a 2D finite-volume model of the conjugate convection, conduction, and radiation heat transfer in a double-glazed window with venetian blinds in its cavity. This numerical model was validated for a range of slat angles using Garnet et al.[62] U-value measurements through a guarded heater plate apparatus. In addition, the authors developed a much simpler model that uses data from convection-only CFD solution and predicted the U-value of the double-glazed/cavity blinds window system within 1.5% when compared to the complete CFD model. Even the geometric effects such as those of offsetting the blind from the center line of the air-filled cavity were explained by this simpler model. Avedissian and Naylor[76] also developed and validated a complete CFD model of the coupled heat transfer for a double-glazed window with inter-pane aluminium blinds in its cavity. The effect of Rayleigh number, aspect ratio, and blind geometry on convective heat transfer in the absence of radiative heat transfer was also investigated. An empirical correlation for average Nusselt number developed from the data was observed to be valid for particular values/range of Rayleigh number, enclosure aspect ratio, width-of-blind to width-of-cavity ratio, slat angle, and axial thermal resistance.

Dalal et al.[77] developed a CFD model of free convection in a double-paned window with a between-panes pleated cloth blind. The authors concluded that for all practical windows, there was no single pleat angle that minimizes the convective heat transfer; thus it was recommended to be set based on aesthetics and cost. A simplified one-dimensional model of the coupled convective and radiative heat transfer for calculating the U-value of this complex fenestration was also presented. This model shows good agreement with the results of a full conjugate CFD solution.

Karmele and Sala[78] have conducted a CFD study to investigate the heat transfer in a rectangular vertical enclosure with an internal louvered blind of high as well as low thermal conductivity. The results of the convective analysis were presented as a set of correlations for both types of materials. For each of the blind materials, the authors expressed the blind efficiency as a bi-quadratic equation in terms of blind material, surface emissivity, and blind angle. The predicted U-values using bi-quadratic curves were compared with the values obtained from WINDOW 6 software, and the discrepancies were explained on the basis of differences in thermal conductivity of aluminium used, the slat width and the boundary conditions considered.

2.5 Studies Related to Heat Transfer through Glazing Systems in Indian Conditions

Bansal and Minke[79] have classified Indian climate into six major zones: cold and sunny, cold and cloudy, warm and humid, hot and dry, composite, and moderate. These climatic zones pose a challenge to understand what glazing systems are expected to deliver in each of the zones so as to improve the comfort as well as the energy conditions of the user. The northern region of India comes under composite climate. Singh and Bansal[80] used the LBNL (Lawrence Berkley National Laboratories) software WINDOW 4.1 to evaluate the U-values, SHGC, and visual transmittance values for different window systems in different climatic zones of the Indian subcontinent. Instead of considering the values of outdoor (h_o) and indoor heat transfer coefficient (h_i) as predicted by the software, the authors used basic calculations to determine h_o and h_i. The authors reported that for a single-glazed window, the U-values in hot and dry Indian climatic conditions were 15% higher than those available in standard books.

Studies involving reduced energy consumption and increased thermal comfort have also been conducted in India. Singh and Bansal[81] investigated the effect of type, climate, size, and orientation of windows on the total energy demand for a building in Indian climatic conditions and determined the optimum window area along with the corresponding specific energy consumption. The authors concluded that optimization of area could be achieved only for buildings with U-values less than 0.6 W/m² K. Kamal[82] presented different passive cooling strategies including the use of shading devices to minimize the incident solar radiation and hence increase thermal comfort in addition to reducing building energy consumption. The author concluded that such strategies helped in reduction of air-conditioner (A/C) size, reducing A/C running time for the same comfort and thus reduced its energy consumption. Singh and Garg[83] analyzed the effect of the different types of glass curtain walls on the heating and cooling load of a building exposed to various climatic conditions in India. The authors concluded that annual energy consumption increased linearly with glazed area. It was reported to be minimum for north orientation and maximum for south orientation irrespective of the types of the glazing systems and the climates. The authors recommended the use of glass curtain walls made of solar control glazing (reflective), which consumed 6–8% lesser energy than the standard window. Manu and Rawal[84] highlighted the ambiguous nature of building regulations in India, particularly for window sizes. The authors conducted a parametric simulation study of cooling/lighting loads for commercial office spaces and reported that for hot and dry Indian climatic conditions, windows with window-to-wall ratio (WWR) less than 20% with proper orientation can save energy during lighting as well as cooling. The authors also concluded that for cases where windows can be provided on only one wall, the minimum window size specified by regulations was inefficient; thus they recommended a revision in existing building regulations.

Parishwad et al.[85] presented a simple procedure for estimating the incidence-angle-dependent solar-optical properties of tinted and coated glass of any thickness. The authors compared the estimated results with measured ones and concluded that up to 70° incidence angle estimated values were within a tolerable relative error of 11.18%. However, from 70°–80° it increased and beyond 80° it finally decreased.

Pal et al.[86] developed mathematical models to simulate the window plane solar radiation and corresponding glazing surface temperature and validated the results with measured values. The developed thermal model was also used to determine the heat flow inside the room through window. The authors concluded that thermal and radiation models accurately predicted the effect of solar radiation on the thermal performance of building interiors.

3. Analysis and Discussion of Review

The analysis from the literature survey of heat transfer through windows shows that

1. Both U-values and SHGC can completely define the heat transfer effects of a glazing system, which can be single-glazed windows, double/multi-glazed windows, or any of these windows along with a shading device such as venetian blinds, draperies, insect screen, etc. Five category heads that may affect the heat transfer rate through glazing systems were identified in general. These category heads along with their concerned characteristics are listed in the Table 1.

2. Various techniques are available for experimentally measuring the U-values of a glazing system. Though a guarded hot plate apparatus can measure only the center-glass U-values, it is preferably used over other apparatuses due to its lower cost and lower time consumed during measurements. Moreover, only center-glass U-values are predominant when frames of lesser thermal conductivity values are considered. In addition, the need of providing reproducible heat coefficient values for U-value determination is eliminated, and the test glazing systems can be tested at any tilt angle.

3. The U-values can be determined using simulation software such as VISION, WINDOW, Window Information System (WIS), and ParaSol. During calculations of U-values using WINDOW (6.3 version), the temperature difference of outer and inner panes can be identified with those of glazing system sample in a guarded hot plate apparatus for comparison or validation purposes.

4. SHGC of the complete glazing system can be determined by knowing the thermal and solar optical properties of the individual layers using a spectrophotometer with an integrating sphere. These values can also be determined using analytical models.

5. From an Indian point of view, many studies associated with windows have been conducted, but most of them were limited to selection of windows direction, the type of glazing to be used, and window area to be kept so as to ensure lower energy consumption and maximizing comfort to occupants for a specific local climate. Due to the lack of research that establishes the credibility of double-glazed windows with cavity blinds to Indian climatic conditions, these devices to date have not been used by the majority of the population of India. The potential of double-glazed windows with inter-pane shading devices as

Table 1. Categories affecting heat transfer and their corresponding characteristics.

Thermal properties	Optical properties	Geometrical properties	Environmental conditions
Thermal resistance of individual layers including that of cavity	Normal and off-normal bi-conical transmittance, reflectance, and absorptance of each layer including that of coating on glass pane as well as shading screen	Window geometry defining length, width, and height of window	Sky radiation
Physical and dynamic properties of fluid in cavity	Index of refraction of each layer	Number of panes and their thickness	Wind speed and direction
Shading coefficient	Emittance (including longwave) of outdoor and indoor surfaces	Cavity depth or pane spacing	Indoor and outdoor temperature
Surface heat transfer coefficient	Beam and diffuse solar optical properties	Shading screen geometry and dimensions	Normal incident solar irradiance on windows
Outdoor–indoor temperature difference	——	Slat angles or openness factors of shade screen	Temperature of fluid in cavity
——	——	Dimensions specifying location of shade screen	Climatic conditions as per latitude and longitude

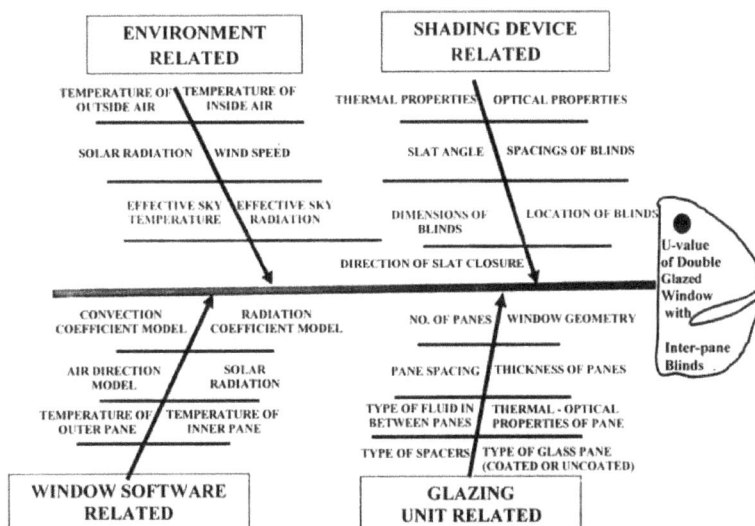

Fig. 2. Ishikawa diagram for factors affecting U-value of double-glazed window with inter-pane blinds using WINDOW software.

applicable to a particular climate can be justified only after proper experimentation. Some of the rarely used but commercially available such devices (whether imported or a replica of an imported one) appear to have been designed based mainly on aesthetics, with little or no consideration to their thermal performance. With a small effort, an improved understanding of the effects of various parameters on the heat transfer through such a glazing unit for the Indian climate can allow its possible usage.

After surveying the literature, the factors that may affect the simulated U-value of a double-glazed window with cavity blinds

in WINDOW 6.3 software are identified and an Ishikawa fishbone diagram is drawn as shown in Figure 2.

4. Gaps and Scope for Future Work

Both SHGC and U-values for this configuration can be measured as well as simulated; however, to start with, the authors intend to simulate the thermal transmittance through double-glazed windows with inter-pane shading devices using composite climatic conditions of India in WINDOW (6.3 version) software and to empirically validate the obtained results. The need of the hour is a low-cost and quick-responding guarded hot plate apparatus for determining thermal transmittance of a double-glazed system with inter-pane shade for temperature levels identified to the Indian climate. Many studies have been performed to investigate the effects of various parameters on the measured U-values; however, quantification of these effects in terms of percentage contribution has never been done. The authors also intend to use design of experiment (DOE) techniques followed by ANOVA to quantify the effect of control parameters.

Appendix: Nomenclature

A	Absorptance
A_{pf}	Total projected area of fenestration (m^2)
G_t	Incident total irradiance (W/m^2)
N	Inward flowing fraction (W/m^2)
q_i	Secondary heat transfer toward inside (W/m^2)
t	Air Temperature (in °C)
T	Transmittance
U	Thermal Transmittance

Greek symbols

α_s	Solar absorptance
θ	Incident angle relative to normal layer
ϕ	Azimuthal angle (in plane of layer about normal)

Superscripts and subscripts

out	outer environment
in	inner environment
dir-sol	direct solar
fH	front hemispherical
k	layer number
L	last layer

References

1. ASHRAE. Fenestrations. In *ASHRAE Handbook—Fundamentals (SI)*; American Society of Heating, Refrigeration and Air-Condition Engineers: Atlanta, GA, 2005; pp 31.1–31.69.
2. Arasteh, D.; Reilly, S.; Rubin, M. A Versatile Procedure for Calculating Heat Transfer Through Windows. *ASHRAE Trans.* 1989, 95, 755–765.
3. Curcija, D. *Trends and Developments in Window Testing Methods.* Baltic Window Conference, Lithuania, April 2000.
4. Elmahdy, A. H.; Bowen, R. P. Laboratory Determination of the Thermal Resistance of Glazing Units. *ASHRAE Trans.* 1988, 94, 1301–1316.
5. Alvarez, G.; Palacios, M. J.; Flores, J. J. A Test Method to Evaluate the Thermal Performance of Window Glazings. *Appl. Thermal Eng.* 2000, 20, 803–812.
6. Christensen, G.; Brown, W. P.; Wilson, A. G. *Thermal Performance of Idealized Double Windows, Unvented.* ASHRAE 71st Annual Meeting, Cleveland, Ohio, June 29–July 1 1964.
7. El Sherbiny, S. M.; Raithby, G. D.; Hollands, K. G. T. Heat Transfer by Natural Convection Across Vertical and Inclined Air Layers. *J. Heat Transfer* 1982, 104, 96–102.
8. Wright, J. L.; Sullivan, H. F. Glazing System U-Value Measurement Using a Guarded Heater Plate Apparatus. *ASHRAE Trans.* 1988, 94, 1325–1337.

9. Wright, J. L.; Sullivan, H. F. Thermal Resistance Measurement of Glazing System Edge-Seals and Seal Materials Using a Guarded Heater Plate Apparatus. *ASHRAE Trans.* 1989, 95, 766–771.
10. J.A. Baker; Sullivan, H. F.; J. L. Wright. *Study of Pane Spacing in Glazing System.* Solar Energy Society of Canada: Penticton, 1989; pp 262–272.
11. Varshney, K.; Rosa, J. E.; Shapiro, I. Method to Diagnose Window Failures and Measure U-Factors on Site. *Int. J. Green Energy* 2012, 9, 280–296.
12. Rubin, M. Calculating Heat Transfer Through Windows. *Energy Res.* 1982, 6, 341–349.
13. Wright, J. L.; H. F. Sullivan. Natural Convection in Sealed Glazing Units: A Review. *ASHRAE Trans.* 1989, 95, 592–603.
14. Baker, J. A.; Sullivan, H. F.; Wright, J. L. Window (Glazing and Frame) Heat Transfer Modelling. *ASHRAE Trans.* 1990, 96, 901–906.
15. Finlayson, E. U.; Arasteh, D. K.; Huizenga, C.; Rubin, M. D.; Reilly, M. S. *WINDOWS 4.0: Documentation of Calculation Procedures;* Lawrence Berkeley Laboratory: Berkeley, CA, 1993.
16. Lawrence Berkeley National Laboratories. WINDOW 6.3. http://windows.lbl.gov/software/window/6/ (accessed Sept 14, 2013).
17. Erhorn, H.; Stricker, R.; Szerman, M. Test Methods for Steady-State Thermal Transmission. In *IEA-Annex XII—Windows and Fenestration Report;* Energy Conservation in Buildings & Community Systems Programme Step 2: Delft, 1987; pp 1–27 (3.3.2).
18. Wright, J. L.; Sullivan, H. F. A 2-D Numerical Model for Natural Convection in a Vertical, Rectangular Window Cavity. *ASHRAE Trans.* 1994, 100, 1193–1206.
19. Wright, J. L.; H. F. Sullivan. A 2-D Numerical Model for Glazing System Thermal Analysis. *ASHRAE Trans.* 1995, 101, 819–831.
20. Wright, J. L. Calculating Window Solar Heat Gain. *ASHRAE J.* 1995, 37, 18–22.
21. J. L. Wright; McGowan, A. Calculating Solar Heat Gain of Window Frames. *ASHRAE Trans.* 1999, 105, 1011–1021.
22. Aydin, O. Determination of Optimum Air-Layer Thickness in Double-Pane Windows. *Energy Build.* 2000, 32, 303–308.
23. J. H. Klems. A New Method for Predicting the Solar Heat Gain of Complex Fenestration Systems: II. Detailed Description of the Matrix Layer Calculation. *ASHRAE Trans.* 1994, 100, 1073–1086.
24. Klems, J. H.; Kelley, G. O. Calorimetric Measurements of Inward-Flowing Fraction for Complex Glazing and Shading Systems. *ASHRAE Trans.* 1996, 102, 947–954.
25. Collins, M. R.; Harrison, S. J. Calorimetric Measurement of the Inward-Flowing Fraction of Absorbed Solar Radiation in Venetian Blinds. *ASHRAE Trans.* 1999, 105, 1022–1030.
26. Collins, M. R. The Effects of Calorimeter Tilt on the Inward-Flowing Fraction of Absorbed Solar Radiation in a Venetian Blind. *ASHRAE Trans.* 2001, 107, 677–683.
27. Collins, M. R.; Harrison, S. J. Test of Measured Solar Heat Gain Variation with Respect to Test Specimen Tilt. *ASHRAE Trans.* 2001, 107, 691–699.
28. Collins, M. R.; Harrison, S. J.; Oosthuizen, P. H.; Naylor, D. Heat Transfer from an Isothermal Vertical Surface with Adjacent Heated Horizontal Louvers; Validation. *ASME J. Heat Transfer* 2002, 124, 1078–1087.
29. Collins, M. R.; Harrison, S. J. Calorimetric Analysis of the Solar and Thermal Performance of Windows with Interior Louvered Blinds. *ASHRAE Trans.* 2004, 110, 474–485.
30. Loutzenhiser, P. G.; Manz, H.; Carl, S.; Simmler, H.; Maxwell, G. M. Empirical Validations of Solar Gain Models for a Glazing Unit with Exterior and Interior Blind Assemblies. *Energy Build.* 2008, 40, 330–340.
31. Simmler, H.; Binder, B. Experimental and Numerical Determination of the Total Solar Energy Transmittance of Glazing with Venetian Blind Shading. *Build. Environ.* 2008, 43, 197–204.
32. Kotey, N. A.; Wright, J. L.; Barnaby, S. C.; Collins, M. R. Solar Gain Through Windows with Shading Devices: Simulation Versus Measurement. *ASRAE Trans.* 2009, 115, 18–30.

33. Rosenfeld, J. L. J.; Platzer, W. J.; Dijk, H. V.; Maccari, A. Modelling the Optical and Thermal Properties of Complex Glazing: Overview of Recent Developments. *Solar Ener.* **2000**, *69 (Suppl.)* (1–6), 1–13.

34. Pfrommer, P.; Lomas, K. J.; Kupke, C. Solar Radiation Transport Through Slat-Type Blinds: A New Model and Its Application for Thermal Simulation of Buildings. *Solar Ener.* **1996**, *57*, 77–91.

35. Datta, G. Effect of Fixed Horizontal Louver Shading Devices on Thermal Performance of Building by TRNSYS Simulation. *Renewable Ener.* **2001**, *23*, 497–507.

36. Collins, M. R.; Wright, J. L. Calculating Centre-Glass Performance Indices of Windows with a Diathermanous Layer. *ASHRAE Trans.* **2006**, *112*, 22–29.

37. Ye, P.; Harrison, S. J.; Oosthuizen, P. H.; Naylor, D. Convective Heat Transfer from a Window with a Venetian Blind: Detailed Modeling. *ASHRAE Trans.* **1999**, *105*, 1031.

38. Collins, M. R.; Harrison, S. J.; Oosthuizen, P. H.; Naylor, D. Sensitivity Analysis of Heat Transfer from an Irradiated Window and Horizontal Louvered Blind Assembly. *ASHRAE Trans.* **2002**, *108*, 503–511.

39. Collins, M. Convective Heat Transfer Coefficients from an Internal Window Surface and Adjacent Sunlit Venetian Blind. *Energy Build.* **2004**, *36*, 309–318.

40. Collins, M. R.; Harrison, S. J. Estimating the Solar Heat and Thermal Gain from a Window with an Interior Venetian Blind. *ASHRAE Trans. J.* **2004**, *110*, 486–500.

41. Shahid, H.; Naylor, D. Energy Performance Assessment of a Window with a Horizontal Venetian Blind. *Ener. Build.* **2005**, *37*, 836–843.

42. Fang, X. D. A Study of the U-Factor of the Window with a High Reflectivity Venetian Blind. *Solar Ener.* **2000**, *68*, 207–214.

43. Wright, J. L.; Kotey, N. A. Solar Absorption by Each Element in a Glazing/Shading Layer Array. *ASHRAE Trans.* **2006**, *112*, 3–12.

44. Edwards, D. K. Solar Absorption by Each Element in an Absorber-Cover Glass Array. Technical Note. *Solar Ener.* **1977**, *19*, 401–402.

45. Papamichael, K.; Winkelmann, F. Solar-Optical Properties of Multilayer Fenestration Systems. In *International Daylighting Conference Proceedings II*, Long Beach, CA, Nov 5–7 1986; Bales, E. J., McCluney, R., Eds.; American Society of Heating, Refrigerating and Air-conditioning Engineers Inc.: Atlanta, Georgia, 1988; 410–418.

46. Klems, J. H.; Warner, J. L. Measurement of Bi-Directional Optical Properties of Complex Shading Devices. *ASHRAE Trans.* **1995**, *101*, 791–801.

47. Klems, J. H. Solar Heat Gain Through Fenestration Systems Containing Shading: Summary of Procedures for Estimating Performance From Minimal Data. *ASHRAE Trans.* **2002**, *108*, 512–524.

48. Milburn, D. I.; Hollands, K. G. T. Solar Transmittance Measurements Using an Integrating Sphere with Broad Area Irradiation. *Solar Ener.* **1994**, *52*, 497–507.

49. Kotey, N. A.; Wright, J. L.; Collins, M. R. *A Simplified Method for Calculating the Effective Solar Optical Properties of a Drapery*. 2nd Canadian Solar Buildings Conference, Calgary, 2007.

50. Collins, M. R.; Jiang, T. Validation of Solar/Optical Models for Louvered Shades Using a Broad Area Illumination Integrating Sphere. *ASHRAE Trans.* **2008**, *114*, 483–490.

51. Yahoda, D. S.; Wright, J. L. Methods for Calculating the Effective Solar-Optical Properties of a Venetian Blind Layer. *ASHRAE Trans.* **2005**, *111*, 572–586.

52. Kotey, N. A.; Wright, J. L.; Collins, M. R. Determination of Angle-Dependent Solar Optical Properties of Insect Screens. *ASHRAE Trans.* **2009**, *115*, 155–164.

53. Kotey, N. A.; Wright, J. L.; Collins, M. R. Determination of Angle-Dependent Solar Optical Properties of Roller Blind Materials. *ASHRAE Trans.* **2009**, *115*, 145–154.

54. Rubin, M. Solar Optical Properties of Windows. *J. Ener. Res.* **1982**, *6*, 123–133.

55. Rubin, M.; Von Rottkay, K.; Powles, R. Window Optics. *Solar Ener.* **1998**, *62*, 149–161.

56. Tzempelikos, A.; Laouadi, A.; Reinhart, C.; Athienitis, A. *Determining the Optical Properties of Shading Devices: Current Modelling*

Approaches and Future Directions. Canadian Solar Buildings Conference, Montreal, 2004.

57. Yahoda, D. S.; Wright, J. L. Heat Transfer Analysis of a Between-Panes Venetian Blind Using Effective Longwave Radiative Properties. *ASHRAE Trans.* **2004**, *110*, 455–462.

58. Kotey, N. A.; Collins, M. R.; Wright, J. L.; Jiang, T. A Simplified Method for Calculating the Effective Solar Optical Properties of a Venetian Blind Layer for Building Energy Simulation. *J. Solar Ener. Eng.* **2009**, *131*, 1–9.

59. Kotey, N. A.; Wright, J. L. Simplified Solar Optical Calculations for Windows with Venetian Blinds. In *Proceedings of the 31st Conference of the Solar Energy Society of Canada, Inc. (SESCI) and 1st Solar Buildings Conference (SBRN)*. Montreal, Quebec, Canada, August 20–24 2006.

60. Laouadi, A.; Parekh, A. Optical Models of Complex Fenestration Systems. *Lighting Res. Tech.* **2007**, *39*, 123–145.

61. Laouadi, A. Thermal Performance Modelling of Complex Fenestration Systems. *Build. Perf. Simulation* **2009**, *2*, 189–207.

62. Garnet, J. M.; Fraser, R. A.; Sullivan, H. F.; L. Wright, J. Effect of Internal Venetian Blinds on Window Centre-Glass U-Values. In *Window Innovations*; Toronto, June 5–6 1995; pp 273–279.

63. Huang, N. Y. T.; Wright, J. L.; Collins, M. R. Thermal Resistance of a Window with an Enclosed Venetian Blind: Guarded Heater Plate Measurements. *ASHRAE Trans.* **2006**, *112*, 13–21.

64. Onur, N.; Sivrioglu, M.; Turgut, O. An Experimental Study on Air Window Collector Having a Vertical Blind for Active Solar Heating. *Solar Ener.* **1996**, *57*, 375–380.

65. Naylor, D.; Lai, B. Y. Experimental Study of Natural Convection in a Window with a Between-Panes Venetian Blind. *Exper. Heat Transfer* **2007**, *20*, 1–17.

66. Breitenbach, J.; Lart, S.; Langle, I.; Rosenfeld, J. L. J. Optical and Thermal Performance of Glazing with Integral Venetian Blinds. *Ener. Build.* **2001**, *33*, 433–442.

67. Rheault, S.; Bilgen, E. Heat Transfer Analysis in an Automated Venetian Blind Window System. *J. Solar Ener. Eng.* **1989**, *111*, 89–95.

68. Zhang, Z.; Bejan, A.; Lage, J. L. Natural Convection in a Vertical Enclosure with Internal Permeable Screen. *J. Heat Transfer* **1991**, *113*, 377–383.

69. Bülow-Hübe, H.; Kvist, H.; Hellström, B. *Estimation of the Performance of Sunshades Using Outdoor Measurements and the Software Tool Parasol v. 2.0*. ISES Solar World Congress, Gothenburg, 2003.

70. Wright, J. L.; Collins, M. R.; Huang, N. Y. T. Thermal Resistance of a Window with an Enclosed Venetian Blind: A Simplified Model. *ASHRAE Trans.* **2008**, *114*, 471–482.

71. Tasnim, S. *Numerical Analysis of Convective Heat Transfer for Horizontal, Between-the-Panes Louvered Blinds*; MASc Thesis, MME, University of Waterloo: Waterloo, Ontario, 2005.

72. Almeida, F.; Naylor, D.; Oosthuizen, P. An Interferometric Study of Free Convection in a Window with a Heated Between-Panes Blind. In *Proc. 3rd SBRN and SESCI 33rd Joint Conference*, Fredericton, 2008; p 8.

73. Collins, M. R.; Tasnim, S.; Wright, J. L. Determination of Convective Heat Transfer for Fenestration with Between-the-Glass Louvered Shades. *Int. J. Heat Mass Transfer* **2008**, *51*, 2742–2751.

74. Collins, M. R.; Tasnim, S. H.; Wright, J. L. Numerical Analysis of Convective Heat Transfer in Fenestration with Between-the-Glass Louvered Shades. *J. Build. Environ.* **2009**, *44*, 2185–2192.

75. Naylor, D.; Collins, M. R. Evaluation of an Approximate Method for Predicting the U-Value of a Window with a Between-Panes Blind. *Numer. Heat Trans. Part A—Appl.* **2005**, *47*, 233–250.

76. Avedissian, T.; Naylor, D. Free Convective Heat Transfer in an Enclosure with an Internal Louvered Blind. *Int. J. Heat Mass Trans.* **2008**, *51*, 283–293.

77. Dalal, R.; Naylor, D.; Roeleveld, D. A CFD Study of Convection in a Double Glazed Window with an Enclosed Pleated Blind. *Ener. Build.* **2009**, *41*, 1256–1262.

78. M. Karmele, U.; Sala, J. M. Heat Transfer Through a Double-Glazed Unit with an Internal Louvered Blind: Determination of the Thermal

Transmittance Using a Biquadratic Equation. *Int. J. Heat Mass Trans.* **2012**, *55*, 1226–1235.

79. Bansal, N. K.; Minke, G. *Climatic Zones and Rural Housing in India;* Forschungszentrum Julich Gmbh: Zentralbibliothek, 1995.

80. Singh, I.; Bansal, N. K. Thermal and Optical Parameters for Different Window Systems in India. *Int. J. Ambient Ener.* **2002**, *23*, 201–211.

81. Singh, I.; Bansal, N. K. Effect of Window Type, Size and Orientation on the Total Energy Demand for a Building in Indian Climatic Conditions. *Int. J. Ener. Tech. Policy* **2004**, *2*, 323–334.

82. Kamal, M. A. A Study on Shading of Buildings as a Preventive Measure for Passive Cooling and Energy Conservation in Buildings. *Int. J. Civil Environ. Eng.* **2010**, *10*, 19–22.

83. Singh, M. C.; Garg, S. N. Suitable Glazing Selection for Glass-Curtain Walls in Tropical Climates of India. *ISRN Renewable Ener.* **2011**, Article 484893. doi: 10.5402/2011/484893.

84. Manu, S.; Rawal, R. In *Impact of Window Design Variants on Lighting and Cooling*. Eleventh International IBPSA Conference, Glasgow, Scotland, July 27–30 2009; pp 286–293.

85. Parishwad, G. V.; Bhardwaj, R. K.; Nema, V. K. Estimation of Solar Optical Properties for Windows. *Arch. Sci. Rev.* **1999**, *42*, 161–167.

86. Pal, S.; Roy, B.; Neogi, S. Heat Transfer Modelling on Windows and Glazing Under the Exposure of Solar Radiation. *Ener. Build.* **2009**, *41*, 654–661.

19

Maximum Power Point Tracking (MPPT) Scheme for Solar Photovoltaic System

AHTESHAMUL HAQUE*

Electrical Engineering, Jamia Millia Islamia University, New Delhi, India

Abstract: Global energy demand is increasing exponentially. This increase in demand causes concern pertaining to the global energy crisis and allied environmental threats. The solution of these issues is seen in renewable energy sources. Solar energy is considered one of the major sources of renewable energy, available in abundance and also free of cost. Solar photovoltaic (PV) cells are used to convert solar energy into unregulated electrical energy. These solar PV cells exhibit nonlinear characteristics and give very low efficiency. Therefore, it becomes essential to extract maximum power from solar PV cells using maximum power point tracking (MPPT). Perturb and observe (P&O) is one of such MPPT schemes. The behavior of MPPT schemes under continually changing atmospheric conditions is critical. It leads to two conditions, i.e., rapid change in solar irradiation and partial shading due to clouds, etc. Also, the behavior of MPPT schemes under changed load condition becomes significant to analyze. This article aims to address the issue of the conventional P&O MPPT scheme under increase solar irradiation condition and its behavior under changed load condition. The modified MPPT scheme is implemented in the control circuit of a DC–DC converter. The simulation study is done using PSIM simulation software. A prototype unit is tested with artificial light setup on a solar PV panel to simulate the changed solar irradiation condition. The results of the modified MPPT scheme are compared with existing schemes. The modified MPPT scheme works fast and gives improved results under change of solar irradiation. Furthermore, the steady state oscillations are also reduced.

Keywords: MPPT, photovoltaic, power electronics, perturb and observe, PSIM

1. Introduction

Global energy crisis and climate change threats are among the major concerns faced by the present civilized world. The limited reservoirs of fossil fuels and emission of greenhouse gases are the major identified reasons for the above concern. Renewable energy sources (RES) such as solar, wind, and tidal are considered the solution to overcome these concerns. Among these RES, solar energy is considered one of the potential sources to solve the crisis as it is available in abundance and free of cost.[1]

Solar photovoltaic (PV) cells are used to convert solar energy into regulated electrical energy using power electronics converter.[2] These solar PV cells exhibit nonlinear characteristics and very low efficiency.[3] The characteristic of solar cells become more complex under changed atmospheric condition such as partial shading.[4] Due to these issues, it becomes essential to

researchers to extract maximum power from solar PV cells under variable atmospheric conditions. Maximum power point tracking (MPPT) scheme is used to extract maximum power from solar PV cells. Various types of MPPT schemes are proposed by researchers,[5–14] namely open circuit, short circuit, perturb and observe (P&O)/hill climbing, incremental conductance, and so forth.

The P&O method is very popular among all these schemes. It is further classified into various types of P&O MPPT schemes and is adopted by the researchers. The nominal value of capacitors connected in parallel with solar PV is taken as a parameter to monitor in extraction of power.[5] However, Xiao and Dunford have proposed an adaptive scheme to extract the power.[6]

The P&O method may give false results, i.e., when solar irradiation is increased, the conventional P&O algorithm moves in the direction of high power. It fixes the operating point, which is not maximum power point (MPP). Also, the other issue is steady state oscillations due to the nature of the P&O MPPT scheme. The limitation of the P&O method has been highlighted under increased solar irradiation condition.[8] However, the MPPT behavior with resistive load change has not been addressed properly. These issues are addressed properly for other MPPT schemes such as incremental conductance.[13,15]

The P&O method is evaluated and the duty cycle is decided by the product of slope (dP/dV) with a constant gain.[12] The drawback of this method is high steady state oscillations

*Address correspondence to: Ahteshamul Haque, Electrical Engineering, Jamia Millia Islamia University, M. M. Ali Jauhar Marg, Okhla, New Delhi 110025, India. Email: ahaque@jmi.ac.in

and longer time to reach steady state under changed power condition.

In this article, the P&O MPPT scheme is modified considering increase in solar irradiation, steady state oscillations, and load changing condition. The results of the modified scheme are compared with the existing P&O MPPT scheme, and the improvements are highlighted. The new MPPT scheme is implemented in the control circuit of a buck DC–DC converter. A prototype hardware model is made and tested in the lab. The microcontroller is used to implement the proposed MPPT algorithm. The solar simulation and its changing irradiation condition have been carried out with flooded artificial lighting on solar PV panel. PSIM simulation software is used for simulation study.

2. Modeling and Characteristics of Solar Photovoltaic (PV) Cell

The basic element of a solar PV system is PV cells. These cells are connected to form modules. It is further expanded in the form of arrays as per the power requirement. These PV cells exhibit nonlinear characteristic. The output of the PV cell varies with solar irradiation and the ambient temperature. The characteristic equations of PV cell are given in Equations 1, 2, and 3.[3]

$$I = I_{ph} - I_{os} \{exp\,[(q/AKT)\,(V + IR_s)] - 1\} - (V + I * R_s)/R_p \tag{1}$$

$$I_{os} = I_{or}\,exp\,[qE_{GO}/Bk\,((1/T_r) - (1/T))]\,[T/T_r]^3 \tag{2}$$

$$I_{ph} = S\,[I_{sc} + K_I\,(T - 25)]\,/100 \tag{3}$$

Where I is the PV cell output current; V, the PV cell output voltage; R_p, the parallel resistor; and R_s, the series resistor. I_{os}, is the PV module reversal saturation current. A and B are ideality factors; T, the temperature (°C); k, the boltzmann's constant; I_{ph}, the light-generated current; q, the electronic charge; and K_I, the short-circuit current temperature coefficient at I_{SC}. S is solar irradiation (W/m²); I_{SC}, the short-circuit current at 25°C and 1000 W/m², E_{GO}, the band gap energy for silicon; T_r, reference temperature and I_{or}, the saturation current at temperature T_r. Figure 1 shows that the power varies nonlinearly with the variation in solar irradiation, and maximum power point (MPP) varies, too. However, the modeling of solar PV under partial shading condition gives the results differently.[4] The MPP point varies with ambient conditions. It is the task of researchers to make this moving point as the operating point to extract the maximum power.

3. Conventional Perturb & Observe/Hill Climbing MPPT Scheme

The other name of the P&O method is hill climbing method. In fact, the hill climbing and P&O methods are different ways to achieve the MPP. Hill climbing MPPT is achieved by perturbing the duty cycle of the power converter. In the P&O method, the perturbation is applied either in the reference voltage or in the

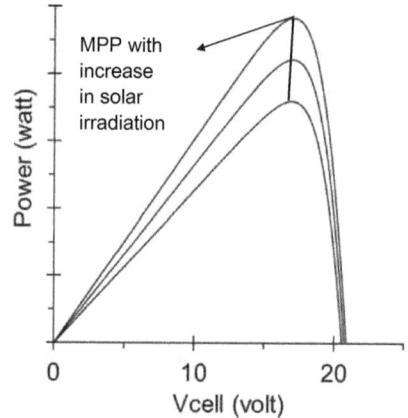

Fig. 1. Solar PV power characteristics with different solar irradiation level.

reference current signal of the solar PV. The flow chart of the conventional P&O method is shown in Figure 2. In this flow chart, Y is shown as the reference signal. It could be either solar PV voltage or current. The main aim is to achieve the MPP. To achieve it, the system operating point is changed by applying a small perturbation (ΔY) in solar PV reference signal. After each perturbation, the power output is measured. If the value of power measured is more than the previous value, then the perturbation in reference signal is continued in the same direction. At any point, if the new value of solar PV power is measured less than the previous one, then the perturbation is applied in the opposite direction. This process is continued till MPP is achieved.[8]

4. Issue Related to the Conventional P&O MPPT Scheme

The main issue with the conventional P&O MPPT scheme is its failure to give the correct MPP under fast changing atmospheric conditions as shown in Figure 3 and is also discussed in Esram and Chapman.[8] Figure 3 shows that the operating point A under one atmospheric condition; a perturbation in reference signal (which is voltage in the figure) brings the operating point at B. This algorithm is reversed back to operating point A due to the decrease in power. However, if the solar irradiation changes, the power curve shifted from P2 to P1 in one sampling period.

The operating point will move from A to C. It is to be noted that C is not the MPP of power curve P1. But the power at C is more than the power at A; however, the perturbation is kept same. Consequently, this phenomenon of divergence from MPP is continued, if solar irradiation increases continuously.

The high ripple content in the power at steady state is the other issue, which may be due to the nature of the P&O method as evaluated.[12]

5. Proposed P&O MPPT Scheme

The P&O MPPT technique depends on the change in power supplied by solar PV. This supplied power level depends on solar

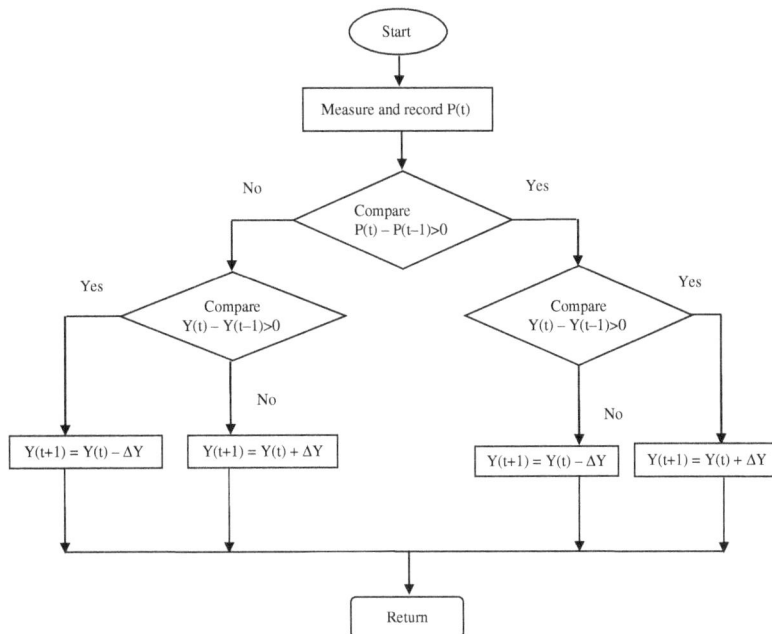

Fig. 2. Flowchart of conventional perturb and observe MPPT method.

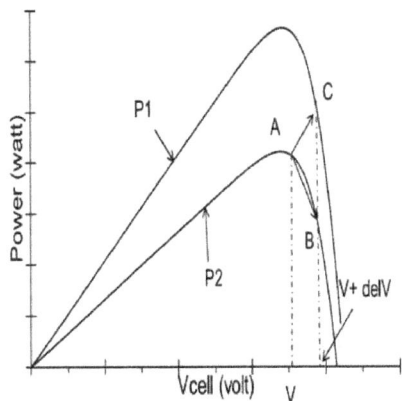

Fig. 3. Issue in P&O MPPT scheme under changing atmospheric condition.

irradiation, ambient temperature, and resistive load. The reference signal is either incremented or decremented periodically, comparing the power obtained in the present cycle. The reference signal is considered as PV voltage in this work. Table 1 presents the summary of the P&O MPPT method, and Table 2 describes the change in power with the change in solar irradiation and load resistance.

Once the solar irradiation is increased, the PV power is increased, and if it decreases, the power is decreased. The PV power varies differently with resistive load.

Table 1. Summary of the perturb and observe MPPT method.

Perturbation	Change in power	Next perturbation
Positive	Positive	Positive
Positive	Negative	Negative
Negative	Positive	Negative
Negative	Negative	Positive

Table 2. Summary of change in power with the change in solar irradiation and load resistance.

Cause	Change	Power	Voltage
Solar irradiation	Increase	Increase	Increase
	Decrease	Decrease	Decrease
Load resistance	Increase	Decrease	Increase
	Decrease	Increase	Decrease

The flowchart of the modified P&O MPPT is shown in Figure 4. The perturbation in the reference is dependent on the slope dP/dV. The slope dP/dV is measured, and the threshold value is shown in Equation 4.

$$\left| \frac{dP}{dV} \right| = 0.05 \tag{4}$$

During the stage, when the slope is high, the perturbation step is more, and when the slope is less, the perturbation step is small. The change in PV current along with PV voltage is also measured. The slope calculation gives the improved and fast MPP,

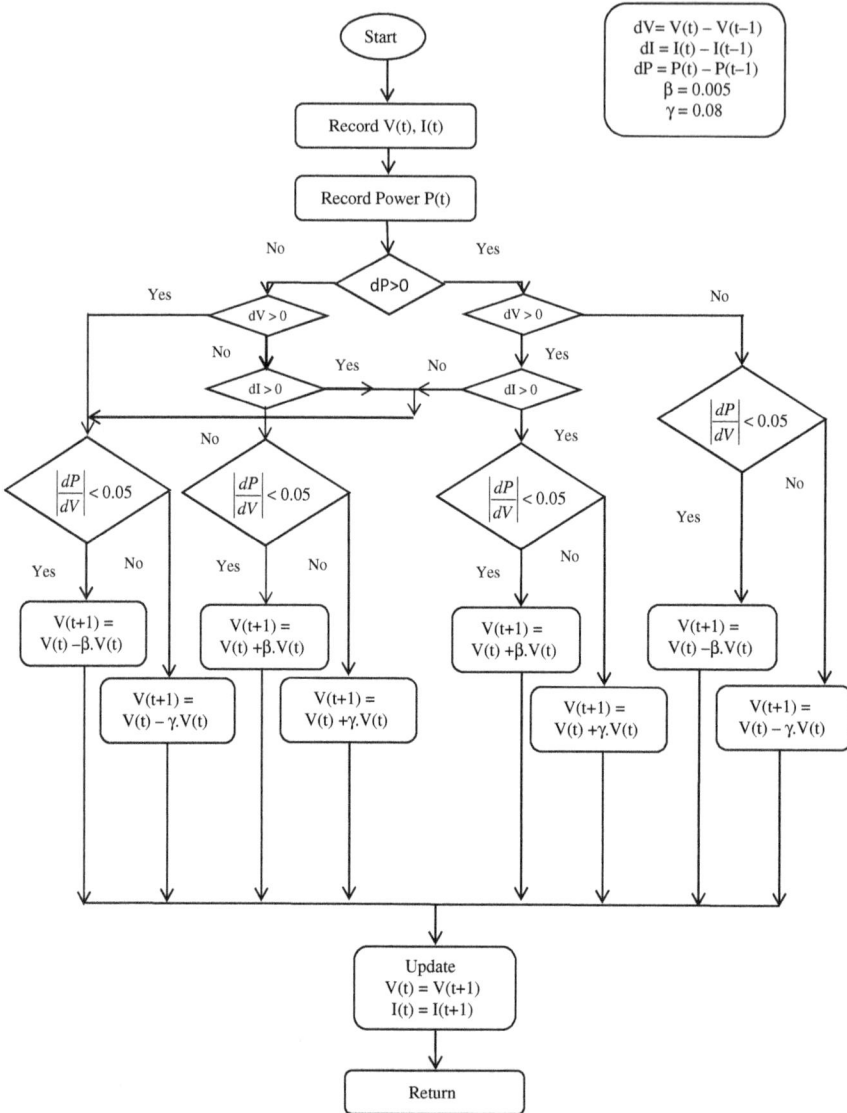

Fig. 4. Flowchart of the proposed perturb and observe MPPT method.

when operating under increased solar irradiation condition. The earlier work is done by multiplying slope with constant gain throughout the operation.[12] Also, the load change is another area where the proposed algorithm gives reliable results.

6. DC–DC Converter Topology and MPPT Working Zone

The MPPT operating zone for solar PV is dependent on DC–DC converter topology and restricts the value of resistive load for which MPPT become effective. The MPPT scheme is

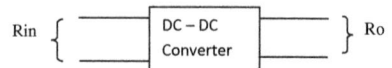

Fig. 5. Block diagram of DC–DC converter.

implemented in the control of DC–DC converter, i.e., it varies the duty cycle. The basic principle of adjusting the duty cycle is to match load impedance with input impedance seen by the DC–DC converter, i.e., impedance of solar PV as shown in Figure 5. Rin (the input impedance seen by the converter) and Ro (the output impedance connected with converter) are related with characteristic equation. This mathematical equation varies with DC–DC converter topologies, which are summarized in Figure 6. Figures 6 and 7 show the operating and non-operating zone of MPPT. The MPPT is working in the entire range of characteristic curve in the case of Buck-Boost and SEPIC (single-ended primary inductor converter) DC–DC converter. Since these converters are complex, they exhibit more cost in comparison to buck or boost converters.

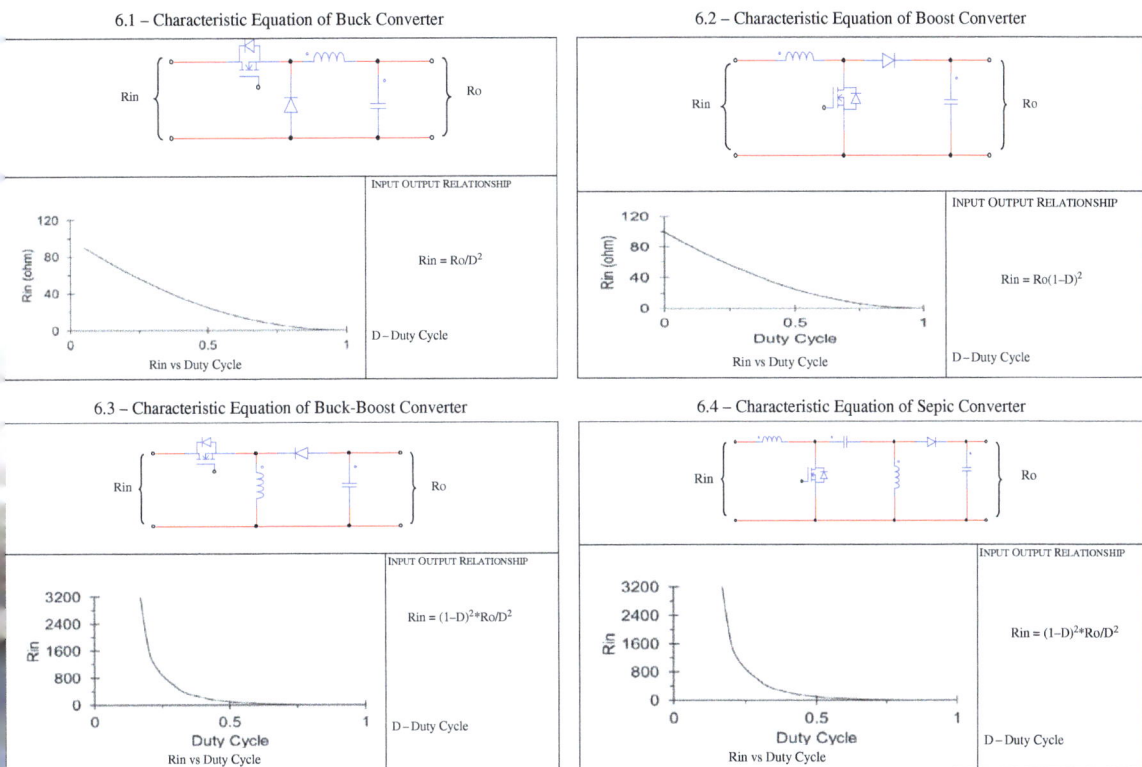

6.1 – Characteristic Equation of Buck Converter

INPUT OUTPUT RELATIONSHIP

$Rin = Ro/D^2$

D – Duty Cycle

Rin vs Duty Cycle

6.2 – Characteristic Equation of Boost Converter

INPUT OUTPUT RELATIONSHIP

$Rin = Ro(1-D)^2$

D – Duty Cycle

Rin vs Duty Cycle

6.3 – Characteristic Equation of Buck-Boost Converter

INPUT OUTPUT RELATIONSHIP

$Rin = (1-D)^2 * Ro/D^2$

D – Duty Cycle

Rin vs Duty Cycle

6.4 – Characteristic Equation of Sepic Converter

INPUT OUTPUT RELATIONSHIP

$Rin = (1-D)^2 * Ro/D^2$

D – Duty Cycle

Rin vs Duty Cycle

Fig. 6. Characteristic equations of commonly used DC–DC converters.

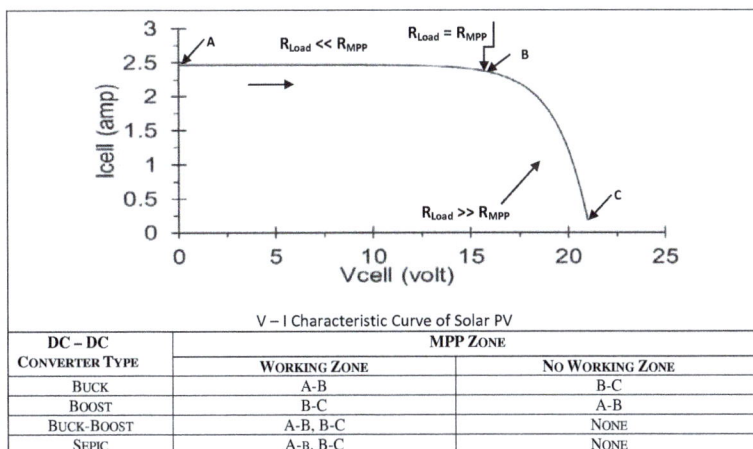

V – I Characteristic Curve of Solar PV

DC – DC CONVERTER TYPE	MPP ZONE	
	WORKING ZONE	NO WORKING ZONE
BUCK	A-B	B-C
BOOST	B-C	A-B
BUCK-BOOST	A-B, B-C	NONE
SEPIC	A-B, B-C	NONE

Fig. 7. MPPT working zone with DC–DC converter topology.

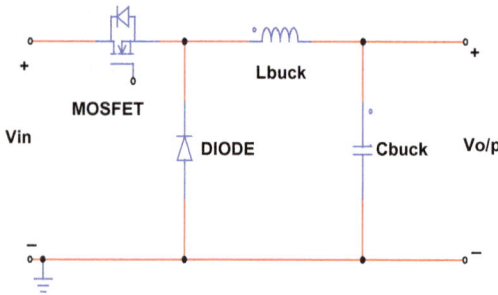

Fig. 8. Schematic: DC–DC buck converter.

7. Design of Buck DC–DC Converter and Solar PV Parameters

The DC–DC buck converter is used to implement the modified MPPT scheme (Fig. 8). The input voltage of DC–DC buck converter is the output voltage of solar PV. Equation 5 gives the relationship between input and output voltage of buck converter.[2]

$$Vo = d.Vin \qquad (5)$$

where Vo is the output voltage, d is the duty cycle and Vin is the input voltage of buck converter. The parameters of buck converter are calculated for the design work (Table 3). The parameters of solar PV model: ELDORA40 poly crystalline is listed in Table 4.

8. Experimental Setup and Results

The block diagram of experimental setup for testing of the modified MPPT algorithm is shown in Figure 9. The solar PV panel in the lab is flooded with artificial light (halogen lamp). The intensity of artificial light is controlled to simulate the fast-changing solar irradiation. The solar PV is connected with the buck DC–DC converter. The DC load is connected at the output of the buck

Table 3. Parameters of DC–DC buck converter.

S. No.	Name of the Parameter	Values
1	Vin	V_{MPP} when MPP is working V_{cell} in NO MPP Zone
2	MOSFET	20A, 600V
3	DIODE	20A, 1000V, Fast Recovery
4	L_{buck}	1mH, 15A Saturation
5	C_{buck}	1000 uF
6	Vo	$d*V_{MPP}$ $d*V_{cell}$
7	Frequency	20 kHz
8	Power Output	35W

Table 4. Parameters of solar PV.

S. No.	Name of the Parameter	Values
1	Name of the Vendor of solar PV	Vikram Solar Model: ELDORA40
2	PV Type	PolyCrystalline
3	No. of Cells: Connected in series	36
4	Rated Maximum Power: P_{MPP}	40W
5	Open Circuit Voltage: V_{OC}	21.9 V
6	Short Circuit Current: I_{SC}	2.45 A
7	Voltage at MPP: V_{MPP}	17.4 V
8	Current at MPP: I_{MPP}	2.3 A

converter. The PV voltage and current is sensed by using sensors and given as input to the microcontroller [Model: PIC- 16F887]. The microcontroller processes the proposed MPPT and gives output as pulse width modulated [PWM] signal to the gate of buck converter. The opto-coupler driver IC [TLP-250] is used to drive the power Mosfet of the converter. The data logger is connected to record the solar PV power, current, and voltage. The gate signal is monitored using digital CRO. A voltmeter and an ammeter are connected at the load end.

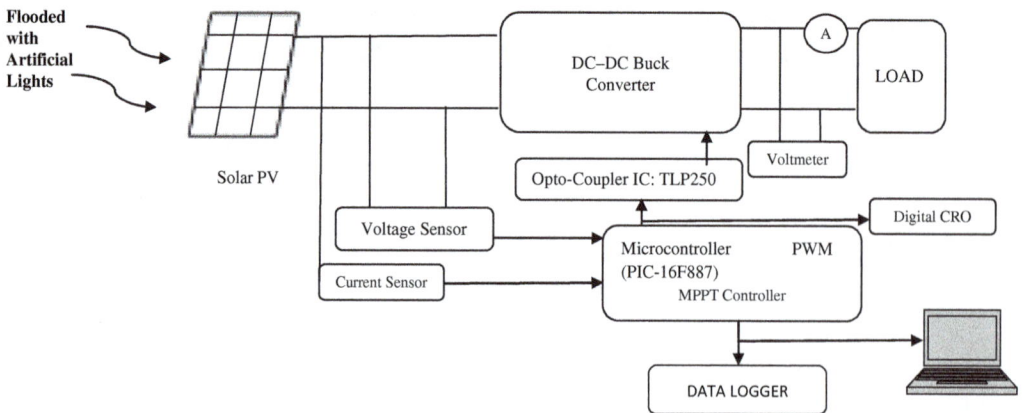

Fig. 9. Block diagram of experimental setup.

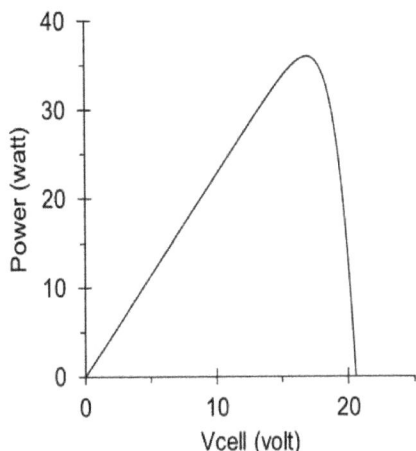

Fig. 10. Simulation results: Power (P)-Voltage (V) curve.

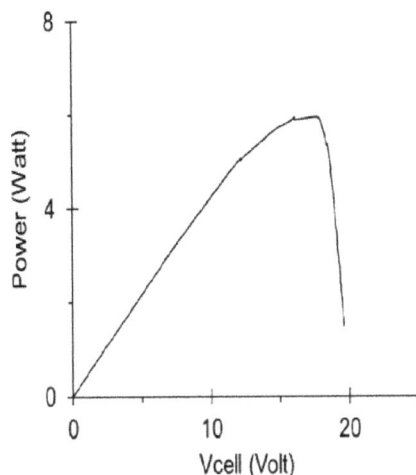

Fig. 11. Experimental results with full artificial light.

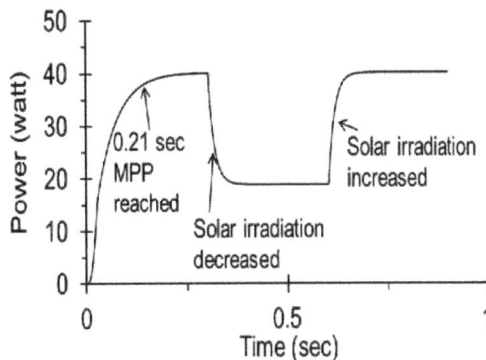

Fig. 12. Simulation results: Power vs. time.

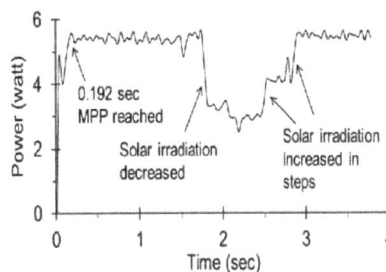

Fig. 13. Experimental results: Power vs. time.

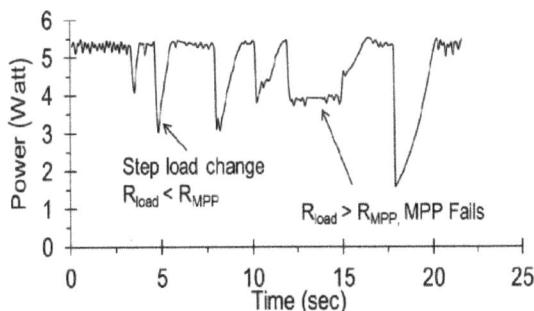

Fig. 14. Experimental results: Solar PV power vs. time at different load conditions.

The rating of solar PV used is 40W @ 1000 w/m² solar irradiation. The simulated Power-Vcell characteristic is shown in Figure 10. Since the PV is flooded with artificial lighting, therefore, power generated is not 40W—rather, it is around 6–7 Watts. But the purpose to test the effectiveness of MPPT scheme is achieved. The Power-Vcell curve of solar PV from the experimental setup with artificial light source, i.e., 1000 W/m², is shown in Figure 11.

Figure 12 shows the experimental results under changed solar irradiation levels. Figure 13 presents the experimental results obtained under similar conditions to that of simulation. It can be seen that the time taken to reach MPP is approximately 0.21 sec, and even if the solar irradiation is changed, the MPP scheme is not giving false results. Also, the experimental and simulation results are almost similar, except the steady state oscillation

is invisible in simulation. It is due to the fact that dP/dV is much less.

Figure 14 shows the experimental results under different load conditions. It is evident that if R_{Load} is less than R_{MPP}, the MPP works and restores the power to MPP level. The moment R_{Load} is greater than R_{MPP}, the proposed MPP fails, which is obviously expected. Figures 15 and 16 shows the experimental results for variable duty cycle.

Table 5 presents a comparison of the result between the gain method and the proposed method. It is evident that time to reach MPP, peak to peak steady state oscillations, and step load change response time is improved.

Fig. 15. Experimental results: PWM gate signal.

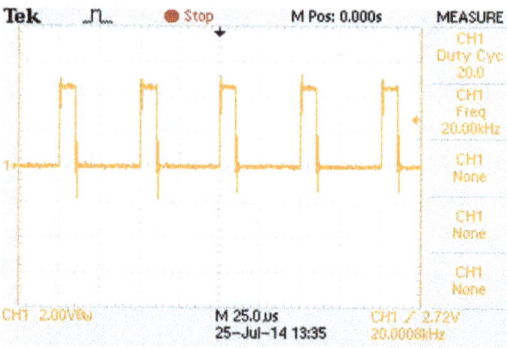

Fig. 16. Experimental results: PWM gate signal.

Table 5. Comparative results of gain and proposed method.

S. No.	Name of the parameter	Gain method	Proposed method
1	Time to reach MPP from zero	0.4 sec	0.192 sec
2	Peak to peak steady state oscillation	2.0 watt	0.500 watt
3	Step load change - response time	2.5 sec	1.800 sec

9. Conclusions

The correct and fast tracking of MPP under change solar irradiation and change load conditions are challenging tasks for researchers. The proposed MPPT scheme provides a solution to improve the existing methods. The proposed scheme may help in achieving accurate and fast response in standalone and grid-connected solar PV energy conversion systema. It can be applied in fast-changing solar irradiation areas where solar PV is used. The limitation of the proposed method is that it is not evaluated under partial shading conditions.

Appendix: Nomenclature

MPPT – Maximum power point tracker
PV – Photovoltaic
Rp – Parallel resistor
Ios – PV module reversal saturation current
T – Temperature in degree Celsius
Isc – Short circuit current
RES – Renewable energy sources
P&O – Perturb and observe
Rs – Series resistor
A, B – Ideality factor
k – Boltzmann's constant
S – Solar irradiation

References

1. Solanki S. Chetan. *Solar Photovoltaics: Fundamentals, Technologies and Applications*, New Delhi, PHI, 2012.
2. Erickson, R. W.; Maksimovic, D. *Fundamentals of Power Electronics*; Kluwer Academic Publishers, 2001.
3. Villalva, M. G.; Gazoli, J. R.; Filho, E.R. Comprehensive Approach to Modelling and Simulation of Photovoltaic Arrays. *IEEE Trans. Power Elec.* **2009**, *5*, 1198–1208.
4. Seyedmahmoudian, M.; Mekhilef, S.; Rahmani, R.; Yusof, R.; Renani, E. T. Analytical Modelling of Partially Shaded Photovoltaic Systems. *MDPI J. Energ.* **2013**, *6*, 128–144.
5. Kasa, N.; Lida, T.; Iwamoto, H. Maximum Power Point Tracking with Capacitor Identifier for Photovoltaic Power System. Proc. of Eighth International Conference on Power Electronics, variable speed drives. **2000**, *147*, 130–35.
6. Xiao, W.; Dunford, W. G. A Modified Adaptive Hill Climbing MPPT method for photovoltaic power systems. Proceedings of 35th Annual IEEE Power Electronics Specific Conference: 1957–63. 2004.
7. Cavalcanti, M.C.; Oliveira, K.C.; Azevedo, G.M.S.; Neves, F.A.S. Comparative Study of Maximum Power Point Tracking Techniques for Photovoltaic Systems. *Electron Potencia* **2007**, *12*, 163–171.
8. Esram, T.; Chapman, L. P. Comparison of Photovoltaic Array Maximum Power Point Tracking Techniques. *IEEE Trans. Energy Conv.* **2007**, *22*, 439–449.
9. Safari, A.; Mekhilef, S. Simulation and Hardware Implementation of Incremental Conductance MPPT with Direct Control Method Using Cuk Converter. *IEEE Trans. Ind. Elec.* **2011**, *58*, 1154–1161.
10. Yang Chih-Yu; Hsieh Chun-Yu; Feng Fu-Kuei; Chen Ke-Horng. Highly Efficient Analog Maximum Power Point Tracking (AMPPT) in a Photovoltaic System. IEEE Transaction on Circuits and Systems-1, 2012, 59.
11. Sayed Khairy; Abdel-Salam Mazen; Ahmed Adel; Ahmed Mahmoud. New High Voltage Gain Dual-boost DC-DC Converter for Photovoltaic Power Systems. *Int. J. Elec. Power Comp. Sys.* **2012**, *40*, 711–728.
12. De Brito, M.; Gomes, A.; Galotto, L.; Poltronieri, L.; Melo Guilherme de Azevedo e Melo, E.; Canesin, C. A. Evaluation of the Main MPPT Techniques for Photovoltaic Applications. *IEEE Trans. Ind. Elec.* **2013**, *60*, 1156–1167.
13. Tey Kok Soon; Saad Mekhilef; Azadeh Safari. Simple and Low Cost Incremental Conductance Maximum Power Point Tracking Using Buck-Boost Converter. *J. Renew. Sustain. Energ.* **2013**, *5*, 023106.
14. Houssam Issam; Locment Fabrice; Sechilariu Manuela. Experimental Analysis of Impact of MPPT Methods on Energy Efficiency for Photovoltaic Power System. *J. Elec. Power Energy Syst.* **2013**, *46*, 98–107.
15. Kok Soon Tey; Saad Mekhilef, Modified Incremental Conductance Algorythm for Photovoltaic System Under Partial Shading Condition & Load Variation. *IEEE Trans. Ind. Elec.* **2014**, *61*, 5384–5392.

20

Feasibility of Concentrated Photovoltaic Systems (CPV) in Various United States Geographic Locations

JIN HO JO*, RYAN WASZAK, and MICHAEL SHAWGO

Illinois State University, Normal, Illinois, USA

Abstract: Renewable energy sources are needed to reduce the country's reliance on greenhouse gas-producing fossil fuels. Many states have mandated renewable portfolio standards (RPS) to begin meeting this need. Photovoltaic (PV) systems can be an option for meeting this requirement. Concentrated photovoltaic (CPV) systems are one of the newest commercially available PV technologies, yielding the highest cell efficiencies. The purpose of this study is to determine the feasibility of a CPV system in relation to a traditional flat-plate PV system in various locations throughout the United States. We conducted a feasibility study concerning the energy production, net present value, and internal rate of return (IRR) for each system. Any available government incentives have been included in the study. Based on the results, we recommend in which states or regions CPV installations should be pursued.

Keywords: Concentrated photovoltaic, energy modeling, renewable energy, renewable portfolio standards, sustainable energy system

1. Introduction

Energy consumption in the United States is a growing concern. While there are many different ways to generate electricity to meet the growing demand, generally the two means of energy production competing are fossil fuels and renewable sources. Many countries around the world have begun to use renewable energy to substitute for fossil fuels, and solar photovoltaic (PV) systems have become a major player in the field.

As of the first quarter of 2014, 6.68 GW of utility PV plants are operating, and the contracted utility PV pipeline secured 12.5 GW.[1] Including all types of photovoltaics, cumulative operating PV capacity stood at 13.4 GW with 482,000 individual systems online at of the end of the first quarter of 2014.[1] It was forecasted that new PV installations will reach 6.6 GW in 2014, up 39% over 2013 and nearly double the market size in 2012.[1] The US EPA's effort to reduce CO_2 emissions from existing power plants by 30% by 2030 will further support the use of clean energy sources in addition to nuclear power and carbon-capturing strategies. Moreover, there are many incentives to produce renewable energy that are available from the federal and state governments. Different states will promote different technologies based on renewable resource availability and types of incentives.

There are multiple ways to produce energy with solar power, but the most widely used are flat-plate photovoltaic systems. Other new forms of solar production are on the rise, such as the use of concentrated photovoltaic (CPV) systems in conjunction with multi-junction PV cells. Typically, the most expensive part of PV modules is the PV cells, made from mono- or polycrystalline silicon. One of the benefits of CPV over a flat-plate PV system is that CPV uses less photovoltaic material and is therefore less expensive in terms of quantity of silicon required for PV module construction. In other words, CPV can reduce costs associated with PV material by concentrating the sunlight into very small cells that are made up of multiple layers of material. The layers associated with multi-junction cells allow the cells to absorb different wavelengths of light. This allows the CPV cells to have higher cell efficiencies than traditional mono- and polycrystalline silicon PV cells. CPV cells are small and are used with a concentrator to focus the light onto the cell. The concentrators can focus the sun by more than 500 times its normal irradiance level. Figure 1 represents a reflective mirror concentrator,[2] which uses two mirrors to concentrate sunlight onto a very small multi-junction solar cell.

Concentrator technology has been around since the mid-1970s but did not make much progress until recently, with the development of multi-junction cells. Multi-junction cells were initially made for space technology, but now can be more efficient on earth because of the concentrating technology. Concentrator solar cells are able to produce more current than conventional PV cells. The sun is converted to electricity in the concentrated cells, but the heat produced must be dispersed by incorporating a heat sink to the receiver to keep temperatures below 80°C. A primary and

*Address correspondence to: Jin Ho Jo, Illinois State University, Campus Box 5100, Normal, IL 61790-5100, USA. Email: jjo@ilstu.edu

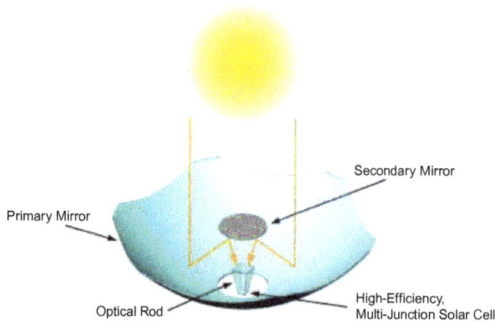

Fig. 1. CPV reflector mirror concentrator.[2]

secondary optic mirror helps to focus as much of the solar radiation as possible. The tracking system allows the array to stay focused on the sun all day long.

Traditional PV cells can absorb almost all the solar radiation that reaches the modules. The drawbacks of CPV modules are that only direct sunlight (80% of total sunlight on average) can be focused on the cells, and the panels must track the sun in order to keep it focused on cells all day long. Although only direct sunlight is used, multi-junction cells can absorb a wider spectrum of light from the sun than silicon cells and are much more efficient because of it. Efficiencies for multi-junction solar cells can reach over 40%.[3] The triple junction cells typically lose 0.15%–0.24%/°C compared to 0.50%/°C for traditional silicon cells, meaning they will have better performance at higher temperatures.[4] This shows that the efficiency of traditional flat-plate PV decreases faster than CPV as temperature rises. Because of the concentrator and the high efficiencies of CPV, it is beneficial to install these systems in places where abundant solar radiation is available. Therefore, the hot, dry climate of the southwestern states makes CPV applications more suitable in the United States.

A number of studies conducted by researchers around the world have investigated several aspects related to the implementation of large-scale solar PV systems, including both traditional flat-plate PV and concentrating solar power systems. Li et al.[5] assess technical and economic feasibility of 100 MW CSP systems to investigate the development potential of three types of CSP systems, including the parabolic trough system, tower system, and sterling dish system at six different locations in China. They report that that the grid parities of the parabolic trough system and tower system will be achieved as early as the next 10–20 years.

Mahtta et al.[6] evaluate the district-wise potential for concentrating solar power (CSP) and centralized solar photovoltaic (SPV) technology-based power plants in India. They utilize advanced techniques based on remotely sensed annual average global horizontal irradiance (GHI) and direct normal irradiance (DNI). The solar irradiation data (GHI and DNI), land-use data, and digital elevation model (DEM) were used in a geographic information system environment while employing land-use criteria and topography to exclude unsuitable sites for harnessing solar energy. They present potential sites for installation of CSP and centralized large-scale SPV plants in India.

Purohit et al.[7] claimed that CSP needs to play an important role to accelerate the decarburization in the Indian power sector. The authors attempt to assess the potential of CSP generation in the northwestern (NW) regions of India. The energy yield was estimated using the system advisor model for four commercially available CSP technologies. They report that with an annual DNI of over 1600 kW h/m^2, it is possible to exploit over 2000 GW CSP in NW India.

Trieb et al.[8] present a strategy utilizing CSP plants in the Middle East and North Africa (MENA). The authors explain the need of MENA countries for sustainable supply of electricity and calculate the cost of electricity for a model case country, and then they calculate the cost of CSP plants.

Although several prior works have examined the feasibility and impact of concentrating solar power into the current power markets, none of these have evaluated the feasibility and impact of implementing two different PV technologies, conventional flat-plate PV and concentrating PV, in various locations. Furthermore, a suitable method for evaluating the specific technology requirement within a state's RPS policy has not been suggested. Therefore, greater understanding is required to evaluate the feasibility and impact of different PV technology options. Throughout this research study, we report feasibility concerning energy production, net present value, and internal rate of return (IRR) for both CPV and conventional flat-plate PV systems in various locations in the United States. Any available government incentives have been included in the study. Based on the results, we recommend in which states or regions CPV installations should be pursued.

2. Methodology

Three main data sources were used in order to carry out the analysis. Solar resource data from the National Renewable Energy Laboratory (NREL) and information from the Database of State Incentives for Renewables and Efficiency (DSIRE) were utilized to select the states for the analysis. A couple of CPV manufacturers were interviewed in order to find pricing and system specifications as inputs in the simulation models. The system advisor model (SAM) from NREL was used to model CPV and PV systems. Figure 2 shows the process of this feasibility study. These data sources and the simulation tool utilized are further explained below.

Solar resource data from NREL were gathered for all 50 states. Both global horizontal irradiance (GHI) and direct normal irradiance (DNI) were collected for the locations selected. DNI is the amount of solar radiation received by a surface that is perpendicular to the rays traveling in a straight path from the sun, and GHI is the total amount of radiation received from the sun on a horizontal surface (both direct and diffuse).[3]

The DSIRE database of state and federal renewable energy incentives were then consulted to find incentives for each location. Table 1 represents the selected state incentives. After reviewing the solar resources along with the state incentives, 12 states were chosen to model to cover a varying range of solar resources. Another location in Texas was also modeled in order to fill in a gap within the range of the solar resources throughout the United States. When states with similar solar resources

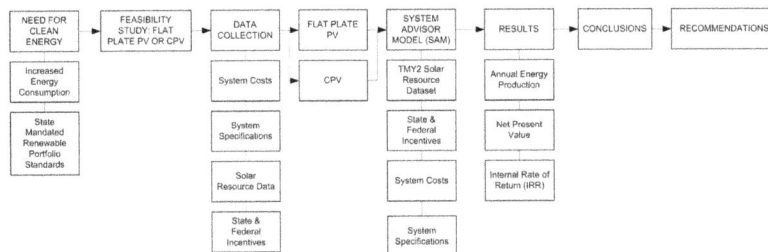

Fig. 2. Methodology flow chart.

Table 1. State incentives.

Location	State incentive
Daggett, CA	$0.05/kWh Production-Based Incentive for 5 years
Tucson, AZ	Production Tax Credit in years 1–10 of first 200,000 MWh as follows: 4¢, 4¢, 3.5¢, 3.5¢, 3¢, 3¢, 2¢, 2¢, 1¢, 1¢ (per kWh)
El Paso, TX	Deduct 10% of amortized cost from franchise tax
Tucumcari, NM	Production Tax Credit in years 1–10 of first 200,000 MWh as follows: 1.5¢, 2¢, 2.5¢, 3¢, 3.5¢, 4¢, 3.5¢, 3¢, 2.5¢, 2¢ (per kWh)
Kahului, HI	35% Investment Tax Credit up to $500,000
San Angelo, TX	Deduct 10% of amortized cost from franchise tax
Oklahoma City, OK	$0.005/kWh Corporate Tax Credit for 10 years
Miles City, MT	35% Corporate Tax Credit (carryover for 7 years)
Cape Hatteras, NC	35% Corporate Tax Credit (carryover for 7 years), capped at $2.5 million
Des Moines, IA	$0.015/kWh Production Tax Credit for 10 years
Miami, FL	$0.01/kWh Production Tax Credit for 4 years
Baltimore, MD	$0.0085/kWh Production Tax Credit for 5 years $0.10/kWh Production-Based Incentive for 20 years (subject to bid)
Caribou, ME	Only facilities <10 MW eligible. Assumed 10 MW eligible plant and 20 MW ineligible

were considered, the one with the best state incentives was selected.

Three CPV companies were contacted in order to find pricing and system specifications. The companies were SolFocus, Amonix, and Soitec. We found that two companies have prices near $2.75 per watt for installed systems.[9–10] The third, Soitec, quoted prices of $2.00–$2.10 for the modules only.[11] SolFocus quoted a balance of system costs near $0.85–$1.00, so assuming Soitec has similar balance of system (BOS) costs, their total price was assumed to be $3 per watt. Given that the other two companies' prices were both near $2.75, this price was used for the simulation models. The SolFocus system was modeled

because there is a published data set for its electrical outputs at a 1 MW facility at Arizona Western College in Yuma, Arizona.[12] For flat-plate PV, the price from Goodrich et al. (2012) was adopted ($3.37 per watt) for flat-plate PV systems utilizing single-axis trackers.[13] The system modeled for flat plate is from SunPower, which is a system made up of 1.5 MW power blocks. The SunPower system was chosen because they are one of the largest flat-plate PV manufacturers and are therefore a good representation of that sector of the PV market.

The SAM was used to model the flat-plate and CPV systems at the selected 13 locations. A 30 MW CPV system was modeled at each location. The SolFocus system has a cell efficiency of 40%, a module efficiency of 30%, and a total system efficiency of 28–29%.[9] The CPV system specifications were manually entered into the SAM and adjusted to give these efficiencies. The total system efficiency of 28.5% was used. The dual-axis tracker module is 15.6 kW, and 1924 trackers were used to model the matching 30 MW system. The total cost of the system was $82.5 million. The SunPower system was then modeled by choosing the 425W SunPower modules (SunPower SPR-425E-WHT-D) preinstalled in the software. A 24.5 MW system was modeled to give the same $82.5 million cost as the CPV system.

Making each system the same total cost allows the systems to be compared as equal investments. The two technologies can be compared using MW size, but the nature of these systems is quite different in terms of efficiency and capacity factor. Therefore, it is more practical to compare these technologies using projects of equal cost. This will provide solar developers with practical insights to evaluate the profits when making investment on such solar projects.

The Siemens Sinvert PVS1401 UL 480 V inverter was used in the model for both systems. This is the largest inverter available in the SAM at 1.4 MW. A 480 V inverter is required for utility power installations, and the largest size available was used because in actuality the system would use an even larger inverter. The SunPower 1.5 MW power blocks, for example, use a 1.5 MW inverter. The SolFocus CPV system utilizes dual-axis trackers because without them, the sunlight would not be focused on the solar cells at all times. The SunPower flat-plate system uses only a single-axis tracker. Flat-plate PV systems typically do not use dual-axis trackers as the increased energy output they give does not justify the added cost. Individual locations for each state were then picked to give the best disbursement of solar resource across the locations. The incentives from the DSIRE database were then entered for each state along with the federal 35% investment tax credit. The models were then simulated for each system at all

13 locations, and the energy production, net present value, and internal rate of return were compiled.

An important consideration when building the models was the plane of array (POA) irradiance at the module efficiency of the multi-junction cells. POA irradiance is the irradiance incident on the plane of the array and is calculated as the direct irradiance component with diffuse component. The reference condition for flat-plate PV cells is 1000 W/m², which is standard for PV calculations. The POA irradiance used for CPV was 850 W/m², which is lower than for flat-plate PV because the diffuse component is removed for CPV.[10] Leaving the POA irradiance at 1000 W/m² for CPV would cause the system's nameplate capacity to read as 15% higher than it actually is.

The other important parameters of the model involve financing and taxes. Each system was modeled with a 50% debt fraction and a 20-year loan at 7%. The analysis period was 30 years. A 2.5% inflation rate and 7.00% real discount rate were used, which gives a nominal discount rate of 9.67%. Federal corporate income taxes of 35% and each individual state's corporate income tax rate was entered into the SAM. Finally, a power purchase agreement (PPA) price of $0.10/kWh was used. There are examples of similar PPAs being granted for PV installations,[15] and this is within the range of current electricity costs paid by the consumer.

Once all the inputs were entered into the SAM, the model was run in all 13 locations for each system. The SAM outputted annual energy production, net present value, and internal rate of return for each system at each location. These results were then analyzed and conclusions drawn.

3. Results and Discussion

The results were derived from modeling a flat-plate PV system on a 1-axis tracker (tilt angle to latitude) versus a CPV system with a 2-axis tracker, with each system amounting to the same initial capital cost. The results show that a DNI of 5.4 or above will favor CPV over a traditional flat-plate PV at the utility scale. The model was run against 13 locations across the United States with DNI ranging from 3.3 to 7.6 kWh/m²/day at fairly even intervals. The global solar radiation, called global horizontal irradiance (GHI), (also measured in kWh/m²/day) spans from 5.4 to 9.1 kWh/m²/day. By modeling areas from low resource to high resource, it is suggested that CPV is best in a dry and hot climate environment and becomes less feasible without substantial incentives as GHI and DNI decrease.

3.1 CPV System with Direct Normal Irradiance

CPV electricity generation is dependent on DNI as it does not absorb any diffuse radiation from the sun. One of the purposes of this study is to determine what locations in the United States are actually suitable for this new technology, and it has been concluded that locations that receive 5.4 peak sun hours or more of direct normal irradiance are suitable for CPV. The location modeled with the highest DNI is Daggett, CA with a DNI of 7.6 kWh/m²/day. This location produced an annual generation of 75 GWh for CPV, which is 12 GWh greater than flat-plate at a generation of 63 GWh in the same location. The difference

between CPV generation and PV generation decreases as DNI decreases until San Angelo, TX with a DNI of 5.4 kWh/m²/day and a CPV annual generation of 52 GWh compared to flat-plate's 5.1 kWh/m²/day.

The next lowest DNI modeled is 5.1 kWh/m²/day in Oklahoma City, OK, which results in flat-plate generating 858 MWh more than CPV (49,827 MWh for the former and 48,969 MWh for the later). As DNI decreases, the difference of flat-plate generation minus CPV generation increases up to a difference of 6,884 MWh in favor of flat-plate in Caribou, ME with a DNI of 3.3 kWh/m²/day.

3.2 Flat-Plate PV System with Global Horizontal Irradiance

The flat-plate 1-axis tilt to latitude system is capable of absorbing the direct beam radiation that CPV utilizes as well as diffuse radiation. These two types of radiation combine to form GHI. The modeled locations range from 3.3 to 7.6 kWh/m²/day for DNI and from 5.4 to 9.1 kWh/m²/day for GHI.[16] While flat-plate receives a higher solar radiation than CPV, the benefits of utilizing multi-junction cells under concentration result in a higher generation for the same cost system in higher DNI resource areas. It is evident that as global radiation decreases, so does DNI; however, further analysis shows that as global irradiance decreases, DNI decreases at a greater rate as a percentage of total radiation. There may be exceptions in locations that are very dry with low GHI where a scenario with a DNI of less than 5.4 hypothetically could produce more energy with a CPV system than a PV system, should the percentage of DNI to GHI be high enough. One exception is Miami, presented in Table 2. As the percentage of diffuse radiation increases with GHI and DNI decreasing, Miami acts as an outlier to the trend due to its high humidity. This is because humid locations contain a greater ratio of diffuse radiation due to more clouds and moisture in the air interfering with the direct beam path from the sun to the collector. Miami's percentage of diffuse radiation to GHI is larger than the general pattern followed in Table 2.

At a resource of 9.1 GHI, 7.6 of the 9.1 are DNI. This results in 84% of total radiation being direct beam. This means that the remaining 16% is diffuse light, which CPV cannot utilize. As GHI decreases, the percentage of diffuse radiation to total GHI increases (meaning the percentage of DNI to GHI decreases). At the lowest resource tested in Caribou, ME, with a total GHI of 5.4 and a DNI of 3.3, 39% is diffuse and 61% is direct beam. This further explains why hot, dry locations are ideal for CPV; the higher the total solar radiation, the higher the percentage of direct beam as long as the area is not excessively humid. Therefore, an increase in global solar radiation is generally directly correlated with an increase in CPV energy generation potential.

3.3 Return on Investment

This study modeled the PV systems based off of a 30 MW CPV plant and a flat-plate PV plant that matches the cost of the CPV in order to compare energy generation and returns from the same initial investment. A positive internal rate of return (IRR) represents a profit adjusted for inflation. A positive value for net present value (NPV) signifies that the system produced a

Table 2. Percentage of diffuse and direct irradiation by location.

Location	Total solar resource (GHI = kWh/m²/day)	Direct solar resource (DNI = kWh/m²/day)	Diffuse (%)	Direct beam (%)
Daggett, CA	9.1	7.6	16.50%	83.50%
Tucson, AZ	8.7	7.2	17.20%	82.80%
El Paso, TX	8.6	6.8	20.90%	79.10%
Tucumcari, NM	8.0	6.3	21.30%	78.80%
Kahului, HI	7.7	6.0	22.10%	77.90%
San Angelo, TX	7.4	5.4	27.00%	73.00%
Oklahoma City, OK	7.0	5.1	27.10%	72.90%
Miles City, MT	6.7	4.8	28.40%	71.60%
Cape Hatteras, NC	6.3	4.4	30.20%	69.80%
Des Moines, IA	6.2	4.3	30.70%	69.40%
Miami, FL	6.5	4.1	36.90%	63.10%
Baltimore, MD	5.9	3.9	33.90%	66.10%
Caribou, ME	5.4	3.3	38.90%	61.10%

greater profit than an alternative investment yielding a 7% annual real return (see Table 3). A negative NPV does not necessarily represent a loss. If a negative NPV has a positive IRR, it still made a profit, but the profit was less than would have been realized from the alternative investment with a 7% real return—thus a negative NPV.

The IRR and NPV are not solely dependent on energy generation but are also greatly affected by government incentives. Applicable state and federal incentives (see Table 1) were entered into the system advisor model and the results show that CPV produced a positive IRR in the nine highest solar resource locations and a negative IRR in the lower four. Flat plate only showed a negative IRR in Baltimore (second lowest GHI). This shows that with sufficient incentives in place, CPV has potential to generate profit in locations where it is not most efficient in terms of annual energy generation. It is also worth noting that in Miles City, MT, the IRR is clearly the highest of the sample group, while its solar resource sits in the middle of the pack due to very favorable incentives.

3.4 Comparison with Real World Outputs

The only publicly available production data for the SolFocus CPV system modeled comes from a research site at Arizona Western College in Yuma, AZ. This site has a 974.4 kW SolFocus CPV system, which in its full year of energy production in 2012 produced 2,469 MWh of electricity.[17] Scaling this production to a 30 MW system would yield annual energy generation of 76,016 MWh. Yuma, AZ has a DNI of 7.2 kWh/m²/day, the same as Tucson, AZ. The SAM output for energy production was 72 GWh for Tucson, which is relatively close to the scaled real-world production of 76 GWh in Yuma; therefore, the SAM outputs appear to be a good approximation for real-world results.

It is worth noting that PV system hardware cost is known to be universal around the world, but what makes the total installed cost different are soft costs, including customer acquisition, financing, contracting, permitting, interconnection, inspection, installation, operations, and maintenance.[18] The US Department of Energy's

SunShot Initiative aims to reduce the total installed cost of solar energy systems to $.06 per kilowatt-hour (kWh) by 2020 without subsidies eventually. Today, SunShot is 60% of its way toward achieving the program's goal, only three years into the program's 10-year timeline. Since SunShot's launch in 2011, the average price per kWh of a utility-scale photovoltaic (PV) project has dropped from about $0.21 to $0.11.[18] Achieving the SunShot Initiative's $.06 per kWh goal would lead to the creation of 390,000 new solar jobs by 2050. Solar energy could meet 14% of U.S. electricity needs by 2030 and 27% by 2050.[14]

4. Conclusions

CPV technology is best suited for hot and dry climates with high solar resources and direct beam incidence. While these locations are ideal, with the aid of incentives they are not the only areas in which this technology is feasible. What makes hot and dry climates optimal for CPV is the high available solar resource coupled with a high DNI percentage of the total GHI. As the ratio of DNI to GHI decreases, the total annual energy output of CPV to PV decreases until the output achieved by PV utilizing GHI in areas with greater diffuse radiation surpasses CPV output. These results closely agree with industry expectations that set the DNI marker for CPV at 6 kWh/m²/day or above. (The study's model suggests 5.4 kWh/m²/day.) And as predicted, these areas are located mainly in the southwestern United States. It has also been proven, however, that CPV can still be profitable given the right incentives in the form of tax credits and/or high PPAs in lower resource areas. Taking a more detailed look at locations across the country to identify locations with a high ratio of DNI to GHI would be helpful in identifying locations that may be suited for CPV outside the Southwest.

Further work is required as CPV installations continue in order to create simulation software that more accurately and readily models real-world CPV technologies. The SAM provided reasonable outputs with traditional flat-plate PV since it has had real-world data to use and company information and specifications preinstalled into the model, while CPV system

Table 3. CPV vs. flat-plate PV SAM outputs.

Location	GHI (kWh/m² /day)	DNI (kWh/m² /day)	System	Annual generation (MWh)	Net present value	IRR
Daggett, CA	9.1	7.6	Flat-Plate PV	63,202	$13,387,778	26.06%
			CPV	75,495	$20,947,600	31.60%
Tucson, AZ	8.7	7.2	Flat-Plate PV	60,532	$11,652,405	23.67%
			CPV	71,960	$18,827,536	28.95%
El Paso, TX	8.6	6.8	Flat-Plate PV	59,455	$7,714,148	20.24%
			CPV	67,679	$11,990,208	23.93%
Tucumcari, NM	8.0	6.3	Flat-Plate PV	56,660	$8,439,092	20.05%
			CPV	61,775	$11,237,051	22.37%
Kahului, HI	7.7	6.0	Flat-Plate PV	53,449	$660,897	10.67%
			CPV	55,969	$1,477,076	11.82%
San Angelo, TX	7.4	5.4	Flat-Plate PV	51,186	$2,766,428	14.64%
			CPV	52,362	$2,825,336	14.78%
Oklahoma City, OK	7.0	5.1	Flat-Plate PV	49,827	($794,083)	8.35%
			CPV	48,969	($1,910,062)	6.15%
Miles City, MT	6.7	4.8	Flat-Plate PV	47,349	$14,466,054	41.31%
			CPV	47,090	$13,713,011	40.85%
Cape Hatteras, NC	6.3	4.4	Flat-Plate PV	45,274	$2,387,225	14.40%
			CPV	42,134	$334,579	8.61%
Des Moines, IA	6.2	4.3	Flat-Plate PV	44,888	($233,523)	9.10%
			CPV	41,899	($2,574,224)	−0.58%
Miami, FL	6.5	4.1	Flat-Plate PV	45,267	($3,581,112)	2.55%
			CPV	38,287	($8,326,899)	−6.54%
Baltimore, MD	5.9	3.9	Flat-Plate PV	42,001	($4,630,447)	−1.23%
			CPV	37,577	($7,779,019)	−7.33%
Caribou, ME	5.4	3.3	Flat-Plate PV	39,427	($284,321)	9.09%
			CPV	31,543	($6,514,553)	−9.72%

model simulation was somewhat restricted. Once CPV systems become more established, and enough field tests have been run on specific CPV modules and technologies, government and industry will greatly benefit from software that has accurate presets for a given CPV technology already entered. This would help choose optimal site locations based on incentives and resources by providing accurate results relative to the technology with minimal error. CPV is still in its early stages at the utility level, and once generation data are made more publicly available through more installations, there will be more resources for research and technology comparisons and analysis.

References

1. Solar Energy Industries Association. U.S. Solar Market Insight Q1 2014. *Solar Energy Industries Association*; Author: Washington, DC, 2014.
2. Practical Environmentalist. Concentrated Photovoltaic Plant Goes Online in Spain. PracticalEnvironmentalist.com: http://www.practical environmentalist.com/alternative-energy/concentrated-photovoltaic-plant-goes-online-in-spain.htm (accessed April 5, 2013).
3. Kurtz, S. *Opportunities and Challenges for Development of a Mature Concentrating Photovoltaic Power Industry*. NREL: Golden, CO, 2012.
4. Seshan, C. CPV: Not Just for Hot Deserts. Photovoltaic Specialists Conference (PVSC), 2010 35th IEEE, 003075-003080, 2010.
5. Li, Y.; Liao, S.; Rao, Z.; Liu, G. A dynamic assessment based feasibility study of concentrating solar power in China. *Renew. Energ.* **2014**, *69*, 34–42.
6. Mahtta, R.; Joshi, P.K.; Jindal, A. K. Solar power potential mapping in India using remote sensing inputs and environmental parameters. *Renew. Energ.* **2014**, *71*, 255–262.
7. Purohit, I.; Purohit, P.; Shekhar, S. Evaluating the potential of concentrating solar power generation in northwestern India. *Energ. Policy* **2013**, *62*, 157–175.
8. Trieb, Franz; Hans Müller-Steinhagen; Jürgen Kern. Financing concentrating solar power in the Middle East and North Africa—Subsidy or investment? *Energ. Policy* **2011**, *39*, 307–317.
9. Desy, K. Phone Interview with Kelly Desy of SolFocus (M. Shawgo, Interviewer), March 19, 2013.
10. Shell, S. Phone Interview with Stephen Shell of Amonix (R. Waszak, Interviewer), March 21, 2013.
11. Stafford, B. Phone Interview with Brent Stafford of Soitec (B. Hogdgon, Interviewer), March 18, 2013.
12. AZWestern. Arizona Western College Solar Installation. Solar with a Purpose: Innovation, Education, Generation. http://www.azwestern. edu/Marketing_and_PR/downloads/solar_infosheet_finallow.pdf (accessed Nov 26, 2013).
13. Goodrich, A.; James, T.; Woodhouse, M. *Residental, Commercial and Utility-Scale Photovoltaic (PV) System Prices in the United States: Current Drivers and Cost-Reduction Opportunities*; NREL: Golden, CO, 2012.
14. Kurtz, S.; Surendran, S.; Lerchenmueller, H. *CPV Industry Converges of Standard Rating Conditions*; NREL: Golden, CO, 2013.

15. Wesoff, E. *PV Project: SunPower's 100MW Henrietta Plant With a PPA Price Below $0.104*. Greentech Media, 2012.

16. National Renewable Energy Laboratory. *The Solar Radiation Data Manual for Flat-Plate and Concentrating Collectors*; Author: Golden, CO, 1994.

17. Mercy, B. *CPV in the Solar Landscape: A Real World Report for Arizona. Solar Power-Gen Conference and Exhibition*; PPA Partners: San Diego, CA, 2013; pp 1–21.

18. Department of Energy. Sunshot Initiative. http://energy.gov/eere/sunshot/sunshot-initiative (accessed March 2013).

21

Synthetic Evaluation of Oilfield Development Plans Based on a Cloud Model

ZHI-BIN LIU[1], LIU-LI LU[1,2]*, ZHI-YONG ZHANG[1], and YI-SHEN CHEN[1,3]

[1]*School of Science, Southwest Petroleum University, Chengdu, Sichuan, China*
[2]*CNPC Chuanqing Drilling Engineering Company Ltd., Chengdu, China*
[3]*CNPC International Engineering and Construction Corp., Beijing, China*

Abstract: In the process of evaluating the oilfield development plans, there always exist uncertainties in data, weight assignment, and scheme grading. In this article, an improved evaluation method is introduced to tackle the uncertainties based on a cloud model. The indicators of a development plan are input into this model and a contribution score cloud is obtained for evaluation. The evaluation clouds of all the development plans are sorted by the Technique for Order Preference by Similarity to an Ideal Solution (TOPSIS), with the weighted hamming distance. Then we can get the optimum development plan according to the order. In this model, the randomness and the fuzziness are both taken into consideration, and the feasibility and validity are verified by an example.

Keywords: Cloud model, synthetic evaluation, oilfield development plan, evaluation indicators

1. Introduction

Oilfield development is a complicated systematic job, which is high-tech, large investment, and high-risk. It plays a crucial role in the development to select an appropriate developing plan.[1,2] However, there always exist uncertainties when evaluating the development plans because of the limit, the subjectivity of human beings, and the uncertainties of the system itself. Thus, it is necessary to find a synthetic method to evaluate and optimize the oilfield development plans, which can deal with the uncertainties. To meet this purpose, several methods have been introduced,[3,4,5,6,7] such as AHP, Fuzzy Evaluating, and Technique for Order Preference by Similarity to Ideal Solution (TOPSIS). To a certain degree, these methods do reach the goal, but the randomness is not discussed synthetically. A new synthetic evaluation method of the oilfield development plans is presented in this article based on a cloud model, which takes the fuzziness and the randomness of the oilfield development both into consideration. The results of this article offer some new ideas for the evaluation and selection of the oilfield development plans.

*Address correspondence to: Liu-Li Lu, School of Science, Southwest Petroleum University, Chengdu, Sichuan 610500, China. Email: luliuly2014@163.com

2. Preliminaries

In this section, we will present some definitions and results that might be utilized in this article. The theory of cloud model was first proposed by Deyi Li,[9] the academician of the Chinese Academy of Engineering. A cloud model can be used to transform a qualitative concept into its quantitative expression.

Definition 2.1. Li and Du.[9] *Suppose that U is a qualitative universe with exact numerical expression, and T is a qualitative concept on U; if there is a mapping*

$$C_T(x) : U \to [0, 1], \forall x \in U, x \to C_T(x)$$

x is a stochastic realization of T, then the distribution of $C_T(x)$ on U is called the Cloud of T, and each x is a cloud droplet.

If $C_T(x)$ is normally distributed, the cloud is called normal. The normal cloud model is a set of random numbers that is normally distributed with a stable tendency. A normal cloud is represented by its numerical characteristics: expectation Ex, entropy En, and hyper entropy He, which is denoted by $C_T(Ex, En, He)$. Here, Ex is the center of the cloud distribution, En is the acceptable range of the linguistic value, and He is the measure of the uncertainty of the entropy. The larger the hyper entropy is, the more dispersed the cloud distribution gets.

2.1 Single Condition Cloud Rule Generator

Given a cloud rule,

$$If\ A, then\ B$$

both of the antecedent A and the consequent B contain uncertainty. We have to construct a cloud rule generator to make sure that the uncertainty can be reflected during the inference. The cloud rule generator is composed of the antecedent cloud generator and the consequent generator. Denote the normal clouds of concepts A and B by $C_A(Ex_A, En_A, He_A)$ and $C_B(Ex_B, En_B, He_B)$, respectively; the algorithm of a single condition cloud rule generator is as follows:[10]

INPUT: $\{Ex_A, En_A, He_A, Ex_B, En_B, He_b\}$, x_0; // x_0 is the initial value in the antecedent universe U_1
OUTPUT: Cloud droplet y in the consequent universe U_2 and its certainty degree μ.

BEGIN

$$
\begin{aligned}
En'_A &= RANDN\,(En_A, He_A) \\
\mu &= e^{\frac{-(x_0 - Ex_A)^2}{2En'^2_A}} \\
En'_B &= RANDN\,(En_B, He_B) \\
if\ x_0 &< Ex\ then \\
y &= Ex_b - \sqrt{-2\ln(\mu)}\,En'_B \\
else \\
y &= Ex_b + \sqrt{-2\ln(\mu)}\,En'_B \\
OUTPUT &:\quad drop\,(y, \mu)
\end{aligned}
$$

END

One can refer to Figure 1 for the procedure of the rule's generation. We should notice that for the same input, the rule generator might output different droplets because of the randomness.

2.2 Backward Cloud Generator

The output of the rule generator is always a droplet, which is quantitative. To transform the quantitative value into a qualitative concept, we employ the backward cloud generator (denoted by CG^{-1}) in Lu and colleagues[11], see Figure 2. Input a group of cloud droplets into the backward generator, we can get a cloud with its numerical characteristics Ex, En, and He. The algorithm is as follows.

Fig. 2. Backward cloud generator.

INPUT: N cloud droplets (y_i, μ_i);
OUTPUT: Ex, En, He.
BEGIN

Step 1: Compute the mean value of the samples y_i: $Ex = mean(y_i)$;
Step 2: Compute the standard deviation of y_i, that is the entropy $En = std(y_i)$;
Step 3: For each pair of (x_i, μ_i), $En'_i = \sqrt{\frac{-(x - Ex)^2}{2\ln \mu_i}}$;
Step 4: Compute the standard deviation of En'_i, that is the hyper entropy $He = std\left(En'_i\right)$.

END

2.3 Floating Clouds and Their Distances

Given n clouds $Y_1(Ex_1, En_1, He_1)$, $Y_2(Ex_2, En_2, He_2)$, . . ., $Y_n(Ex_n, En_n, He_n)$, called the basis clouds, to represent the null linguistic values between them, we need to employ the concept of floating cloud.[12] Denote the generated floating cloud by $Y(Ex, En, He)$, then its numerical characteristics are

$$Ex = \omega_1 Ex_1 + \omega_2 Ex_2 + \ldots + \omega_n Ex_n \tag{1}$$

$$En = \frac{\omega_1^2 En_1 + \omega_2^2 En_2 + \ldots + \omega_n^2 En_n}{\omega_1^2 + \omega_2^2 + \ldots + \omega_n^2} \tag{2}$$

$$He = \frac{\omega_1^2 He_1 + \omega_2^2 He_2 + \ldots + \omega_n^2 He_n}{\omega_1^2 + \omega_2^2 + \ldots + \omega_n^2} \tag{3}$$

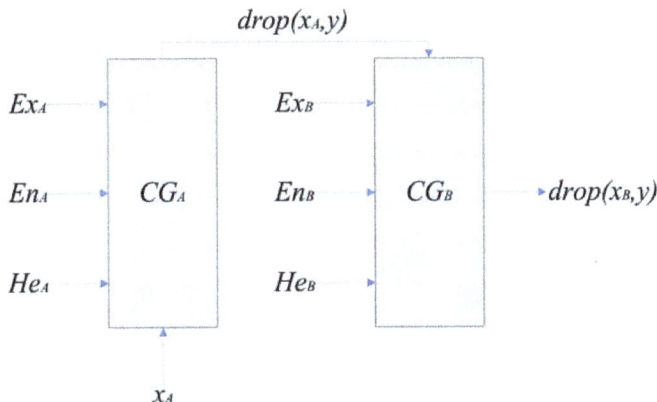

Fig. 1. Single condition rule generator.

Definition 2.3.1. *Suppose that F is the set of all the floating clouds, $y_1 = (Ex_1, En_1, He_1)$ and $y_2 = (Ex_2, En_2, He_2) \in F$. If a mapping $d : F \times F \to R$ satisfies the following condition:*

$$d(y_1, y_2) = d(y_2, y_1) \geq 0$$

and for a given floating cloud y_3,

$$d(y_1, y_3) \leq d(y_1, y_2) + d(y_2, y_3)$$

then $d(y_1, y_2)$ is the distance between y_1 and y_2.

In this article, we employ the weighted Hamming formula to define the distance between two floating clouds,

$$d(y_1, y_2) = \beta_1 |Ex_1 - Ex_2| + \beta_2 |En_1 - En_2| + \beta_3 |He_1 - He_2| \tag{4}$$

Here, $\beta_1, \beta_2, \beta_3 \geq 0$ are the weights, and $\sum_{i=1}^{3} \beta_i = 1$.

3. Synthetic Evaluation of Oilfield Developing Plan Based on a Cloud Model

A synthetic evaluation model based on the cloud theory contains 4 sets: the set of indicators $U = \{U_1, U_2, \ldots, U_m\}$, the set of comments of the indicators $V_1 = \{A_1, A_i, \ldots, A_m\}$, the set of the contribution scores of the indicators $V_2 = \{B_1, B_2, \ldots, B_m\}$ and the set of weights $W = \{W_1, W_2, \ldots, W_m\}$. To evaluate a development plan, we need to get the contribution scores of all the indicators and combine them with weighted average, i.e., a floating cloud. Then we can sort multiple development plans according to the scores and choose the best one.

3.1 The Indicator System and the Evaluation of the Indicators Based on a Cloud

In this section, the indicator system of a development plan is divided into 2 levels. Let U be a family of sets, denoted by $\{U_1, U_2, U_3\}$, where U_1, U_2, U_3 are the first level indicators.

U_1 is a set of the development indicators (the second level), $U_1 = \{u_{11}, u_{12}, u_{13}\}$, where u_{11} is the final recovery efficiency, u_{12} is the composite decline rate, and u_{13} is the oil recovery rate.

U_2 is the set of dynamic indicators $U_2 = \{u_{21}, u_{22}, u_{23}\}$, where u_{21} is the dynamic investment return period, u_{22} is the net value, and u_{23} is the internal rate of return.

U_3 is the set of static indicators $U_3 = \{u_{31}, u_{32}, u_{33}\}$, where u_{31} is the total investment, u_{32} is the total profit, and u_{33} is the average unit cost.

The comments set $V_1 = \{A_1, A_2, A_3, A_4, A_5\}$ represents the levels of the quantitative indicators, which are "excellent", "better", "common", "poor" and "very poor," respectively.

Each element of the set of contribution scores $V_2 = \{B_1, B_2, B_3, B_4, B_5\}$ stands for "highest", "higher", "common", "low" and "very low."

Thus, given an indicator u_{ij}, we say that the grade of u_{ij} is A_k, it is equivalent to construct a cloud $C_{A_k} (Ex, En, He)$, u_{ij} is a droplet in this cloud with its certainty degree $C_{A_k} (Ex_{uij}, En_{uij}, He_{uij})$.

Table 1. The cloud parameters of the comment set and the contribution score set.

Comment set	Cloud parameters	Contribution score set	Cloud parameters
Excellent	(1,0.103,0.013)	Highest	(100,10.31,0.26)
Better	(0.691,0.064,0.008)	Higher	(69.1,6.37,0.16)
Common	(0.5,0.039,0.005)	Common	(50,3.93,0.10)
Poor	(0.309,0.064,0.008)	Low	(30.9,6.37,0.16)
Very poor	(0,0.103,0.013)	Very low	(0,10.31,0.26)

Similarly, to demonstrate that the contribution of u_{ij} in the evaluation system is B_k, we generate another cloud $C_{B_k} (Ex', En', He')$.

According to the expert experience and the model-driven method based on golden section in Wang and colleagues[13], we can get the value of the parameters of the clouds of first-level indicators; see Table 1.

3.2 Compute the Contribution Score of the Second Level Indicators

The method of uncertainty clouds control[9] is applied to obtain the contribution scores. If we input the values of the indicators of a development plan, we will generalize the virtual clouds of the contribution scores through the rule generator. Then the contribution scores can be obtained through the backward cloud generator. Take the final recovery efficiency u_{11}, for example; we simply construct five single condition rule generators as follows:

If u_{11} is *excellent* (A_1), then the contribution score of u_{11} is *highest* (B_1);

If u_{11} is *better* (A_2), then the contribution score of u_{11} is *higher* (B_2);

If u_{11} is *common* (A_3), then the contribution score of u_{11} is *common* (B_3);

If u_{11} is *poor* (A_4), then the contribution score of u_{11} is *low* (B_4);

If u_{11} is *very poor* (A_5), then the contribution score of u_{11} is *very low* (B_5).

Input the value of u_{11} into the five antecedents of this generator, we select the rule with the antecedent of maximum certainty degrees, and then generate the droplets of the contribution score cloud C_{B_k} repeatedly. Put the generated droplets into the backward cloud generator, and we can get the contribution score cloud of u_{11}.

3.3 Compute the Weights of the Indicators

The weights of the indicators play a key role in the evaluation results. There are many methods to determine the indicator weights, such as expert evaluation method, AHP[14] and grey correlation analysis.[15] To eliminate the effect of the subjectivity of human beings, we employ the method in Hu and colleagues[16] which combines the cloud model and the correlation analysis together. The result might be more consistent with actual conditions, see Table 2.

Table 2. The weights of the indicators.

First level	U_1				U_2			U_3	
		0.40			0.27			0.33	

Second level	u_{11}	u_{12}	u_{13}	u_{21}	u_{22}	u_{23}	u_{31}	u_{32}	u_{33}
	0.40	0.30	0.30	0.33	0.56	0.11	0.55	0.36	0.09

3.4 Synthesize and Sort the Evaluation Clouds

By using floating clouds, we synthesize the contribution score clouds of the second-level indicators with the weights in the above subsection. The results are the synthetic score clouds of first-level indicators. Repeat this routine, and one can get the synthetic score clouds of all the development plans. Then, we can obtain the positive/negative ideal solution for each synthetic score cloud of the development plans through TOPSIS method, which can be applied to sort the plans and get the optimized one. The whole procedure of this evaluation process can be described in Figure 3.

4. Application

Take 8 development plans of some oilfield, for example, all the evaluation indicators have been normalized; see Table 3. One can refer to the literature[7] for the detail of this oilfield.

With these data, we can compute the contribution score of each plan. By inputting the data into the cloud rule generator repeatedly, we can get a group of droplets, then put these droplets in the backward cloud generator, and the contribution score cloud is obtained. In this research, we take the amount of the cloud droplets to be 1000. See Table 4 for the contribution score clouds of all the plan's indicators.

According to the weights in Table 3, we can get the synthesized score clouds of the development plans by using floating clouds algorithm. The result is in Table 5.

Generally, the score clouds of the plans are sorted by graphical methods, which is very intuitive. Figure 4 is the graph of the standard evaluation clouds. Figure 5 is the synthetic contribution score clouds of the plans. Comparing the score clouds with the standard evaluation clouds in Figure 5, we can directly select the optimal one.

However, as is shown in Figure 5, the clouds of plan 2, 3, and 7 are too close to tell the difference. According to the idea of Wang,[17] we employ TOPSIS method to sort the plans with weighted Hamming distance, which can emphasize the importance of the expectation of the clouds and sort the plans numerically.

Let the positive ideal plan be

$$y^+ = (\max Ex_i, \min En_i, \min He_i)$$

and the negative ideal plan be

$$y^- = (\min Ex_i, \max En_i, \max He_i)$$

We first compute the weighted Hamming distance ($\beta_1 = 0.5$, $\beta_2 = 0.3$, $\beta_3 = 0.2$) from the plans to the positive/negative ideal plans, respectively, by formula (4).

Denote the Hamming distance from the score cloud y_i of the ith plan to the ideal plans by d_i^+ and d_i^-, one can sort the programs by the nearness degree $d_i^* = \frac{d_i^+}{d_i^+ + d_i^-}$. The program of smaller d_i^* is the better one. See Table 6 for the results. Obviously, the Program 4 is the best one.

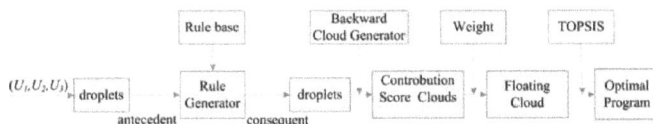

Fig. 3. The procedure of evaluating the development plans.

Table 3. The normalized data of the development plans.

	Development indicators U_1			Dynamic indicators U_2			Static indicators U_3		
	u_{11}	u_{12}	u_{13}	u_{21}	u_{22}	u_{23}	u_{31}	u_{32}	u_{33}
1	0.4397	1.0000	0.0803	0.4947	0.2099	0.1442	0.8333	0.2471	0.4688
2	0.0993	0.4921	0.6569	0.8842	0.6711	0.5096	0.8222	0.6988	0.8125
3	0.0567	0.3665	0.7883	0.9579	0.8538	0.7276	0.4111	0.8996	0.8203
4	1.0000	0.8038	0.5474	0.8842	0.7361	0.5865	0.4111	0.9266	1.0000
5	0.0922	0.0000	1.0000	1.0000	1.0000	1.0000	0.0000	1.0000	0.8750
6	0.1418	0.9351	0.0000	0.0000	0.0000	0.0000	0.5889	0.0000	0.0000
7	0.0000	0.5541	0.5109	0.8842	0.6874	0.5801	1.0000	0.6873	0.9922
8	0.2270	0.5931	0.5766	0.7789	0.4925	0.3526	0.3889	0.5251	0.5469

Table 4. The contribution score clouds of the indicators.

Plan	u_{11}	u_{12}	u_{13}	u_{21}	u_{22}	u_{23}	u_{31}	u_{32}	u_{33}
1	(44.68,0.89,0.39)	(100,0,0)	(8.18,1.04,0.76)	(49.54,0.07,0.27)	(20.85,1.34,0.57)	(14.72,1.94,0.8)	(83.25,1.94,0.73)	(24.64,0.81,0.49)	(47.3,0.35,0.25)
2	(10.13,1.35,0.8)	(49.31,0.1,0.27)	(65.46,0.48,0.5)	(88.2,1.55,0.79)	(67,0.27,0.52)	(50.84,0.11,0.25)	(82.44,1.86,0.68)	(69.93,0.11,0.51)	(81.58,1.67,0.64)
3	(5.74,0.76,0.78)	(36.73,0.79,0.54)	(79.46,1.32,0.54)	(95.71,0.57,0.82)	(85.08,1.92,0.78)	(72.98,0.51,0.52)	(41.23,1.28,0.47)	(89.8,1.4,0.82)	(82.27,1.86,0.69)
4	(100,0,0)	(80.89,1.55,0.62)	(54.12,0.58,0.28)	(88.15,1.63,0.81)	(73.9,0.62,0.49)	(58.12,1.36,0.49)	(41.2,1.29,0.48)	(92.53,1,0.83)	(100,0,0)
5	(9.4,1.24,0.82)	(0,0,0)	(100,0,0)	(100,0,0)	(100,0,0)	(100,0,0)	(0,0,0)	(100,0,0)	(87.32,1.64,0.79)
6	(14.51,1.96,0.79)	(93.4,0.87,0.81)	(0,0,0)	(0,0,0)	(0,0,0)	(0,0,0)	(58.3,1.34,0.49)	(100,0,0)	(0,0,0)
7	(0,0,0)	(54.72,0.69,0.31)	(50.95,0.13,0.26)	(88.22,1.55,0.8)	(68.72,0.05,0.5)	(57.4,1.29,0.47)	(100,0,0)	(68.7,0.05,0.53)	(99.21,0.11,0.81)
8	(22.62,1.06,0.48)	(58.78,1.33,0.51)	(57.16,1.29,0.47)	(78.4,1.17,0.5)	(49.35,0.09,0.26)	(35.31,0.57,0.48)	(39.05,1.14,0.51)	(52.19,0.3,0.28)	(54.1,0.53,0.26)

Table 5. The synthesized score clouds of the plans.

Plan	1	2	3	4	5	6	7	8
Synthesized score cloud	(45.60,0.94,1.62)	(60.61,1.30,2.02)	(60.27,0.88,1.70)	(74.43,0.82,1.59)	(57.86,0.10,1.14)	(22.02,1.19,0.53)	(61.40,0.36,1.48)	(48.27,0.88,1.29)

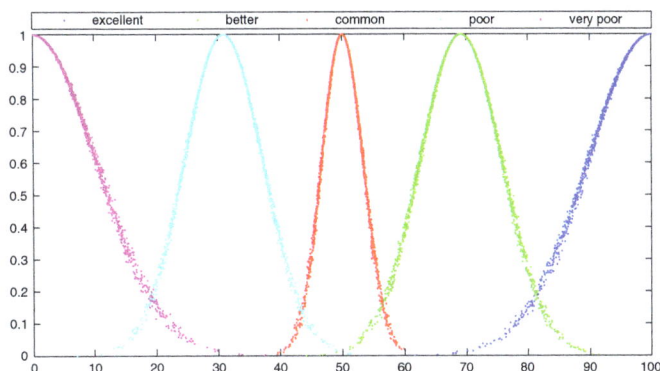

Fig. 4. The standard evaluation clouds.

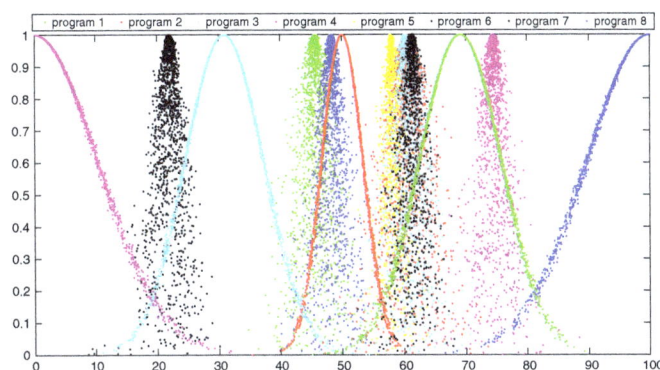

Fig. 5. The synthetic score clouds of the development plans.

Table 6. The nearness degrees of the plans.

y^+			(74.40,0.11,1.11)					
y^-			(24.11,1.28,2.00)					
plan	4	7	3	2	5	8	1	6
d_i^*	0.0114	0.2490	0.2780	0.2785	0.3099	0.4991	0.5522	0.9929

5. Conclusion

The synthetic evaluation method based on a cloud model presented in this article takes the randomness and the fuzziness of the development program's indicators both into consideration. First, the evaluation system based on uncertain inference is able to avoid the information loss of the traditional methods. Second, the weighted Hamming distance can help tell the difference between the score clouds that are too close graphically. The computation results show that this method is efficient and reliable. It is also applicable in the synthetic evaluation of complicated systems.

Funding

This article is supported by the fund from the doctoral program of China (20095121110003) and the fund from the Sichuan Provincial Department of Education (13ZB0203).

References

1. Kasap, E.; Sanza, G. J.; Izat Ali, M.; et al. *Field Development Plan by Optioneering Process Sensitive to Reservoir and Operational Constraints and Uncertainties*, Abu Dhabi International Petroleum Exhibition and Conference, SPE-101556-MS, 2006.

2. Vilela, M. J.; Marin, P. R.; Rodrigo, M.; et al. Building the Field Development Plan for a New Gas Field, Located in Algeria, Reggane Trend, Asia Pacific Oil and Gas Conference and Exhibition, SPE-108831-MS, 2007.

3. Hu, J.; Liu, Z.-B. Nonlinear Fuzzy and Comprehensive Evaluation Model of Oilfield Development Program. *J. Oil Gas Tech.* **2011**, *10*, 132–135.

4. Zhao, M.-C.; Chen, Y.-M.; Yuan, S.-B. Application of TOPSIS Method to Optimization of Oil-Gas Field Development Scheme. *Xinjiang Petrol. Geol.* **2006**, *4*, 484–486.

5. Lin, W.; Shi, Y.-W.; Zhang, X.-L. Application of Grey Correlation Analysis in the Preferred Field Development Scheme. *J. Oil Gas Tech.* **2006**, *3*, 419–421.

6. Li, Shun; Wang, H.-Q., Chen, J.-J. Study on Two Levels Fuzzy Comprehensive EIA in Oil Field Exploitation. *Hydrogeol. Eng. Geol.*, **2002**, 2, 38–41.

7. Chen, Z.-G.; Lang, Z.-X. An Application of Fuzzy Comprehensive Judgement to Determination of Oilfield Development Projects. *J. Univ. Petrol.* **1988**, 1, 59–64.

8. Sun, F.-J.; Cheng, L.-S.; Li, X.-S. An Application of Hierarchic Analysis to Comprehensive Evaluation and Optimization of Oilfield Development Plan. *China Offshore Oil Gas* **2002**, 5, 328–332

9. Li, D.-Y.; Du, Y. Artificial Intelligence with Uncertainty. National Defence Industry Press, 2005.

10. Chen, H.; Li, B. Approach to Uncertain Reasoning Based on Cloud Model. *J. Chinese Comp. Syst.* **2011**, 32, 2451–2455.

11. Lu, H.-J.; Wang, Y.; Li, D.-Y.; Liu, C.-Y. The Application of Backward Cloud in Qualitative Evaluation. *Chinese J. Comp.* **2003**, 8, 1009–1014.

12. Di, K.-C.; Li, D.-Y.; Li, D.-R. Cloud Theory and Its Applications in Spatial Data Mining and Knowledge Discovery. *J. Image Graphics* **1999**, 14, 930–935.

13. Wang, J.; Xiao, W.-J.; Zhang, J.-L.; Deng, Bin. An Analysis of an Improved Effectiveness Evaluation Based on Cloud Model. *Fire Control Comm. Control* **2010**, 5, 97–99.

14. Fan, M.; Liu, Y.-L.; Wu, Y.-J.; Yang, X.-Y. Ecological Impact Evaluation for Land Consolidation Based on Cloud Model. *Geomatics Info. Sci. Wuhan Univ.* **2008**, 9, 986–989.

15. Xu, F.-Y.; Zhu, X.-S.; Yan, Qi-B.; Chen, D.-L. An Approach to Define Weight of Indexes in the Quantitative Evaluation of Reservoirs. *Acta Petrolei Sinica* **1996**, 2, 30–35.

16. Hu, S.-Y.; Li, D.-R.; Liu, Y.-L.; Li, D.-Y. Mining Weights of Land Evaluation Factors Based on Cloud Model and Correlation Analysis. *Geomatics Info. Sci. Wuhan Univ.* **2006**, 5, 423–427.

17. Wang, J.-Q.; Liu, T. Uncertain Linguistic Multi-criteria Group Decision-making Approach Based on Integrated Cloud. *Control Decision* **2012**, 8, 1185–1190.

22

Structuring Procurement to Improve Sustainability Outcomes of Power Plant Projects

GIGIH UDI ATMO ⓘ, COLIN FRASER DUFFIELD ⓘ, and DAVID WILSON ⓘ

Department of Infrastructure Engineering, The University of Melbourne, Melbourne, Australia

Abstract: Electricity consumption throughout Asia and the Pacific is projected to more than double from 2010 to 2035, reaching 16,169 terawatt-hours in 2035. While environmental factors are a pressing issue internationally, governments from developing countries in Asia also have a priority to deliver adequate power supplies to sustain their desired level of economic growth. The purpose of this article is to compare the implications of delivering power plant projects via either public private partnerships (PPPs) or traditional public procurement. A mixed method approach was used to evaluate four Indonesian power plant case studies. The article compares project performance based on project finance availability, construction and commissioning timelines, and operational reliability. It also investigates carbon emission factors from different power plant combustion technologies in relation to project financial structuring. The results show that (1) power plant projects that are procured through PPPs appear to be delivered in a more timely manner, and they have substantially better performance during the first years of operation than those of traditional public procurement; and (2) availability of project finance is influenced by a careful consideration of environmental and sustainability factors such as the selection of fuel type and the combustion technology.

Keywords: Procurement strategies, project finance, power plant technologies, project outcomes

1. Introduction

Developing countries in Asia require substantial investment in the energy sector to sustain their rapid economic growth, especially in emerging economies such as China, India, and Indonesia. For instance, it is estimated that Indonesian power plants will require US$97 billion of investment between 2015 and 2024 to meet economic growth projections.[1] The Indonesian public sector budget is simply not adequate to finance all the required infrastructure.[2,3,4]

Many Asian governments have undertaken regulatory reforms, aimed to improve productivity and attract private-sector investment for such infrastructure developments.[5] These reforms provide greater opportunities for private participation in energy infrastructure projects and in the commercialization of natural energy resources.[6] However, this also raises a concern as to whether private-sector investment in the energy sector

Address correspondence to: Gigih Udi Atmo, The University of Melbourne, Infrastructure Engineering, Victoria, Melbourne, 3010 Australia. Email: gigih.atmo@gmail.com

benefits a nation's energy supply security.[7] Countries dependent on importing energy have an additional concern from the adverse impact of energy price fluctuations in international markets. For example, Turkey (an energy-import-dependent country) had a foreign trade deficit of approximately US$44 billion in 2011, due largely to importation of coal and natural gas for electricity generation.[8]

A secure power supply needs significant capital investment to develop adequate capacity and to mitigate against exposure to the volatility in the price for fuel.[9] Some countries have adopted the single buyer electricity market model where governments invite private investment in electricity generation as independent power producers.[10] In this market model, the contractual relationships between the public and private sector is governed through a power purchase agreement that typically assigns the private sector to arrange finance, design, construction, and operation of a power plant facility in return for performance-based payments from the public-sector contracting party. This structure of contractual relationships is consistent with the general principles of public–private partnerships (PPPs) procurement. Accordingly, governments that adopt the single buyer electricity model have to strategically select between PPPs and traditional public procurement to develop a power plant project.

In this article, traditional public procurement is defined to cover the situation where a government raises project finance (typically via a blend of government budget and international loans) and undertakes the engineering and operation of the facility through a state electric company. The selection process needs to consider whether the chosen procurement method produces better project outcomes to create a sustainable energy system.[11]

Fossil fuel–based power plants are subjected to international scrutiny as to their global carbon emissions.[12] The Asian Development Bank projected that coal consumption in Asia and the Pacific economies is forecasted to increase by more than 50% between 2010 and 2035 because electricity demand in this period is projected to double, reaching 16,169 terawatt-hours in 2035.[13] Environmental sustainability issues in power plant projects have become a major concern for international finance institutions.

The structure of project finance in energy projects in emerging economies contains a mixture of finance from international development banks, export credit agencies, and commercial finance institutions. International development agencies and many international commercial finance institutions have now incorporated social and environmental sustainability into their lending due diligence processes.[14,15,16] These requirements have increased support for renewable energy projects and more efficient thermal-based power technologies. It is worth noting that pre-committed offtake agreements, sometimes called feed-in tariff policies, that aim to promote the deployment of renewable-based power plants may increase government budget exposure, although such policies create greater certainty of financial returns for the private sector.[17] Accordingly, consideration of fuel and technology selection can shape the structure of finance in a power project.

It is evident that the link between project finance structuring, performance targets, and environmental sustainability for power plant projects still needs to be quantified. Globally, project developers from the United States and European countries gained extensive experience with PPP power projects in the 1990s.[4] Regional emerging economies such as China and India have expanded their involvement in international infrastructure projects through a focus on promoting outlets for equipment supplies, project development services, and finance with support from their export credit agencies.[18,19] Governments from emerging economies can benefit from enhanced market competition between project developers for power plant projects. The choice of procurement strategy between PPPs and traditional public procurement can assist these governments to achieve the expected outcomes from power plant projects.

Albeit power projects face increasing pressure to achieve environmental sustainability, there is limited research that compares the implications of procurement strategies, fuel types, and combustion technologies toward project outcomes. Therefore, the aim of this article is to compare the outcomes achieved through the use of either a PPP mechanism or traditional public procurement for delivery of power plant projects in emerging economies. The article compares project outcomes of time and operational quality based on power projects delivered via either a PPP or traditional procurement in Indonesia. Appropriate outcomes from a power project are based on the assessment of project financial viability, construction performance, power supply availability, and environmental performance during project operation.

2. Methodology

A mixed methods approach was used to investigate four Indonesian power project case studies. Case study investigation is an appropriate framework to investigate a complex phenomenon that is specific and where there is little information known about the subject of investigation.[20,21] Specific measures of relative performance were used to compare project outcomes. This research then undertook within-case analyses to explain factors that led to different project outcomes. The within-case analysis can reveal the key explanatory variables that link to an outcome.[22]

This article compares the performance of the case study projects based on the two key project stages: the period between contract award and project commercial operation date (stage 1) and between commercial operation date and the first two years of project operation (stage 2), as presented in Figure 1. The commercial operation date (COD) is a substantial milestone that represents the completion of project construction and commissioning and indicates the commencement of the project operation phase.

Stage 1 of the analysis concentrates on time performance during project construction and commissioning. In Indonesia, on-time completion of the project construction and commissioning is essential if energy supply deficits are to be overcome.[23] The project financing process has been analyzed to reveal the typical structure adopted by project financiers and the commercial terms of their associated loan agreement. Stage 2 investigates project operating performance during the first two years of operation. The first operating years are considered the most critical because there may be technical defects that have not previously been identified during project commissioning and thus reduce the power plant operating performance.[24]

2.1 Project Performance Metrics

Time has been adopted as the metric to assess the performance during construction and commissioning. The project time performance was measured at two different project milestones between the time of government contract approval and the time of completion of project construction and commissioning. The project operation performance measures the power plant availability factor and carbon emission factor. The detailed performance metrics are discussed follows.

2.1.1 Project Time Performance

Analysis of the period between contract award and the completion of commissioning has been used to test any performance differences between PPPs and traditional procurement projects.[25] For traditional procurement, the time of contract signoff between the Indonesian government and the main contractor of the engineering procurement and construction (EPC) contract was selected as the time of project contract award. The PPP procurement used the time when the public and private parties signed off the power purchase agreement (PPA) as the time of project contract award. The completion of project construction and commissioning is a major milestone that indicates the commencement of commercial operation of a new power plant.

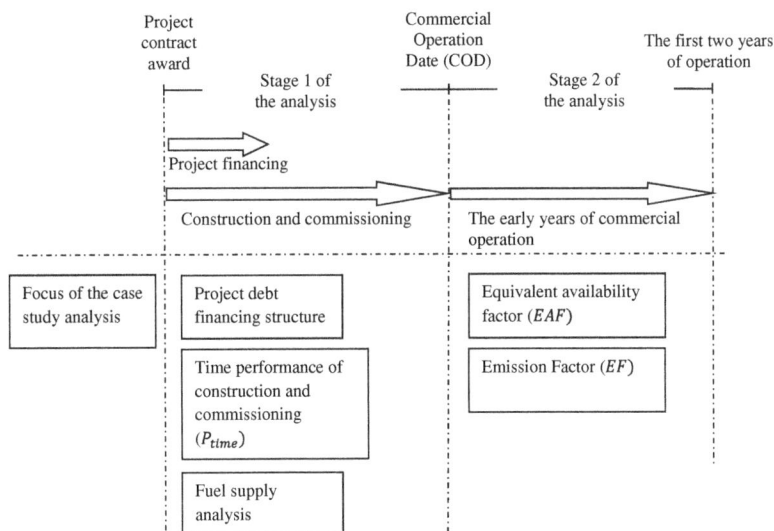

Fig. 1. Project stages that are investigated in the case study comparison.

The metric for measuring project time performance in PPPs and traditional power projects is expressed in equation (1).

$$P_{time} = \frac{T_{AC} - T_{CS}}{T_{CC} - T_{CS}} \quad (1)$$

Definitions:

P_{time}: Project time performance

T_{AC}: Actual finish date of project construction and commissioning

T_{CC}: Contractual finish date of project construction and commissioning

T_{CS}: Contractual start date of project construction and commissioning

2.1.2 Project Quality Performance

Assessment of project quality performance has been based on the availability of a power plant to operate according to its design capacity and carbon emission level. The equivalent availability factor (EAF) is used to measure project operation reliability, while an emission factor (EF) is adopted to investigate carbon emissions per megawatt hour (MWh) of electricity generated by each power plant unit. Early years of operational life of a power plant are essential to measure power plant performance, which is influenced by the commissioning process, potential random failures, and operating procedures.[26] The EAF index is derived from the international standard of IEEE number 762-2006, which measures the percentage of maximum electric generation available over a period of time.[27] The EAF index is considered to be one of the most important performance indices of power plant operating performance.[28] This measurement method has been recognized internationally and used by many international power utilities and organizations, such as the North American Electric Reliability Council (NERC).[29] The EAF index is measured in equation (2).

$$EAF = \frac{AG}{MG} \times 100\% \quad (2)$$

where

EAF: Equivalent availability factor on power plant *(i)* and unit number *(m)* thathas been procured using procurement method *(proc)*

AG: Available electricity generation from a power plant unit

MG: Maximum rated electricity generation from a power plant unit

This study also investigates the carbon emission levels in the power plant operation that is measured in the carbon emission factor (EF). The EF equation is adapted from United Nations Framework Convention on Climate Change (UNFCCC) emission factor formulas that are used for calculating electric power grid emissions, and it follows equation (3).[30]

$$EF_{EL,m,y} = \frac{\sum_i FC_{i,m,y} x NCV_{i,y} x EF_{CO_2,i,y}}{EG_{m,y}} \quad (3)$$

where

$EF_{EL,m,y}$: CO_2 emission factor of a power plant unit m in year y (tCO_2/MWh)

$FC_{i,m,y}$: Amount of fossil fuel type i consumed by a power plant unit m in year y (mass or volume unit)

$NCV_{i,y}$: Net calorific value of fossil fuel type i in year y (GJ/mass)

$EF_{CO_2,i,y}$: CO_2 emission factor of fossil fuel type i in year y (tCO_2/GJ)

$EG_{m,y}$: Net quantity of electricity generated to the grid by power unit m in year y (MWh)

m: case studies of power plant units connected in year y

i: fossil fuel types combusted in power plant unit m in year y

y: year of the calculation (the first and second year operation)

It is worth noting that the case study projects utilize different fuel combustion technologies. Therefore, the EF results are not intended to be a direct comparison between PPPs and traditional public procurement as these projects typically utilize different power plant combustion technologies. It has been highlighted in the literature review that project finance markets are now moving toward more environmentally sustainable projects, and thus power projects that require financial support from the broader international finance communities need to adhere to environmental requirements. Therefore, it is important to evaluate the different emission factors from power plant technologies in relation to their respective project financial structuring.

2.2 Case Study of Indonesian Power Projects

One of the important processes in the case study analysis was the selection of the case studies.[20] Four representative Indonesian power plant projects were selected for this study, namely the Wayang Windu geothermal fired power plant (GTPP), the Cirebon coal-fired power plant (CFPP), the Cilegon combined cycle gas fired turbine (CCGT), and the Paiton-9 CFPP. The first of the two projects was procured through PPPs, while the other two projects were delivered via traditional public procurement. The key project information of the four power plant projects are summarized in Table 1.

The data source used to conduct the analysis was gathered from historical data on the power plant reliability operation from the Java-Bali Dispatch Centre; data on the calculation of power plant carbon emissions from the Ministry of Energy and Mineral Resources of Indonesia; public disclosure information regarding project capital costs and construction performance on traditional power projects; and official press releases from the project companies participating in the case study projects. Figure 2 illustrates indicative locations of the Indonesian case study projects.

3. Comparative Project Performance Analyses

The comparative performance analysis for the case study projects is divided into three sections. The first section investigates the financial structure of the case studies and then compares project time performance for both construction and commissioning. It also analyzes the respective power plant operating availability and carbon emissions.

Table 1. Case study data of Indonesian power plant projects.

Key project information	Public private partnerships (PPPs)		Traditional public procurement	
	Wayang Windu GTPP[31,32]	Cirebon Electric CFPP[33,34,35,36]	Cilegon CCGT[37,38,39]	Paiton-9 CFPP[40,41]
Fuel type	Geothermal	Coal	Natural gas	Coal
Combustion technology	Dry steam saturated geothermal technology	Supercritical pulverized technology	Combine-cycle gas turbine	Subcritical pulverized technology
Power capacity (MW)	117	660	740	660
Estimated project cost	US$300 m	US$850 m	US$345 m	US$ 460 m
Contractual date (T_{CS})	21 Nov 2006	20 Aug 2007	10 Feb 2004	12 Mar 2007
Date of financial close	13 June 2007	8 March 2010	5 November 2004	30-Jan-08 (USD) 18-Apr-08 (IDR)
Duration to reach financing closure	6.7 months	30.6 months	8.8 months	10.7 months (USD) 13.2 months (IDR)
Commercial operation date (T_{AC})	2 March 2009	27 July 2012	11 May 2006	9 June 2012
Contractual finish time (T_{AC})	13 March 2009	31 August 2011	31 October 2005	12 March 2010
Project sponsors/ Engineering Procurement and Construction (EPC) contractor	Star Energy	Marubeni (32.5%) KOMIPO (27.5%) Samtan (20%) Indika Energy (20%)	Mitsubishi Heavy Industries and Truba Jurong Engineering (EPC contractors)	Harbin Power Engineering and Mitra Selaras Hutama Energi (EPC contractors)
Fuel supply arrangement	a Joint-venture for geothermal resources development	Two energy companies, the Samtan and Indika Energy, participated in the project ownership structure	Natural gas supply contracts with a state gas distribution company and a natural gas producer	Annual coal supply contracts with domestic coal suppliers

Notes: GTPP: geothermal power plant; CFPP: coal-fired power plant; CCGT: cmbined cycle gas turbine power plant.

Fig. 2. Indicative location of the four case study projects on the Indonesian island of Java.

3.1 Financial Structure

Investigation of project financing structures is an important step in understanding the interests and commercial strategies of debt providers. It includes an analysis of project debt structures in both PPPs and traditionally procured projects.

There is a significant variation in project financing duration between the four case study projects examined (Table 1). The financing process in a PPP is typically more complex than traditional procurement. The commercial terms of the international creditors in the PPP projects requested government guarantees that contributed to a longer loan agreement process (e.g., approximately 30 months in the Cirebon Electric PPP project). The Wayang Windu GTPP obtained project financing in less than 7 months. A standard Indonesian PPP contract stipulates a period of 12 months for the private sector to reach financing closure from the date of contractual agreement. In traditional procurement, the public sector should complete the needed finance before it selects project developers.

The case study of the Cirebon Electric CFPP project (a PPP) received debt facilities totaling US\$595 million from two international export credit agencies, namely the Japanese Eximbank (JBIC) and the Korean Eximbank, and international commercial banks. They included the Bank of Tokyo-Mitsubishi, Mizuho Corporate Bank, and the ING Bank N.V.[33,34] The two export credit agencies also provided political risk guarantees on loan portions provided by the commercial banks.

The Wayang Windu project received credit facilities of US\$300 million from the loan syndication arranged by the Standard Chartered Bank of Singapore.[31] The financial institutions that participated in the loan syndication included the General Electric Energy Financial Services and two Indonesian state-owned banks, namely, the Bank Nasional Indonesia and the Bank Jawa Barat (Jabar).[42,43] This project sets a benchmark for Indonesian PPPs in that it did not require government guarantees to secure project loans because it has stable revenues from the existing power plant operation and the international carbon credit transaction.

Interestingly, project financing delays also occurred in traditional power projects. The Indonesian government required around 9 months for the Cilegon CCGT, and a year was taken to finalize bilateral loan agreements in the Paiton-9 CFPP from its international trading partners. In the Cilegon CCGT project, the Japan Bank for International Cooperation (JBIC) provided a loan facility amounting in total to ¥30.4 billion.[38] This loan was co-financed with international commercial banks, including the Bank of Tokyo-Mitsubishi, the Hong Kong and Shanghai Banking Corporation Limited, and Standard Chartered Bank.[38] The Paiton-9 CFPP received loan facilities from the Export Import Bank of China and PT Bank Mega, an Indonesian private bank, for US\$331 million and IDR 601 billion, respectively.[40]

It can be seen that the geothermal power project and fossil-fuel-based power plants that utilize enhanced combustion technologies attract a broader range of project financiers. Loan syndications of these projects consisted of a mixture of international finance institutions. These institutions publicly supported more environmentally sustainable energy projects. Some were signatories to the equator principles, an international framework for socio-environmentally responsible lending practices.[44] It appears that the Cirebon Electric project that utilizes supercritical pulverized boiler technology was accepted by broader international finance communities, while the Paiton-9 project that adopted conventional subcritical pulverized boiler technology attracted only a limited pool of potential lenders.

This study shows that the international pressure on environmental sustainability has influenced the availability of project finance from international markets. However, Indonesian banks have maintained their flexibility in their lending policies. Project financing has a critical role in the development of a power plant project, and it may take a considerable time to finalize loan agreements. The international financial institutions often require project loan guarantees if long-term project revenues are not financially sustainable. Accordingly, the anticipated duration of the project financing process may affect the project completion schedule.

3.2 Time Performance of Construction and Commissioning

The results of the time performance of the case studies are presented in Figure 3 and show that the time performance ratio for three of the case study projects is above unity, while the

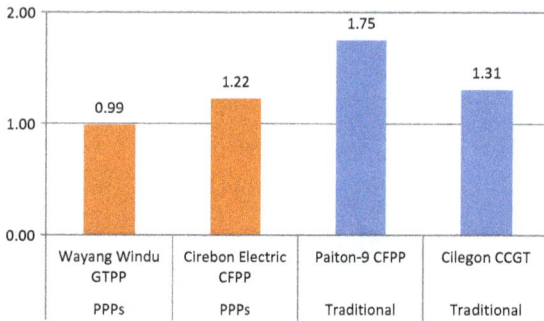

Fig. 3. Project time performance ratio for construction and commissioning.

Wayang Windu GTPP has a time ratio of just less than one. The Wayang Windu GTPP was the only project that was completed ahead of the original contractual schedule. In contrast, the other three projects had experienced completion delays. The project sponsors of the Wayang Windu GTPP contracted the Sumitomo Corporation as the main Engineering Procurement and Construction (EPC) contractor and Fuji Electric as the equipment manufacturer. These companies had developed the first unit of the Wayang Windu GTPP, which had a track record of high operational reliability.[45] The other PPP project, the Cirebon Electric CFPP, experienced the shortest ratio of completion delay of 22% over the original contractual schedule.

The private sponsors of these two projects utilized their internal cash flow to finance construction activities until debt finance was obtained from project financiers. This financing strategy specifically assisted the Cirebon Electric project to reduce the impact of delays from the debt financing process and avoided construction cost overruns.

The Cilegon CCGT project had a completion time overrun of 31% over the original contractual schedule, while the Paiton-9 CFPP was 75% over. These two projects were traditionally procured. The sources of delays were identified from fuel supply shortages, prolonged debt financing, and a power equipment breakdown. The Cilegon project was constructed in conjunction with a mega project valued at US$1.36 billion of the South Sumatera-West Java transmission gas pipeline.[37] The inadequacy of the natural gas supply from this pipeline project had prevented the EPC contractor of the Cilegon CCGT to complete the commissioning on schedule. The Paiton-9 project required around one year to finalize foreign and local loan agreements. This also delayed client progress payments to the EPC contractor which

adversely affected their performance due to a lack of cash flow. The project was also delayed by 6 months because of a power transformer failure that required replacement.[41]

Based on these four case study projects, it appears that PPPs are completed in a more timely manner than the traditional procurement projects. The case study projects also indicate that construction performance can be managed by combined financial and technical decisions. Project sponsors of PPPs were incentivized to utilize their capital resources to finance construction albeit that debt facilities were not completely finalized. The availability of fuel supply is a risk that also needs to be anticipated to prevent delays in project commissioning. It appears that technologically complex projects like a power plant require the involvement of an experienced EPC contractor and reliable power equipment manufacturers to manage the project.

3.3 Operation Performance Analysis

Investigation of operating performance compares the operational reliability of the power plant and is measured as the equivalent availability factor (EAF), as previously defined. The study also evaluates greenhouse gas emissions using an emission factor (EF). Table 2 summarizes calculations of the EAF and EF. The input data used for the EF computations is provided in Appendices 1 and 2. The EAF indices from the case study projects were summarized from six annual power plant performance reports from the Java-Bali Dispatch Centre from 2007 to 2013.[46,47,48,49,50,51] The Java-Bali Dispatch Centre automatically records the operational performance of all power plants that are connected in the Java and Bali electric grid, including the case study projects.

3.3.1 The Equivalent Availability Factor (EAF) Analysis

The average equivalent availability factor (EAF) detailed in Table 2 leads to the conclusion that PPP power plants had much higher reliability than traditional power plants. Of the two traditional projects, the Paiton-9 CFPP had the lowest EAF index relative to the other case study projects, with the average EAF index of just below 44%.

The EAF indices for the PPPs (detailed in Table 2) immediately reached above 80% in their first year of commercial operation. The power projects that were procured through traditional procurement had a much lower operating performance in their first year of operation—e.g., the Cilegon project and Paiton-9 project had EAF indexes of 32.4% and 26.7%, respectively.

The analysis of EAF indices in the second year of operation reveals that the two PPP projects had EAF indices that were

Table 2. Summary of project operation performance of the case study.

Power plant project	EAF (%)			EF (tCO$_2$/MWh)		
	1st yr	2nd yr	Average	1st yr	2nd yr	Average
Wayang Windu (PPP)	98.0	97.0	97.5	0.03	0.03	0.03
Cirebon Electric (PPP)	97.1	89.2	93.2	0.91	0.94	0.92
Cilegon (Traditional)	32.4	91.9	62.1	0.50	0.72	0.61
Paiton-9 (Traditional)	26.7	60.6	43.7	1.00	1.07	1.04

slightly lower than those in the first year of power plant operation. The Wayang Windu GTPP and Cirebon Electric CFPP projects had the second year's EAF index of 97% and 89.2%, respectively. On the other hand, the two traditional procurement projects had improved EAF indices of 91.9% and 60.6% in their second year of operation due to an increased reliability of the natural gas supply for the Cilegon CCGT and reduced technical issues in the Paiton-9 CFPP projects.[47,51]

Even though the Paiton-9 CFPP performance improved in the second year of operation, its reliability remained substantially lower than the other projects considered. These findings from the traditionally procured projects also reveal that energy supplies critically influenced operating performance.

3.3.2 Fuel Supply Arrangements

Investigation of the fuel supply arrangements in the two PPP case study projects in Table 1 revealed that they included an energy company in the project consortium. In the Wayang Windu project, there was strong service integration between the development of the geothermal resource and power plant operations. In the case of the Cirebon PPP coal-fired power project, two private coal companies participated in the ownership structure of the project. They utilized the project to secure a market for coal with a low ranking calorie from Indonesian mining operations. It provides these coal companies with comparative cost efficiency rather than transporting the coal for international buyers. The power plant has also been designed to consume coal that matches the specifications of low rank coal.

The fuel supply for traditionally procured projects is sourced via long-term contracts arranged separately by the Indonesian government with energy suppliers that are not directly connected to specific power plants. The low reliability in the performance of these projects is in part attributed to fuel supply problems. The Paiton-9 CFPP project had a mismatch between coal supply availability and power plant specifications that caused lower thermal operating efficiency.[52] The Cilegon gas-fired power plant also encountered fuel supply problems that caused delays in project commissioning and resulted in a low availability factor in the first year of operation. The project was also disadvantaged by delays in the construction of the South Sumatera–West Java transmission gas pipeline and inadequate natural gas supplies that were 45% lower in the supply volume than contracted.[53] It would appear there are advantages in directly involving fuel supply companies in the development and operation of power plants. This has been achieved in the PPP arrangements.

3.3.3 The Emission Factor Analysis

Table 2 also shows the emission factor (EF) from the four Indonesian power projects. The EF value depends on the fuel type and power plant fuel efficiency. The analysis of emission factors also demonstrates that the geothermal power plant is the most carbon-neutral power plant, while gas-fired power plant utilizing combined cycle technology may provide a better alternative than coal-based power plants in terms of carbon emission reductions. The Wayang Windu geothermal project had a very low EF of approximately $0.03\,tCO_2/MWh$, and the Cilegon CCGT that utilizes combined cycle gas technology had an average EF level of $0.61tCO_2/MWh$. Although the EF of the Cilegon CCGT is much

higher than that of the Wayang Windu, it is substantially lower than those in the coal-based power projects. The Cirebon Electric CFPP that utilizes supercritical boiler technology had an average EF of $0.92tCO_2/MWh$, while the Paiton-9 CFPP that adopts a conventional subcritical boiler technology had an average EF of $1.04tCO_2/MWh$. Interestingly, the deviation of average EF level between these two coal-fired power projects was just above 10%, although the investment cost differences were US$390 million. The stated capital costs of the Cirebon Electric and the Paiton-9 project were US$850 million and US$460 million, respectively.

4. Discussion

The broad objectives of this article were to understand the implications of procurement strategies on the outcomes achieved from either coal or cleaner energy sources. It has been established from the literature that many emerging economies in Asia have undertaken regulatory reforms to increase private participations in energy sector development. These include private power plants in an electricity market under a single buyer electricity market model. Governments in emerging economies need to strategically select procurement methods between PPPs and traditional public procurement in order to deliver power plant infrastructure. The adoption of an in-depth case method has enabled new understandings of the complex interaction that exists between infrastructure development, finance, procurement methods, and environmental sustainability. These are discussed next under the headings of financing structure, construction and operation performance, and environmental performance. A summary of key performance criteria between the four case study projects is shown in Table 3.

4.1 Financing Structure

The investigation into the project financing structure adopted for the four Indonesian power projects demonstrated that finance was a critical issue, especially in PPPs where choice of the power plant combustion technology influences whether international and domestic lending markets are interested in financing the project. International financiers are attracted more to power projects that utilize renewable energy and cleaner technologies such as geothermal power, combined cycle gas turbine technology, and supercritical pulverized coal-fired technology. In contrast, conventional coal-fired power projects utilizing subcritical pulverized technology attract limited interest from international commercial banks. Therefore, financing for these projects would need to rely on bilateral credit facilities and domestic finance sources. The international finance institutions often request government guarantees when revenues generated from a power plant operation are not adequate to cover debt service payments.

4.2 Construction and Operating Performance

The case studies show that for project construction and commissioning PPPs have better time performance than traditional procurement. The Wayang Windu GTPP was completed just ahead of the contract schedule, while the Cirebon Electric,

Table 3. Summary of project performance evaluation.

Key performance criteria	Wayang Windu GTPP	Cirebon Electric CFPP	Paiton-9 CFPP	Cilegon CCGT
Broader access to project financiers	Yes	yes	no	Yes
Request loan guarantees	No	yes	yes	Yes
Ranking of timelines of construction and commissioning	1st	2nd	4th	3rd
Reliable power plant equipment	Yes	Yes	No	Yes
Competent EPC Contractor	Yes	Yes	Yes	Yes
Fuel supply security	Yes	Yes	No	No
Ranking of average power plant availability operation	1st	2nd	4th	3rd

Paiton-9, and Cilegon projects were 22%, 75%, and 31% over the contractual schedules, respectively. Delays in obtaining debt finance were the cause of time slippage in both the Cirebon Electric PPP project and the traditional Paiton-9 project. The Paiton-9 project also suffered from equipment failure that caused an additional 6 months delay. The Wayang Windu project raised debt finance from commercial finance institutions that did not include commercial terms that utilized project developers from a specific country provider. Accordingly, the project consortium had greater flexibility to choose experienced EPC contractors and select technically proven power plant equipment to develop the Wayang Windu power plant. The project has been operating at the highest availability factor when compared with the other three case studies.

The Wayang Windu PPP and the Cirebon Electric PPP had an average equivalent availability factor of 97.5% and 93.2%, respectively. In contrast, the traditional procurement approach of the Cilegon and Paiton-9 projects had an average equivalent availability factor of 62.1% and 43.7%, respectively. While use of emerging project developers is required to raise full project finance, it is critical that such developers are chosen wisely so that they do not cause a substantial drop in project performance. Evidence from the Paiton-9 project shows that it suffered from substantial completion delays and had a low level of power plant availability during the first two years of operation.

The evidence from the two PPP projects (Wayang Windu geothermal and Cirebon Electric coal-fired power project) suggests that project sponsors aligned their business interests to finance construction and ensure fuel supply security during project operation. However, traditional procurement projects suffered from a low level of sustained energy supply that led to completion delay and operation underperformance. The case studies suggest that encouraging energy company participation through a PPP mechanism offers enhanced project development and operation outcomes.

4.3 Environmental Performance

The case projects utilized different fuel and combustion technologies. It is obvious that the Wayang Windu geothermal project produces the least level of carbon emissions than the other three projects. The Cilegon project that adopted a combined-cycle gas turbine had average emission factors that were lower than the

coal-fired power plant projects. The average emission factor for the Wayang Windu geothermal project was just 0.03 tCO_2/MWh, while the Paiton-9 had the highest average emission factor of 1.04 tCO_2/MWh. Interestingly, Cirebon Electric, which utilizes supercritical coal-fired technology, had a higher average emission factor (0.92 tCO_2/MWh) when compared to the average emission factors of the Cilegon gas-fired project. There were slight differences seen in the emission factors between the conventional subcritical power plant and that of the supercritical pulverized boiler technology. The former produces around 10% higher carbon emissions than the latter coal technology. These environmental gains do not appear commensurate with the additional capital cost for the supercritical coal power plant. It was US$390 million more expensive than the conventional coal power plant with the same level of 660 MW power output. The benefit–cost tradeoff for the choice of the enhanced thermal efficiency of coal-fired power plant warrants review. A geothermal power plant or combined cycle gas turbine power plant may provide better options to substantially reduce carbon emission levels from electricity generation.

It is acknowledged that a limitation of the approach taken here is that these results are based on a relatively small sample of projects in Indonesia. But they are nonetheless major projects and point the way to future best practice. Similar studies need to be conducted in other developing countries to confirm our findings.

5. Conclusions

This article has reviewed the international literature relating to mechanisms to encourage investment in Asian power generation. An in-depth analysis was then conducted of four Indonesian case study projects to ascertain the relationship between finance, fuel types, choice of technology, and environmental expectations. It has shown that PPPs can produce appropriate outcomes based on the assessment of project construction and commissioning time performance, project financing structure, operational reliability, and environmental performance.

The case study investigation found that power projects in an emerging economy such as Indonesia generally rely on a significant proportion of finance from international markets to be viable, and attracting such finance is often predicated on the selection of appropriate power plant technologies and fuel type. Many international financiers incorporate environmental sustainability as a consideration for making investment decisions. The case studies examined have also shown that, for PPPs, time performance and operating reliability is consistently better than for the traditionally procured power plants.

International financiers accepted the environmental performance from the supercritical pulverized boiler technology even though it was only marginally better than conventional coal technology (0.1 tCO_2/MWh). The supercritical pulverized boiler technology carries a high cost premium (plus US$390 million for 660 MW) compared to conventional subcritical pulverized boiler technology. Further consideration of geothermal and natural gas power technologies is warranted to achieve an optimal environmental/financing outcome.

In summary, it is concluded that PPPs are an appropriate method to procure power plant projects and that they appear to be delivered in a timelier manner than traditional projects. More importantly, PPPs have substantially better performance than traditional projects in terms of availability during the first years of operation. Attracting finance to Asian power plant projects is enhanced by careful consideration of environmental factors including the choice of fuel type and the combustion technology.

References

1. Perusahaan Listrik Negara. *Rencana usaha penyediaan tenaga listrik PT PLN (Persero) 2015–2024*; Perusahaan Listrik Negara: Jakarta, 2014.
2. Chan, A. P. C.; Lam, P. T. I.; Chan, D. W. M.; Cheung, E.; Yongjian, K. Critical Success Factors for PPPs in Infrastructure Developments: Chinese Perspective. *J. Construction Eng. Manag.* **2010**, *136*, 484–494.
3. Cheung, E.; Chan, A. P. C. Risk Factors of Public Private Partnership Projects in China: A Comparison Between the Water, Power and Transportation Sectors. *J. Urban Plan. Dev.* **2011**, *1*, 54–54.
4. Eberhard, A.; Gratwick, K. N. IPPs in Sub-Saharan Africa: Determinants of Success. *Energ. Policy* **2011**, *39*, 5541–5549.
5. Cheung, E.; Chan, A. P. C. Evaluation Model for Assessing the Suitability of Public–Private Partnership Projects. *J. Manag. Eng.* **2011**, *27*, 80–89.
6. Vedavalli, R. *Energy for Development: Twenty-first Century Challenges of Reform and Liberalization in Developing Countries*; Anthem Press: London, 2007.
7. Banerjee, P. Energy Security through Privatisation: Policy Insights from Hydroelectric Power Projects (HEPs) in India's Northeast. *Perceptions: J. Int. Affairs* **2014**, *19*, 39–53.
8. Şenel, B.; Şenel, M.; Bilir, L. Role of Wind Power in the Energy Policy of Turkey. *Energ. Tech. Policy* **2014**, *1*, 123–130.
9. Varley, C.; Lammers, G. *Electricity Market Reform: An International Energy Agency Handbook*; OECD: Paris, 1999.
10. Jamasb, T. Between the State and Market: Electricity Sector Reform in Developing Countries. *Util. Policy* **2006**, *14*, 14–30.
11. Atmo, G.; Duffield, C. Improving Investment Sustainability for PPP Power Projects in Emerging Economies. *Built Environ. Proj. Asset Manag.* **2014**, *4*, 335–351.
12. International Energy Agency. *Fossil Fuel-Fired Power Generation: Case Studies of Recently Constructed Coal- and Gas-Fired Power Plants*; International Energy Agency: Paris, 2007.
13. Asian Development Bank: *Energy Outlook for Asia and the Pacific*; Asian Development Bank: Philippines, 2013.
14. World Bank: *Toward a Sustainable Energy Future for All: Directions for the World Bank Group's Energy Sector*; World Bank: Washington, 2013.
15. Andrew, J. The Equator Principles: Project Finance and the Challenge of Social and Environmental Responsibility. *Issues Soc. Environ. Account.* **2007**, *1*, 40–53.
16. Sarro, D. Do Lenders Make Effective Regulators? An Assessment of the Equator Principles on Project Finance. *German Law J.* **2012**, *13*, 1522–1555.
17. Ahmad, S.; bin Mat Tahar, R. Feedback-Rich Model for Assessing Feed-In Tariff Policy. *Energ. Tech. Policy* **2014**, *1*, 45–51.
18. Hazard, E.; De Vries, L.; Barry, M. A.; Anouan, A. A.; Pinaud, N. The Developmental Impact of the Asian Drivers on Senegal. *World Econ.* **2009**, *32*, 1563–1585;
19. McDonald, K.; Bosshard, P.; Brewer, N. Exporting Dams: China's Hydropower Industry Goes Global. *J. Environ. Manag.* **2009**, *90*, S294–S302.
20. Eisenhardt, K. M. Building Theories from Case Study Research. *Acad. Manag. Rev.* **1989**, *14*, 532–550.
21. Yin, R. K. *Case Study Research: Design and Methods*, 4th ed.; Thousand Oaks, CA, 2009.
22. Mahoney, J. Qualitative Methodology and Comparative Politics. *Comparative Political Studies* **2007**, *40*, 122–144.
23. Ministry of Energy and Mineral Resources of Indonesia. *Rancangan Umum Ketenagalistrikan Nasional: 2012–2021*; Directorate General of Electricity: Jakarta, 2013.
24. Perez-Canto, S.; Rubio-Romero, J. C. A Model for the Preventive Maintenance Scheduling of Power Plants Including Wind Farms. *Reliability Eng. Sys. Safety* **2013**, *119*, 67–75.
25. Raisbeck, P.; Duffield, C.; Ming, X. U. Comparative Performance of PPPs and Traditional Procurement in Australia. *Construction Manag. Econ.* **2010**, *28*, 345–359.
26. de Souza, G.; Carazas, F.; Guimarães, L.; Rodriguez, C. Combined-Cycle Gas and Steam Turbine Power Plant Reliability Analysis. In *Thermal Power Plant Performance Analysis*, de Souza, G. F. M., Ed. Springer: London, 2012; pp 221–247.
27. IEEE-SA Standards Board. IEEE Standard Definitions for Use in Reporting Electric Generating Unit Reliability, Availability, and Productivity. *IEEE Std 762-2006 (Revision of IEEE Std 762-1987)* 2007, C1–66.
28. Carazas, F.; de Souza, G. Reliability Analysis of Gas Turbine. In *Thermal Power Plant Performance Analysis*, de Souza, G.; Martha, G. F., Ed. Springer: London, 2012; pp 189–220.
29. North American Electric Reliability Corporation. *Generating Availability Data System (GADS): What It Is, What It Can Do, and Why You Should Care*; North American Electric Reliability Corporation: Atlanta, 2012.
30. Clean Development Mechanism. *Methodological Tool: Tool to Calculate the Emission Factor for an Electricity System (Version 3.0.0)*; United Nations Framework Convention on Climate Change: Bonn, 2012.
31. CDM-Executive Board. *Clean Development Mechanim Project Design Document Form (CDM-PDD) version 3*; United Nations Framework Convention on Climate Change: Bonn, 2006.
32. Syah, Z.; Kajo, J.; Antro, Z.; Raharjo, M. *Project Development of the Wayang Windu Unit 2 Geothermal Power Plant*; The World Geothermal Congress: Bali, Indonesia, 2010.
33. Japan Bank for International Cooperation. *JBIC Provides Project Finance and Political Risk Guarantee for Cirebon Thermal Power Plant Project in Indonesia: Supporting Japanese Business Development in Indonesia's Power Infrastructure under Framework of Policy Dialogue*; Press and External Affairs Division, Ed. Japan Bank for International Cooperation: Tokyo, 2010. https://www.jbic.go.jp/en/information/press/press-2009/0308-7152 (accessed 11 February 2015).

34. Korea Eximbank. *Korea Eximbank Provides USD 238 Million Through Project Financing for Indonesia Cirebon Independent Power Producer*; Korea Eximbank: Seoul, 2010. http://www.koreaexim.go.kr/en/bbs/noti/view.jsp?no=9606&bbs_code_id=1316753474007&bbs_code_tp=BBS_2 (accessed 11 February 2015).

35. Marubeni Corporation. *Marubeni Secured the Power Generation Development Right for Coal Fired Steam Power Plant in Cirebon, Indonesia Promotion of the 1st Large IPP Project in Indonesia after Asian Economic Crisis;* Marubeni Corporation: Tokyo, 2007. https://www.marubeni.com/dbps_data/news/2007/070820e.html (accessed 13 Feb. 2015)

36. Indika Energy. *Cirebon Electric Power Statement of Commercial Operation Date*; SVP Corporate Finance & Investor Relations, Indika Energy: Jakarta, 2012. http://www.indikaenergy.co.id/wp-content/uploads/2014/05/14-AGS-2012-Cirebon-Electric-Power-Statement-of-Commercial-Operation-Date.pdf (accessed 13 Feb. 2015).

37. Directorate General of Electricity of Indonesia. *PLTGU Cilegon pasok 740 MW*; Division of Information Management, Ed. Directorate General of Electricity of Indonesia: Jakarta, 2007. portal.djlpe.esdm.go.id/modules/news/index.php?_act=detail&sub=news_media&news_id=475 (accessed 26 September 2014).

38. Japan Bank for International Cooperation. *An export loan to Indonesia: supporting Japanese power plant exports*. Press and External Affairs Division, Ed. Japan Bank for International Cooperation: Tokyo, 2004. https://www.jbic.go.jp/en/information/press/press-2004/1105-6886 (accessed 24 April 2014).

39. Mitsubishi Heavy Industry. *Full-Turnkey Order for 740 MW Combined-Cycle Power Plant;* Daiya Public Relation, Ed. Mitsubishi Heavy Industry: Tokyo, 2004. https://www.mhi-global.com/news/sec1/e_0981.html (accessed 16 February 2014).

40. Ministry of Finance of Indonesia. *Laporan kajian risiko fiskal atas proyek percepatan pembangunan pembangkit 10.000 MW*; Badan Kebijakan Fiskal Jakarta: Jakarta, 2011. http://www.fiskal.depkeu.go.id/2010/adoku/2012/kajian/pprf/Laporan_Kajian_Risiko_Fiskal_atas_Proyek_Percepatan_Pembangunan_Pembangkit_10.000MW.pdf (accessed 15 July 2013).

41. Marison, E.; Doka, E.; Pangidoan, J.; Ramayuni, L.; Suhartanto, T. *Laporan inspeksi teknik PLTU 2 Jatim-Paiton Unit 9 (1 × 600 MW)*; Ministry of Energy and Mineral Resources of Indonesia: Jakarta, 2011.

42. General Electric Energy Financial. *GE Taps Asia, Renewables Growth, Makes Loan to Indonesia's Biggest Geothermal Power Plant*; General Electric Energy Financial: Jakarta, Indonesia, 2013. http://geenergyfinancialservices.com/files/press-releases/09_11_04_Wayang_Windu_FINAL.pdf (accessed 11 February 2015).

43. Bank Jabar Banten; Standard Chartered Bank. *Bank Jabar Banten and Standard Chartered Bank Establish Cooperation on Trade Finance, Corporate Banking and Treasury;* Standard Chartered Bank: Bandung, Indonesia, 2008. https://www.sc.com/id/_documents/press-releases/en/11-PReleaseWBBankJabarEng20080407.pdf (accessed 11 February 2015).

44. The Equator Principles Association. *The Equator Principles: Members and Reporting*; The Equator Principles Association, 2011. http://www.equator-principles.com/index.php/members-reporting (accessed 11 February 2015).

45. H. Murakami; Y. Kato; N. Akutsu. In *Construction of the Largest Geothermal Power Plant for Wayang Windu Project Indonesia*, The World Geothermal Congress, Kyushu-Tohoku, Japan, 2000; pp 3239–3244.

46. Perusahaan Listrik Negara. *Evaluasi operasi sistem Jawa Bali 2006*; PT PLN Penyaluran dan Pusat Pengatur Beban Jawa Bali: Jakarta, 2007.

47. Perusahaan Listrik Negara. *Evaluasi operasi sistem Jawa Bali 2007*; PT PLN Penyaluran dan Pusat Pengatur Beban Jawa Bali: Jakarta, 2008.

48. Perusahaan Listrik Negara. *Evaluasi operasi sistem Jawa Bali 2008*; PT PLN Penyaluran dan Pusat Pengatur Beban Jawa Bali: Jakarta, 2009.

49. Perusahaan Listrik Negara. *Evaluasi operasi sistem Jawa Bali 2009*; PT PLN Penyaluran dan Pusat Pengatur Beban Jawa Bali: Jakarta, 2010.

50. Perusahaan Listrik Negara. *Evaluasi operasi sistem Jawa Bali 2010*; PT PLN Penyaluran dan Pusat Pengatur Beban Jawa Bali: Jakarta, 2011.

51. Perusahaan Listrik Negara. *Realisasi indek kinerja pembangkit sistem Jawa Bali tahun 2012 and 2013*; PT PLN Penyaluran dan Pusat Pengatur Beban Jawa Bali: Jakarta, 2014.

52. Directorate General of Electricity of Indonesia. *Inventarisasi permasalahan program 10.000 MW tahap I*; Division of Information Management, Ed. Directorate General of Electricity of Indonesia: Jakarta, 2013.

53. Cahyadi. Kajian teknis pembangkit listrik berbahan bakar fossil. *Jurnal Ilmiah Teknologi Energi* **2011**, *1*, 21–32.

54. Febijanto, I. *Jawa Madura Bali (Jamali) grid system ver. 02*. Badan Pengkajian dan Penerapan Teknologi, Jakarta, 2006.

Appendix 1: Net Electricity Outputs and Fuel Consumption on Each Power Plant Case Study Projects

Year	Fuel type	Fuel specification	Wayang Windu	Cirebon Electric	Cilegon unit 1	Cilegon unit 2	Paiton-9
1st year		Net electricity output (MWh)	952,870	2,559,566	*	*	1,510,407
	High Speed Diesel oil	Consumption (liter)	0	0	0	0	7,329,338
		Net Calorific Value (TJ/Gg)	–	–	–	–	42.42
	Coal	Consumption (ton)	–	1,353,120	–	–	903,495
		Net Calorific Value (TJ/Gg)	–	18.50	–	–	17.87
	Natural gas	Consumption (MMBtu)	N/A	N/A	*	*	N/A
		Net Calorific Value (TJ/Gg)	–	–	–	–	–
	Geothermal**	Steam (tons)***	7,221,067	N/A	–	–	N/A
2nd year		Net electricity output (MWh)	960,802	4,482,863	595,275	453,997	1,846,750
	High Speed Diesel oil	Consumption (liter)	0	0	0	0	1,395,502.2
		Net Calorific Value (TJ/Gg)	–	–	–	–	37.67
	Coal	Consumption (tons)	N/A	2,381,923	N/A	N/A	1,194,847.3
		Net Calorific Value (TJ/Gg)	–	19.08	–	–	17.87
	Natural gas	Consumption (MMBtu)	N/A	N/A	8,046,533	5,250,234	N/A
		Net Calorific Value (TJ/Gg)	–	–	34.06	29.14	–
	Geothermal**	Steam (tons)***	7,281,178	N/A	N/A	N/A	N/A

Notes: *No publicly available information regarding fuel consumption for the first year operation of the two units of the Cilegon CCGT project was discovered. Instead, the calculation of the power plant emission factor was based on the public disclosure of total carbon emission and net electricity output that was 371.9 tons of CO_2 emission and 742 MWh, respectively.[54]
**Geothermal CO_2 emissions from the steam production is calculated based on the Wayang Windu project document for Clean Development Mechanism.[31]
***Estimated value is based on steam production per MWh in the Wayang Windu project document for Clean Development Mechanism.[31]

Appendix 2: Default CO_2 Emission Factor for Each Combustion Type Power Plant*

Fuel Type	Effective CO_2 emission factor (kg/TJ)
HSD (High Speed Diesel)	72,600
Natural gas	54,300
Coal	92,800

Note: *Source: IPCC 2006 Volume 2 Energy, table 1.4, pp 1.23–1.24.

23

Optimal Energy Utilization of Photovoltaic Systems Using the Non-Binary Genetic Algorithm

SYAFARUDDIN[1]*, H. NARIMATSU[2], and HAJIME MIYAUCHI[2]

[1]*Department of Electrical Engineering, Universitas Hasanuddin, Tamalanrea-Makassar, Indonesia*
[2]*Department of Computer Science and Electrical Engineering of Kumamoto University, Kumamoto, Japan*

Abstract: The article presents the energy model development based on the real-time output power of PV system installations at Kumamoto University, Japan. There are four sites of installation inside the university, which consist of two locations with thin-film silicon (5.76 kW and 7.56 kW) in the Faculty of Engineering and in the Faculty of Engineering Research Building, respectively, monocrystalline silicon (50 kW) in the Research Laboratory, and CIGS (30 kW) on the rooftop of the Forico Building. From the real-time output power information in one week continuous measurement during winter and summer in 2013, the accumulative energy and capacity factor are calculated. These data are then used as the raw information for developing an energy model in order to estimate the energy production based on capacity factor. The optimal capacity factor is obtained using the non-binary genetic algorithm (NB-GA). Obtaining the optimal capacity factor is important for the prediction of accumulative energy output following the forecasted meteorological information.

Keywords: Real-time output power, accumulative energy, capacity factor, NB-GA, PV system

1. Introduction

Photovoltaic generation systems have been the fastest-growing renewable energy sources worldwide since the last decade. There has been recently estimated a capacity of around 30% of new installation of PV systems, which are contributed from small solar cell applications to giant solar parks connecting to the power grid. The positive trend is due to the mature PV systems' technology level in terms of increasing in power conversion efficiency, low cost material, and other new innovations to support the system development. In a symposium held in Kumamoto in 2010 about community base photovoltaic systems, the Japanese government has incredibly targeted new PV system installation to about 50 GW by 2030. The output policy from this meeting mentioned that all new homes or apartments should be attached with photovoltaic panels on the rooftop of the building. The policy is independently promoted by the government, but they receive constant support from infrastructure, financial, business, and society sectors.

*Address correspondence to: Syafaruddin, Department of Electrical Engineering of Universitas Hasanuddin, 90245 Tamalanrea-Makassar, Indonesia. Email: syafaruddin@unhas.ac.id

Like other countries, the current situation of PV system practice in Japan is showing a positive trend after about 3 years of the policy implementation. People may easily find new homes or apartments with PV panels on the rooftop of the building. This scenery is becoming clearer in the locations of the smart city grid project, such as in Higashida, Kitakyushu Japan. The smart community project considers PV systems one of the renewable energy sources to supply the neighborhood. Also, as part of green energy campaign, the PV system is commonly installed in campus areas. For instance, Kumamoto University has four sites of PV systems installation with thin-film silicon (5.76 kW and 7.56 kW), monocrystalline silicon (50 kW), and CIGS (30 kW). In the large-scale capacity of PV systems, the Kyushu Electric Company has built 3-MW giant solar power plants in Omuta, Fukuoka Prefecture. In addition, island huge solar power plants of 53,000 photovoltaic modules have been installed in the former airport in Kagoshima, the southern part of Kyushu, with a total annual energy production of 9.8 million kWh. The complete information of technology and manufacture of solar energy in Japan can be found online.[1]

One of the challenging problems in PV system utilization is the output power fluctuation highly affected by weather conditions. Since weather conditions are sometimes unpredictable and highly variable, the output power changes drastically. The only way to obtain the output power of PV systems is by estimation techniques either by considering the system behavior or meteorological information at the ground level. Some parameters affect the sunlight intensity arriving on the module surface—for instance, the cloud cover variability, atmospheric aerosol levels, and indirectly the extent of participating gases in the atmosphere.[2] In the forecasting output power of PV systems, some methods are successfully confirmed in terms of their simple algorithms or fast computational techniques. The combination between the statistical method and artificial intelligence technique improves the forecasting accuracy using past historical output power and meteorological information.[3] The combination of artificial neural network and time series information for daily prediction of global radiation is claimed better than any conventional prediction methods, such as ARIMA techniques, Bayesian inference, Markov chains, and k-Nearest-Neighbors.[4] The artificial intelligence method by means of the artificial neural network is the most common method for prediction of daily global radiation in PV system sites.[5–9] A new statistical method for short-term forecasting in a grid-connected PV system considering weather and artificial neural network models has been proposed for the PV system participation in the one-day ahead market.[10] The confirmation of a real-time model of output power and energy prediction is effectively demonstrated for practitioners in PV systems sites.[11]

The variable output power consequently has an impact on the energy production of PV generation systems. Knowing prior information about the next day of energy production in a PV system may improve the operation planning, especially for the grid-connected system. In general, it may enhance the cost of reserves generation, dispatchable and ancillary generations, and grid reliability performance. Therefore, developing an energy model for PV system behavior is not less important for energy output prediction. A real-time model for small-scale grid-connected PV systems was used to predict the real-time AC output power using 30 minutes data within one year measurement.[12] The other proposed energy model considered the statistical analysis to describe the temperature, efficiency, and accuracy of output power prediction models, and this model is claimed suitable for smart metering. Another energy model was proposed by considering the reference evaluation of solar transmittance for the solar irradiance estimation and so the output power calculation.[13] However, the static model sometimes cannot be justified for high accurate prediction because of the dynamic nature surrounding the PV system installation. Therefore, the dynamic thermal model based on the total energy balance was proposed.[14] In this study, the environmental factors related to the thermal mechanism between PV panels and their environments are assessed. Several researchers were concerned with the variability and probability solar radiation pattern in order to calculate the energy production of PV systems. However, the complexity of solar radiation pattern in time resolution is not giving significant impact to energy estimation results.[15] To enhance the energy output estimation including optimizing other balance of system components, the stochastic simulation model for statistical analysis of the site solar radiation are performed.[16] The energy estimation method is in very wide perspectives, after the artificial neural network method is claimed provides better results than other conventional methods considering minor secondary effects of low irradiance, angular, and spectral effects.[17]

In our current study, we are not predicting the solar irradiation or considering dynamic environmental factors in order to estimate the energy generation of PV systems. The energy model by means of the objective function is generated following the real-time output behavior of PV systems during certain periods in winter and summer. The idea for this research is that there will be certain and optimal capacity factors for each PV technology to contribute to the total output power and energy according to sky conditions. The minimum and maximum output energy corresponds to the cloudy and clear sky conditions, respectively. The optimal capacity factor is determined using the non-binary genetic algorithm. Since the forecasted meteorological information is highly accurate in Japan, then the obtained optimal capacity factor can be simply used to calculate the estimated energy output the next day or one week ahead according to the weather forecast. The detailed explanation of our proposed method is presented as follows.

2. Case Study System

The real-time and continuous output monitoring system for the large scale of PV systems has been installed on the Kurokami campus of Kumamoto University. The measured output power is the AC power, which can be directly used to supply the parts of daily load inside the university. The PV system installation does not have a battery and charge controller because the total output capacity is quite low compared to the electricity demand requirement. There are four sites of installation inside the university, which consists of two locations with thin-film silicon (5.76 kW and 7.56 kW) in the Faculty of Engineering (35) and in the Faculty of Engineering Research Building (33), respectively, monocrystalline silicon (50 kW) in the Research Laboratory (48) and CIGS (30 kW) in the rooftop of the Forico Building (54). The PV system location can be found through the map of Kumamoto University shown in Figure 1.

The real-time output power measurement including irradiance and temperature data is recorded in the database system in 1-minute intervals. These data profiles are simultaneously displayed in the monitor located in the entrance gate of Faculty Building No. 2 (35). The data have been recorded since the first installation in 2010; therefore, it requires high-capacity memory space for the database system. The data logger records the information in four fields. The first field is the type of data; the second field is the time stamp in the server when data is recorded. Meanwhile, the third field is the time stamp in the power conditioner, and, finally, the fourth field is the data categories, i.e., the data of output power, irradiation, and temperature.

The typical data in the first field can be designated as the coding for the location of the PV system including the output measurement. The classification of this data type can be found in Table 1. The time in power conditioning when data recorded and measured is exactly similar to the time of data recorded for thin-film silicon and CIGS PV location. There is a slight difference in the time processing for crystalline silicon PV technology. There is a 23-minute time delay between the parameter measured in the PV site and the time when the data is recorded. In comparison, the output power measurement of CIGS PV starts being recorded at 4 a.m. and is terminated at 12 midnight every day. However, in the actual practice the CIGS PV system continues the operation; therefore, these data are not typical

Fig. 1. Overview of Kumamoto University map for the location of PV system.

Table 1. The classification of data type of each site of PV systems.

PV technology/locations	Measured parameters
Thin-film silicon (solar arc) Faculty of Engineering (35)	irradiation of solar-arc [kW/m^2] temperature of solar-arc [°C] power output of solar-arc [kW]
Thin film Silicon (Wall-stuck) Faculty of Engineering Research Building (33)	irradiation of wall PV [kW/m^2] temperature of wall PV [°C] power output of wall PV [kW]
Monocrystalline Silicon Research Laboratory (48)	irradiation of 50 kW PV [kW/m^2] temperature of 50 kW PV [°C] power output of 50 kW PV [kW]
CIGS Forico (54)	irradiation of 30 kW PV [kW/m^2] temperature of 30 kW PV [°C] power output of 30 kW PV [kW]

with some other reasons. The measured irradiance and temperature in Figure 2 are, respectively, obtained from phyranometer and temperature sensors that are embedded in the PV panel. It means that the irradiance measurement results might be considered as the irradiance measured in the plane of array (POA), while the temperature is the ambient temperature. Figure 2a and Figure 2b show the profile of irradiance and temperature in a PV location in one day during winter and summer seasons, respectively. Both input parameters are variable and changed overtime along the day. The irradiance is unpredictable with high potential of cloudy conditions that cause sunlight intensity changing rapidly. The irradiance level is also affected by seasons; it is less bright in winter, only reaching about 500 W/m^2 and fully bright with a maximum of around 1,000 W/m^2 in summer. On the other hand, the low temperature may go to below 0°C before 9 a.m. during winter and over 30°C after midday during summer. These fluctuations, of course, influence the output power of PV systems.

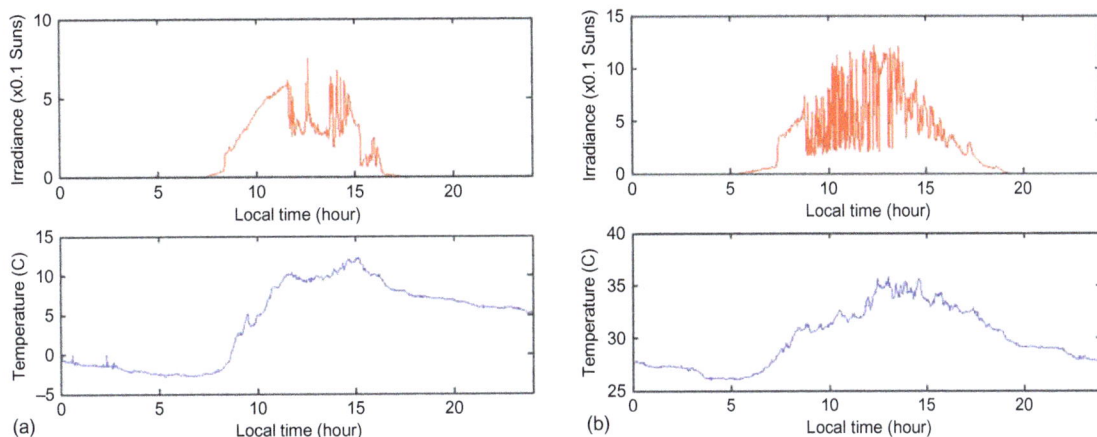

Fig. 2. Profile of irradiance and temperature during winter (a) and summer (b).

3. Real-time Output Power for Energy Model

For the purpose of energy analysis, a CSV file of 10 MB is taken from the server database system. The output power data obtained from real-time database system consists of one week during peak winter (7 January 2013–13 January 2013) and another one week during peak summer (8 July 2013–14 July 2013). We can utilize much data, but we are limited to the computer memory for data processing and classification. Typical real-time one-day output power measurement during winter (7 January 2013) and one-day data in summer (14 July 2013) are shown in Figures 3 and 4, consecutively. Generally, in these figures the output power measurement is highly dependent on the irradiance profile and temperature location. Also, the time range for capturing output power during summer is about 4 hours wider than in wintertime because the daytime period is longer than the nighttime period.

In wintertime (Fig. 3), the weather is not only designated with low temperature but also the quality of sunlight intensity decreases. The minimum temperature may reach below 0°C before 9 a.m.; the maximum temperature during the day is 10°C, and the irradiance level may decrease to 500 W/m². As a result, percentage output power to the rated power reduced to around 52%, 13.2%, and 70% for thin-film silicon (solar arc and wall-stuck types) and monocrystalline silicon, respectively, as shown in Figure 3a–3c. Conversely, good performance is shown for CIGS PV technology with increasing output power of 66% to the rated output power (Fig. 3d). The results indicate output power of silicon-based technology is highly affected by the irradiance level, although it may reduce as well as the temperature increases for silicon PV technology. In comparison, the CIGS PV technology can produce much higher output under low temperature although the irradiance level reduces 50% than the normal one.

In summertime (Fig. 4a–Fig.4d), the performance of all PV system technology is, of course, very good, with a percentage of 86%, 19.8%, 90%, and 53% of rated output for both thin film Si, monocrystalline silicon PV, and CIGS PV systems, respectively. The installation of thin-film silicon is categorized with solar-arc

and wall-stuck configurations. The solar-arc system may follow the light direction better than the wall-stuck system. That is the way the irradiance profile received by wall-stuck system is totally changed after 12 noon; the irradiance level is lower, which may reach only 30% from normal irradiance. But this is not typical data by other reasons. By the other analysis, the efficiency of the wall-stuck type in winter is better than the summer because the angle of the sun irradiation is better. The problems may come from the analyzed output power data, which is not sufficient compared to the sun irradiation. There may be some troubles for PCS or data logger. The remaining silicon and CIGS PV technology give excellent output due to their locations on the top of the building and exposure to the sunlight direction along the day. An installation place of CIGS might be in shadow during the evening. But the CIGS PV system is strong for the partial shadowing; therefore, their output does not change.

4. Capacity Factor for Energy Model

In terms of energy production, the accumulative energy is calculated in four locations of PV systems. The PV systems designated with TF1 and TF2 are for thin-film with solar-arc and wall-stuck, respectively, monocrystalline Si and CIGS PV systems are with just Si and CIGS, respectively. The accumulative energy is defined as the total submission of energy production in 24-hour operations, which can be formulated as follows:

$$E_a = \int_0^t P_{PV} dt \qquad (1)$$

where E_a is the accumulative energy in kWh, t is the time operation (24 hours), and P_{PV} is the output power produced by each PV site, which is variable from hour to hour. According to (1), the accumulative energy results are shown in Figure 5a and Figure 6a for 7 days measurement in winter and summer, respectively. In these results, the thin-film PV system with wall-stuck configuration did not produce any output power from 10–13 January

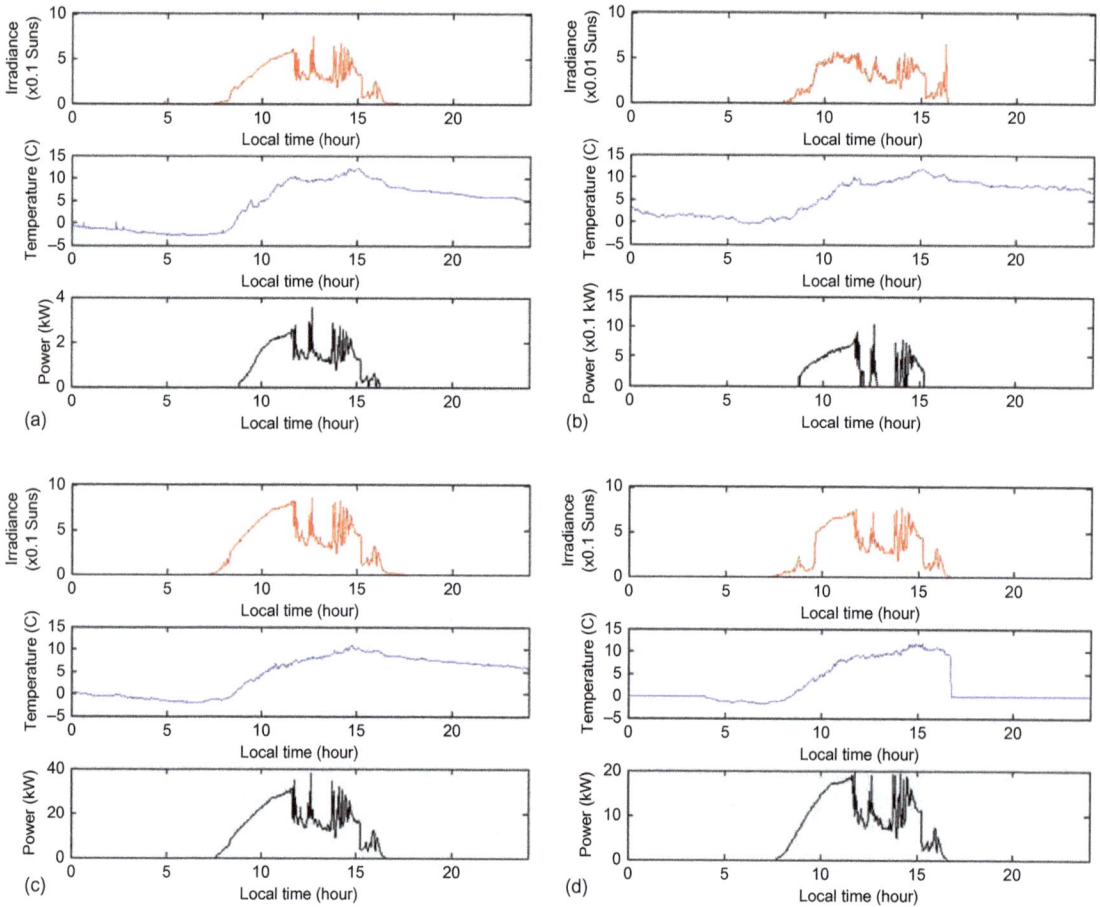

Fig. 3. Typical real-time one-day data measurement during winter (7 January 2013): (a) Thin-film Si (solar arc); (b) Thin-film Si (wall-stuck); (c) Monocrystalline Si; (d) CIGS.

2013, and this condition is similar for the CIGS PV system on 9 July 2013. This condition does not mean that the PV system is not properly working; the high potential problem is from the sensor circuit and data logger. However, such data with zero accumulative energy is still considered for the energy analysis of the system.

Another interesting point to be discussed is related to the energy capacity factor, which might be defined as follows:

$$c_i = \frac{E_a}{E_{max}} \times 100\% \qquad (2)$$

where c_i is the energy capacity factor in (%) for each PV system, E_a is the accumulative energy in (kWh) that was calculated in (1), and E_{max} is the maximum energy production of PV system in one day operation. The E_{max} is simply obtained from the multiplication between the rated maximum power of PV systems and 24 hours operation. For instance, the thin film with solar arc (TF1) has the maximum capacity of 5.76 kW; this rated power is

multiplied with 24 h in order to obtain the E_{max} of 138.24 kWh. The similar procedure is applied for other PV sites—then the E_{max} is obtained as 181.44, 1200, and 720 for TF2, Si, and CIGS, respectively. Later, these E_{max} values become the coefficient for energy model development, as expressed in (3).

The measurement results for capacity factor are shown in Figure 5b and Figure 6b for one week in winter and summer operations, respectively. The results are very interesting because the high output power of the PV system in one location does not indicate directly to the high capacity factor. The trend values are different for different PV locations. For instance, the silicon PV system is always producing much higher than other PV systems due to the total install capacity, which is 50 kW. However, CIGS has slight higher capacity factor during winter, and thin film with solar-arc configuration (TF1) has better capacity factor during summer than other types of PV systems. The minimum and maximum values of capacity factor as shown in Table 2 represent the PV system participation in total energy supply system in winter and summer seasons, respectively.

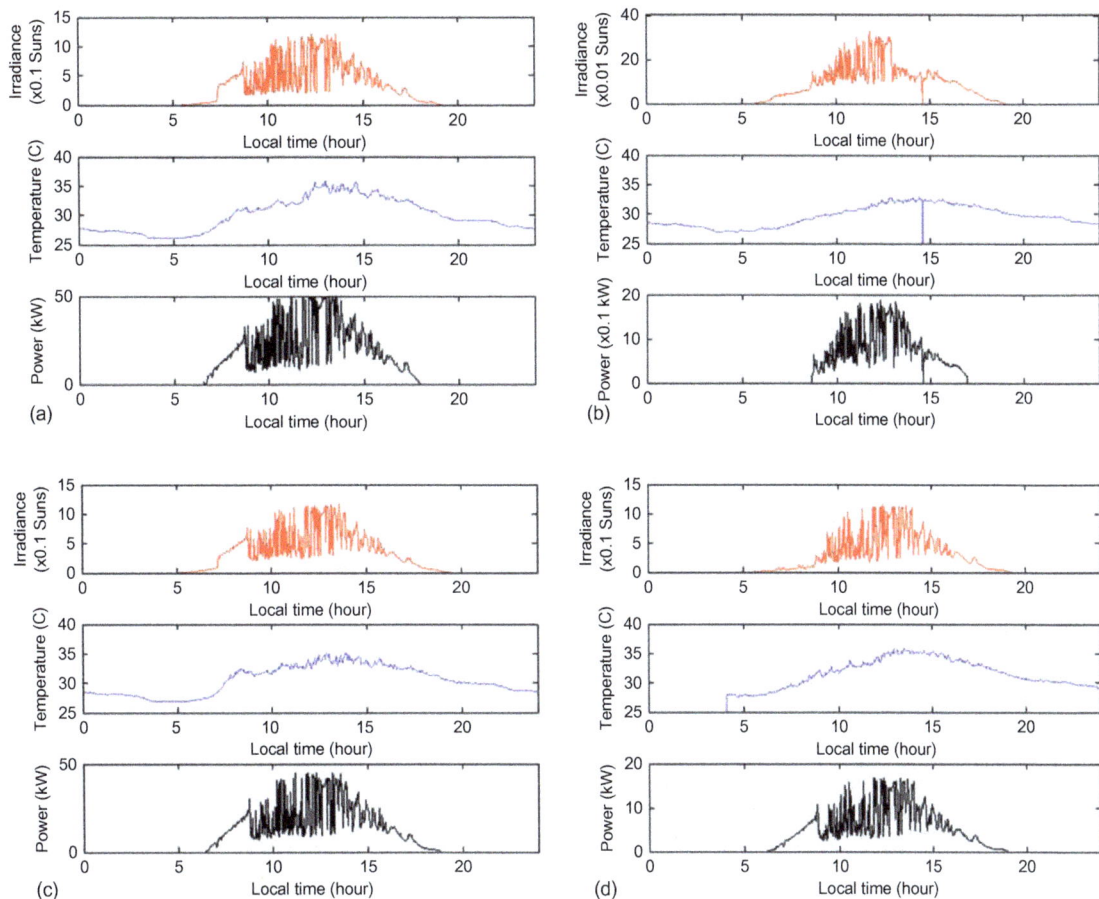

Fig. 4. Typical real-time one-day data measurement during summer (14 July 2013): (a) Thin-film Si (solar arc); (b) Thin-film Si (wall-stuck); (c) Monocrystalline Si; (d) CIGS.

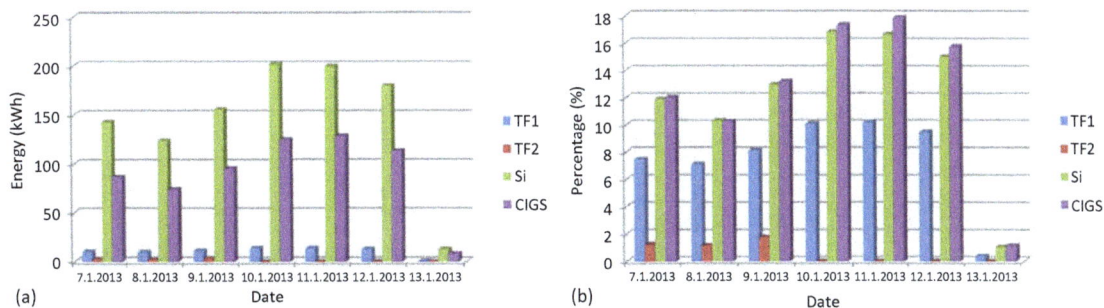

Fig. 5. Accumulative energy and capacity factor during winter measurement: (a) Accumulative energy; (b) Capacity factor.

The capacity factor is changing daily according to the environmental factors; therefore, the optimal value of capacity factor is important to be determined. The optimal capacity factor is the main consideration in the energy analysis model. The prediction of optimal capacity factor is inevitable since the weather forecast in Japan—such as cloudy, partially cloudy, rainy, and sunny conditions—can be predicted one week ahead with high accuracy. It means that when the weather is already forecasted for the

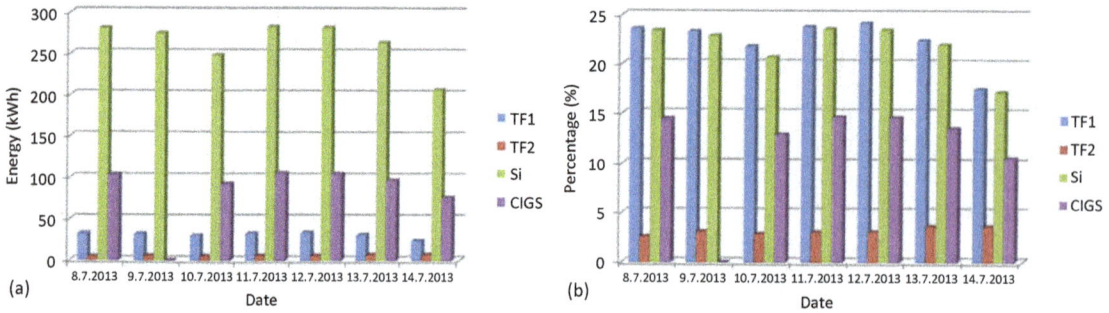

Fig. 6. Accumulative energy and capacity factor during summer measurement: (a) Accumulative energy; (b) Capacity factor.

Table 2. Minimum and maximum values of capacity factor.

PV systems	Capacity factor (min-max)	
	Winter (%)	Summer (%)
TF1 (c_1)	0.376–10.21	17.44–24.14
TF2 (c_2)	0.000–1.773	2.564–3.623
Si (c_3)	1.054–16.81	17.17–23.53
CIGS (c_4)	1.133–17.85	0.000–14.62

next week, the total minimum and maximum energy output can be estimated using the optimal values of capacity factor. The minimum and maximum energy output may correlate with the cloudy and sunny weather conditions, respectively. The current PV system installation is not accompanied with a battery for energy storage systems; therefore, it seems the energy output prediction is less important in this location. However, for further expansion of the system or learning from the current PV system model for new installation in other PV system locations, such prediction is very necessary.

5. Non-Binary Genetic Algorithm Based Optimal Energy Production

For the purpose of prediction of energy production, a single mathematical model is developed to represent the four PV system installations. The mathematical equation for the energy model is expressed as follows:

$$138.24c_1^* + 181.44c_2^* + 1200c_3^* + 720c_4^* = total\ energy\ output \tag{3}$$

where c_1^* to C_4^* are the optimal capacity factor, which will be lately determined by the non-binary genetic algorithm method. For prediction purposes, we may consider the optimal capacity factor for maximum total energy production in the case of a sunny weather forecast and minimum total energy production for cloudy weather energy production. Meanwhile, the total energy output refers to the minimum and maximum energy output for one week continuous measurement during winter are 21.33 kWh and 342.31 kWh, respectively. Meanwhile, they are

about 312.31 kWh and 425.99 kWh during summer, respectively. The range value of capacity factor is quite varied; however, the values in between optimal capacity factor during winter and summer can be used for energy output prediction during spring and autumn seasons.

There are four unknown variables of capacity factor in the objective function that need to be calculated. Conventionally, to solve such a mathematical equation, it is necessary to have four equations. It is possible to have four equations from four days of measurement; however, it may not represent the whole operating time. In addition, the constants will be basically the same in four equations that make singular matrix, which is impossible to be solved with conventional algebraic equations. On the other hand, solving single linear mathematical equations is not so simple, because the capacity factor may have an infinite number of solutions. Therefore, we try to find solutions with nonconventional techniques by using the non-binary genetic algorithm. The method is well recognized as a powerful method to find the optimal solution for a single linear equation with more coefficients without solving rigorous mathematical equations.

This artice utilizes the non-binary genetic algorithm to find the optimal capacity factor during minimum and maximum daily accumulative energy output of PV systems during winter and summer. For this purpose, the objective function is defined as follows:

$$maximum:\quad \left| F\left(c_1^*, c_2^*, c_3^*, c_4^*\right) \right| \equiv 138.24c_1^* + 181.44c_2^* + 1200c_3^* + 720c_4^* \tag{4}$$

Within the constraints of

$$\left| F\left(c_1^*, c_2^*, c_3^*, c_4^*\right) \right| = total\ energy\ output \tag{5}$$

The range of capacity factor in winter:

$$0.376 \leq c_1^* \leq 10.21$$
$$0 \leq c_2^* \leq 1.773$$
$$1.054 \leq c_3^* \leq 16.81 \tag{6}$$
$$1.133 \leq c_4^* \leq 17.85$$

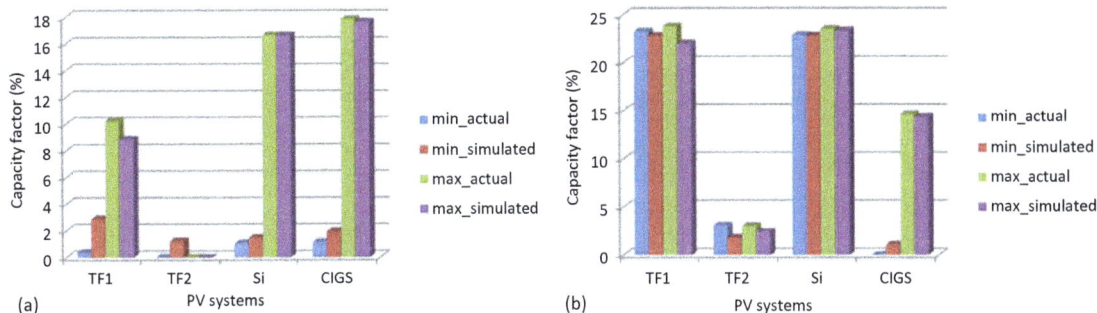

Fig. 7. The actual and simulated capacity factors: (a) Winter season; (b) Summer season.

The range of capacity factor in summer:

$$17.44 \leq c_1^* \leq 24.14$$

$$2.564 \leq c_2^* \leq 3.623$$

$$17.17 \leq c_3^* \leq 23.53 \tag{7}$$

$$0 \leq c_4^* \leq 14.62$$

We are not going deeply into the explanation of the genetic algorithm (GA) process because this method is commonly known as one of the powerful optimization techniques in engineering problems. In our simulation model, the initialization of the GA process is designated with total generation of 100, population size of 10, and mutation rate of 0.8 with roulette-wheel selection. The initial chromosome by means of the capacity factor of each PV system was selected by random number within the constraints. The best chromosome will be the solution for the optimal capacity factors. To confirm the accuracy level in capacity factor measurements, the performance index is designated by the absolute error between the actual (c_i) and simulation results (c_i^*). The equation for the performance index is shown as follows:

$$|e| = |c_i - c_i^*|, \; where \; i = 1 \; to \; 4 \tag{8}$$

The comparison of capacity factor between the actual value (real-time measurement) and the simulated value using the non-binary genetic algorithm for both minimum and maximum accumulative energy output during winter and summer seasons are shown in Figure 7a and Figure 7b, respectively. The results are very close designated by the absolute error. In the winter season, the absolute errors for minimum and maximum accumulative energy output are about 16.01 kWh and 3.55 kWh, respectively. Meanwhile, they are about 3.46 kWh and 7.43 kWh, respectively, during summer season. The simulation result may be forced to converge very close to the actual value of accumulative energy output by increasing the total generation to 1000 and the population size to 100. However, the capacity factor may reach 0 or 1, meaning the PV system is totally not producing any output power or that the PV system reaches the maximum capacity. Such results are impractical since the current PV systems are continuing to produce output power within specified capacity factor.

The simulation results about the optimal or simulated capacity factor imply to be used based on the meteorological forecast. As previously mentioned, the weather forecast in Japan is highly accurate and provides for one week ahead. Therefore, the optimal capacity factor is possibly used to predict the daily accumulative energy output according to the weather forecast. The optimal capacity factor for minimum output may represent cloudy sky conditions. Conversely, the maximum one is used for sunny sky conditions. Because the optimal capacity factor is in the interval value, they are possible to estimate the energy output for the partially cloudy condition. Finally, the energy model developed in this article is not limited to the number of data; therefore, more accurate prediction can be obtained if the output power during the spring and autumn seasons are also brought online.

6. Conclusion

The study has presented the energy model for PV systems installation at Kumamoto University in Japan. Parameters of accumulative energy and capacity factor are calculated from the real-time output power of thin-film silicon (solar arc and wall-stuck), monocrystalline silicon, and CIGS PV systems. The minimum and maximum accumulative energy corresponds to cloudy and clear sky conditions, respectively. Meanwhile, for the prediction of total energy production, there will be certain and optimal capacity factor for each PV technology to contribute to the total output power and energy according to sky conditions. Therefore, the optimal capacity factor is determined using the non-binary genetic algorithm. The proposed method may confirm the accuracy of optimal capacity factor to the actual measurement indicated by performance index of absolute error. The proposed method can be used to estimate seasonal energy output for the next day or one week ahead according to the forecasted meteorological information. To be more accurate in estimation results, further data in the spring and autumn seasons should be brought online as well.

References

1. http://factsanddetails.com/japan.php?itemid=1795&catid=23&subcat id=152 (accessed on 21 August 2013).
2. Inman, R. H.; Pedro, H. T. C.; Coimbra, C. F. M. Solar Forecasting Methods for Renewable Energy Integration. *Prog. Energ. Combust. Sci.* **2013**, *39*, 535–576.

3. Chen, C.; Duan, S.; Cai, T.; Liu, B. Online 24-h Solar Power Forecasting Based on Weather Type Classification Using Artificial Neural Network. *Solar Energ.* **2011**, *85*, 2856–2870.

4. Paoli, C.; Voyant, C.; Muselli, M.; Nivet, M.-L. Forecasting of preprocessed daily solar radiation time series using neural networks. *Solar Energ.* **2010**, *84*, 2146–2160.

5. Mellit, A.; Kalogirou, S. A. Artificial Intelligence Techniques for Photovoltaic Applications: A Review. *Prog. Energ. Combust. Sci.* **2008**, *34*, 574–632.

6. Voyant, C.; Muselli, M.; Paoli, C.; Nivet, M.-L. Optimization of an Artificial Neural Network Dedicated to the Multivariate Forecasting of Daily Global Radiation. *Energy* **2011**, *36*, 348–359.

7. Ding, M.; Wang, L.; Bi, R. An ANN-based Approach for Forecasting the Power Output of Photovoltaic System. *Procedia Environ. Sci.* **2011**, *11*, 1308–1315.

8. Mellit, A.; Benghanem, M.; Kalogirou, S.A. An Adaptive Wavelet-Network Model for Forecasting Daily Total Solar-Radiation. *Appl. Energ.* **2006**, *83*, 705–722.

9. İzgi, E.; Öztopal, A.; Yerli, B.; Kaymak, M.K.; Şahin, A.D. Short–Mid-term Solar Power Prediction by Using Artificial Neural Networks. *Solar Energ.* **2012**, *86*, 725–733.

10. Fernandez-Jimenez, L. A.; Muñoz-Jimenez, A.; Falces, A.; Mendoza-Villena, M.; Garcia-Garrido, E.; Lara-Santillan, P. M.; Zorzano-Alba, E.; Zorzano-Santamaria, P. J. Short-term Power Forecasting System for Photovoltaic Plants. *Renew. Energ.* **2012**, *44*, 311–317.

11. Su, Y.; Chan, L.-C.; Shu, L.; Tsui, K.-L. Real-time Prediction Models for Output Power and Efficiency of Grid-connected Solar Photovoltaic Systems. *Appl. Energ.* **2012**, *93*, 319–326.

12. Ayompe, L.M.; Duffy, A.; McCormack, S.J.; Conlon, M. Validated Real-Time Energy Models for Small-Scale Grid-Connected PV-systems. *Energy* **2010**, *35*, 4086–4091.

13. Rizwan, M.; Jamil, M.; Kothari, D.P. Solar Energy Estimation Using REST Model for PV-ECS Based Distributed Power Generating System. *Solar Energ. Mater. Solar Cells* **2010**, *94*, 1324–1328.

14. Lobera, D. T.; Valkealahti, S. Dynamic Thermal Model of Solar PV Systems Under Varying Climatic Conditions. *Solar Energ.* **2013**, *93*, 183–194.

15. Perpiñan, O.; Lorenzo, E.; Castro, M.A.; Eyras, R. On the Complexity of Radiation Models for PV Energy Production Calculation. *Solar Energ.* **2008**, *82*, 125–131.

16. Kaplani, E.; Kaplanis, S. A Stochastic Simulation Model for Reliable PV System Sizing Providing for Solar Radiation Fluctuations. *Appl. Energ.* **2012**, *97*, 970–981.

17. Almonacid, F.; Rus, C.; Pérez-Higueras, P.; Hontoria, L. Calculation of the Energy Provided by a PV Generator. Comparative Study: Conventional Methods vs. Artificial Neural Networks. *Energy* **2011**, *36*, 375–384.

PERMISSIONS

The contributors of this book come from diverse backgrounds, making this book a truly international effort. This book will bring forth new frontiers with its revolutionizing research information and detailed analysis of the nascent developments around the world.

We would like to thank all the contributing authors for lending their expertise to make the book truly unique. They have played a crucial role in the development of this book. Without their invaluable contributions this book wouldn't have been possible. They have made vital efforts to compile up to date information on the varied aspects of this subject to make this book a valuable addition to the collection of many professionals and students.

This book was conceptualized with the vision of imparting up-to-date information and advanced data in this field. To ensure the same, a matchless editorial board was set up. Every individual on the board went through rigorous rounds of assessment to prove their worth. After which they invested a large part of their time researching and compiling the most relevant data for our readers.

The editorial board has been involved in producing this book since its inception. They have spent rigorous hours researching and exploring the diverse topics which have resulted in the successful publishing of this book. They have passed on their knowledge of decades through this book. To expedite this challenging task, the publisher supported the team at every step. A small team of assistant editors was also appointed to further simplify the editing procedure and attain best results for the readers.

Apart from the editorial board, the designing team has also invested a significant amount of their time in understanding the subject and creating the most relevant covers. They scrutinized every image to scout for the most suitable representation of the subject and create an appropriate cover for the book.

The publishing team has been an ardent support to the editorial, designing and production team. Their endless efforts to recruit the best for this project, has resulted in the accomplishment of this book. They are a veteran in the field of academics and their pool of knowledge is as vast as their experience in printing. Their expertise and guidance has proved useful at every step. Their uncompromising quality standards have made this book an exceptional effort. Their encouragement from time to time has been an inspiration for everyone.

The publisher and the editorial board hope that this book will prove to be a valuable piece of knowledge for researchers, students, practitioners and scholars across the globe.

LIST OF CONTRIBUTORS

Zaheeruddin and Munish Manas
Department of Electrical Engineering, Jamia Millia Islamia (A Central University), New Delhi, India

A. A. Adeyanju
Mechanical Engineering Department, Ekiti State University, Ado-Ekiti, Ekiti, Nigeria

Paris A. Fokaides, Angeliki Kylili and Andri Pyrgou
School of Engineering and Applied Sciences, Frederick University, Nicosia, Cyprus

Christopher J. Koroneos
Laboratory of Heat Transfer and Environmental Engineering, Aristotle University of Thessaloniki, Thessaloniki, Greece

Anu K. Gupta and Shelley M. Rinehart
Faculty of Business, University of New Brunswick, Saint John, New Brunswick, Canada

Dale C. Roach
Department of Engineering, University of New Brunswick, Saint John, New Brunswick, Canada

Lisa A. Best
Department of Psychology, University of New Brunswick, Saint John, New Brunswick, Canada

Andreas Poullikkas
Department of Electrical Engineering, Cyprus University of Technology, Limassol, Cyprus

Islam F. Mohamed and Ahmed M. Ibrahim
Faculty of Engineering, Cairo University, Giza, Cairo, Egypt

Shady H. E. Abdel Aleem
15th of May Higher Institute of Engineering, 15th of May City, Helwan, Cairo, Egypt

Ahmed F. Zobaa
School of Engineering and Design, Brunel University, Uxbridge, Middlesex, United Kingdom

Amir Erfani, Milad Muhammadi, Soheil Asgari Neshat, Mohammad Masoud Shalchi and Farshad Varaminian
School of Chemical, Gas and Petroleum Engineering, Semnan University, Semnan, Iran

Mini S. Thomas, Ikbal Ali and Nitin Gupta
Department of Electrical Engineering, Faculty of Engineering and Technology, Jamia Millia Islamia, New Delhi, India

Bilgin Şenel and Mine Şenel
Industrial Engineering Department, Tunceli University, Tunceli, Turkey

Levent Bilir
Energy Systems Engineering Department, Atılım University, Ankara, Turkey

Ravindra B. Sholapurkar and Yogesh S. Mahajan
Department of Chemical Engineering, Babasaheb Ambedkar Technological University, Lonere, Tal Mangaon, District Raigad, Maharashtra, India

Ikbal Ali, Mini S. Thomas, Sunil Gupta and S. M. Suhail Hussain
Department of Electrical Engineering, Faculty of Engineering & Technology, Jamia Millia Islamia, New Delhi, India

Ambarish Datta and Bijan Kumar Mandal
Department of Mechanical Engineering, Bengal Engineering and Science University, Howrah, West Bengal, India

R.S. KEMP
Department of Nuclear Science and Engineering, Massachusetts Institute of Technology

Seyyed Mohsen Mousavi Ehteshami and Siew Hwa Chan
Nanyang Technological University, Singapore, Singapore

Amir Hossein Fakehi Khorasani and Somayeh Ahmadi
Department of Energy Engineering, Sharif University of Technology, Tehran, Iran

Mohammad Ali Moradi
Faculty of Mechanical Engineering, K. N. Toosi University of Technology, Tehran, Iran

Deepali Sharma
Guru Tegh Bahadur Institute of Technology, New Delhi, India

Prerna Gaur and A. P. Mittal
Netaji Subhas Institute of Technology, New Delhi, India

Tamer Khatib, Andrea Monacchi, Wilfried Elmenreich and Dominik Egarter
Institute of Networked & Embedded Systems/ Lakeside Labs, Alpen-Adria-Universität Klagenfurt, Klagenfurt, Austria

Salvatore D'alessandro and Andrea M. Tonello
WiTiKee s.r.l., Udine, Italy

Anuranjan Sharda
Department of Mechanical Engineering, Rayat and Bahra Institute of Engineering and Bio-Technology, Punjab, India

Sudhir Kumar
Department of Mechanical Engineering, National Institute of Technology, Kurukshetra, Haryana, India

Ahteshamul Haque
Electrical Engineering, Jamia Millia Islamia University, New Delhi, India

Jin Ho Jo, Ryan Waszak and Michael Shawgo
Illinois State University, Normal, Illinois, USA

Zhi-Bin Liu and Zhi-Yong Zhang
School of Science, Southwest Petroleum University, Chengdu, Sichuan, China

Liu-Li Lu
School of Science, Southwest Petroleum University, Chengdu, Sichuan, China
CNPC Chuanqing Drilling Engineering Company Ltd., Chengdu, China

Yi-Shen Chen
School of Science, Southwest Petroleum University, Chengdu, Sichuan, China
CNPC International Engineering and Construction Corp., Beijing, China

Gigih Udi Atmo, Colin Fraser Duffield and Davidwilson
Department of Infrastructure Engineering, The University of Melbourne, Melbourne, Australia

Syafaruddin
Department of Electrical Engineering, Universitas Hasanuddin, Tamalanrea-Makassar, Indonesia

H. Narimatsu and Hajime Miyauchi
Department of Computer Science and Electrical Engineering of Kumamoto University, Kumamoto, Japan

Index